MW00634870

HANSEN SOLUBILITY PARAMETERS

A USER'S HANDBOOK

HANSEN SOLUBILITY PARAMETERS

A USER'S HANDBOOK

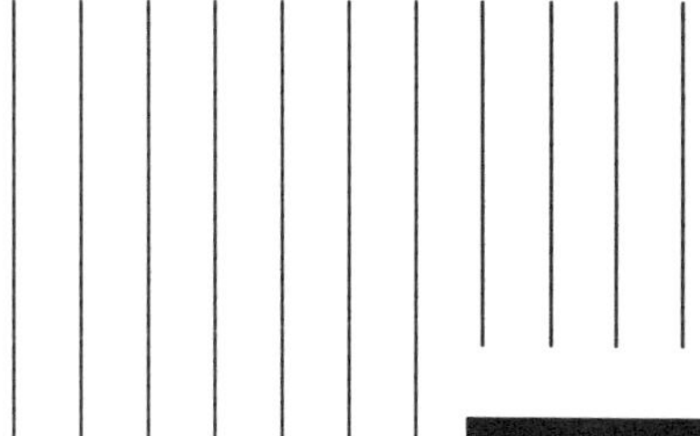

Charles M. Hansen

CRC Press
Boca Raton London New York Washington, D.C.

Acquiring Editor: Tim Pletscher
Project Editor: Naomi Lynch
Cover design: Shayna Murry

Library of Congress Cataloging-in-Publication Data

Hansen, Charles M.
Hansen solubility parameters: a user's handbook / Charles M. Hansen
p. cm.
Includes bibliographical references and index.
ISBN 0-8493-1525-5 (alk. paper)
1. Solution (Chemistry) 2. Polymers--solubility. 3. Thin films.
I. Title.
[DNLM: 1. Hepatitis B virus. QW 710 G289h]
QD543.H258 1999
547'.70454—dc21
DNLM/DLC
for Library of Congress 99-26234
CIP

International Standard Book Number 0-8493-1525-5
Library of Congress Card Number 99-26234
Printed in the United States of America 1 2 3 4 5 6 7 8 9 0
Printed on acid-free paper

Preface

My work with solvents started in Denmark in 1962 when I was a graduate student. The major results of this work were the realization that polymer film formation by solvent evaporation took place in two distinct phases and the development of what has come to be called Hansen solubility (or cohesion) parameters, abbreviated in the following by HSP. The first phase of film formation by solvent evaporation is controlled by surface phenomena such as solvent vapor pressure, wind velocity, heat transfer, etc., and the second phase is controlled by concentration-dependent diffusion of solvent molecules from within the film to the air surface. It is not controlled by the binding of solvent molecules to polymer molecules by hydrogen bonding as was previously thought. My solubility parameter work was actually started to define affinities between solvent and polymer to help predict the degree of this binding which was thought to control solvent retention. This was clearly a futile endeavor since there was absolutely no correlation. The solvents with smaller and more linear molecular structure diffused out of the films more quickly than those with larger and more branched molecular structure. HSP were developed in the process, however.

HSP have been used widely since 1967 to accomplish correlations and to make systematic comparisons which one would not have thought possible earlier. The effects of hydrogen bonding, for example, are accounted for quantitatively. Many of these correlations are discussed in the following, including polymer solubility, swelling, and permeation; surface wetting and dewetting; solubility of inorganic salts; and biological applications including wood, cholesterol, etc. The experimental limits on this seemingly universal ability to predict molecular affinities are apparently governed by the limits represented by energies of the liquid test solvents themselves. There had/has to be a more satisfactory explanation of this universality than just "**semi**empirical" correlations.

I decided to try to collect my experience for the purpose of a reference book, both for myself and for others. At the same time, a search of the major theories of polymer solution thermodynamics was undertaken to explore how the approaches compared. A key element in this was to explain why the correlations all seemed to fit with an apparently "universal" 4 (or ¼ depending on which reference is used). This is described in more detail in Chapter 2 (Equations 2.5 and 2.6). My present view is that the "4" is the result of the validity of the geometric mean rule to describe not only dispersion interactions, but also permanent dipole–permanent dipole and hydrogen bonding (electron interchange) interactions in mixtures of unlike molecules. The Hildebrand approach uses this and was the basis of my earliest approach. The Prigogine corresponding states theory yields the "4" in the appropriate manner when the geometric mean rule is adopted (Chapter 2, Equation 2.11). Any other kind of averaging gives the wrong result. Considering these facts and the massive amount of data which has been correlated using the "4" in the following, it appears proven beyond a reasonable doubt that the geometric mean assumption is valid not only for dispersion-type interactions (or perhaps more correctly in the present context those interactions typical of aliphatic hydrocarbons), but also for permanent dipole–permanent dipole and hydrogen bonding as well.

For those who wish to try to understand the Prigogine theory, I recommend starting with an article by Donald Patterson.[1] This article explains the corresponding states/free volume theory of Prigogine and co-workers in a much simpler form than in the original source. Patterson[2] has also reviewed in understandable language the progression of developments in polymer solution thermodynamics from the Flory-Huggins theory, through that of Prigogine and co-workers, to the so-called "New Flory Theory".[3] Patterson also has been so kind as to aid me in the representations of the earlier theories as they are presented here (especially Chapter 2). All of the previous theories and their extensions also can be found in a more recent book.[4] For this reason, these more classical theories are not treated extensively as such in this book. The striking thing about all of this previous work is that no one has dared to enter into the topic of hydrogen bonding. The present quantitative treatment of permanent dipole–permanent dipole interactions and hydrogen bonding is central to the results reported in every chapter in this book. An attempt to relate this back to the previous theories is given briefly here and more extensively in Chapter 2. This attempt has been directed through Patterson,[1] which may be called the Prigogine-Patterson approach, rather than through the Flory theory, since the relations with the former are more obvious.

I strongly recommend that studies be undertaken to confirm the usefulness of the "structural parameters" in the Prigogine theory (or the Flory theory). It is recognized that the effects of solvent molecular size, segment size, and polymer molecular size (and shapes) are not fully accounted for at the present time. There is hope that this can be done with structural parameters.

The material presented here corresponds to my knowledge and experience at the time of writing, with all due respect to confidentiality agreements and the like.

I am greatly indebted to many colleagues and supporters who have understood that at times one can be so preoccupied and lost in deep thought that the present just seems not to exist.

CMH
October 19, 1998

REFERENCES

1. Patterson, D., Role of Free Volume Changes in Polymer Solution Thermodynamics, *J. Polym. Sci. Part C,* 16, 3379–3389, 1968.
2. Patterson, D., Free Volume and Polymer Solubility. A Qualitative View, *Macromolecules*, 2(6), 672–677, 1969.
3. Flory, P. J., Thermodynamics of Polymer Solutions, *Discussions of the Faraday Society*, 49, 7–29, 1970.
4. Lipatov, Y. S. and Nesterov, A. E., *Polymer Thermodynamics Library, Vol. 1, Thermodynamics of Polymer Blends*, Technomic Publishing Co., Inc., Lancaster, PA, 1997.

The Author

Charles M. Hansen is currently Senior Scientist at FORCE Institute in Broendby (Copenhagen) Denmark. He received a BChE from the University of Louisville and an MS degree from the University of Wisconsin. After being awarded the Dr. Techn. degree from the Technical University of Denmark in 1967, he held leading positions with PPG Industries in Pittsburgh, and as Director of the Scandinavian Paint and Printing Ink Research Institute in Hoersholm, Denmark.

Dr. Hansen is perhaps best known for his extension of the Hildebrand solubility parameter to what are now called Hansen solubility parameters. These have been found mutually confirming with the I. Prigogine corresponding states theory of polymer solutions and can be used to directly calculate the Flory-Huggins interaction coefficient. He has published widely in the fields of polymer solubility, diffusion and permeation in polymers and films, surface science, and coatings science.

He is currently vice president of the Danish Society for Polymer Technology, on the editorial board of Progress in Organic Coatings, and a member of the Danish Academy of Technical Sciences (ATV).

Key to Symbols

A_{12}	Energy difference defined by Chapter 2, Equation 2.12
c	Dispersion cohesion energy from Chapter 1, Figure 1.2 or Figure 1.3
D	Diffusion coefficient in Chapter 8.
D	Dispersion cohesion (solubility) parameter — in tables and computer printouts
DM	Dipole moment — debyes
E_D	Dispersion cohesion energy
E_P	Polar cohesion energy
E_H	Hydrogen bonding cohesion energy
ΔE_v	Energy of vaporization (=) cohesion energy
G	Number of "good" solvents in a correlation, used in tables of correlations
ΔG^M	Molar free energy of mixing
$\Delta G^M_{noncomb}$	Noncombinatorial molar free energy of mixing
H	Hydrogen bonding cohesion (solubility) parameter — in tables and computer printouts
ΔH_v	Molar heat of vaporization
ΔH^M	Molar heat of mixing
P	Permeation coefficient in Chapter 8
P	Polar cohesion (solubility) parameter — in tables and computer printouts
P*	Total pressure, atm. (Chapter 8, Figures 8.4 and 8.5)
R	Gas constant (1.987 cal/mol K)
Ra	Distance in Hansen space, see Chapter 1, Equation 1.9 or Chapter 2, Equation 2.5
RA	Distance in Hansen space, see Chapter 2, Equation 2.7
R_M	Maximum distance in Hansen space allowing solubility (or other "good" interaction)
Ro	Radius of interaction sphere in Hansen space
RED	Relative energy difference (Chapter 1, Equation 1.10)
S	Solubility coefficient in Chapter 8
ΔS^M	Molar entropy of mixing
T	Absolute temperature
T	"Total" number of solvents used in a correlation as given in tables
T_b	(Normal) boiling point, degrees K
T_c	Critical temperature, degrees K
T_r	Reduced temperature, Chapter 1, Equation 1.12
V	Molar volume, cm^3/gram molecular weight
V_M	Volume of mixture
f	Fractional solubility parameters, defined by Chapter 3, Equations 3.1 to 3.3
i	Component "i" in a mixture
k	Constant in Equation 4.1
r	Number of segments in a given molecule, Chapter 2
t_s	Sedimentation time, see Chapter 5, Equation 5.1

x	Mole fraction in liquid phase (Chapter 8, Figures 8.4 and 8.5)
y	Mole fraction in vapor phase (Chapter 8, Figures 8.4 and 8.5)
H	Ratio of cohesive energy densities; Chapter 2, Equation 2.6
Σ	Summation
Δ_T	Lydersen critical temperature group contribution
α	Thermal expansion coefficient
β	Constant in Chapter 2, Equation 2.1
δ_D	Dispersion cohesion (solubility) parameter
δ_H	Hydrogen bonding cohesion (solubility) parameter
δ_P	Polar cohesion (solubility) parameter
δ_t	Total (Hildebrand) cohesion (solubility) parameter
d	Prigogine normalized interaction parameter, Chapter 2, Equation 2.8
ε	Cohesive energy for a polymer segment or solvent
γ	Surface free energy of a liquid in air or its own vapor
η	Viscosity of solvent, Chapter 5, Equation 5.1
η_s	Viscosity of solution
η_o	Viscosity of solvent
$[\eta]$	Intrinsic viscosity, see Chapter 6, Equation 6.4
φ_i	Volume fraction of component "i"
ν	Interaction parameter, see Chapter 2, Equation 2.11
θ	Contact angle between liquid and surface
θ_a	Advancing contact angle
θ_r	Receding contact angle
ρ	Prigogine parameter for differences is size in polymer segments and solvent, Chapter 2, Equation 2.10
ρ	Density in Chapter 5, Equation 5.1
ρ_p	Particle density in Chapter 5, Equation 5.1
ρ_s	Solvent density in Chapter 5, Equation 5.1
σ	Prigogine segmental distance parameter, Chapter 2, Equation 2.10
χ	Polymer–liquid interaction parameter (Flory-Huggins), Chapter 2
χ_{12}	Interaction parameter — "New Flory Theory"
χ_c	Critical polymer–liquid interaction parameter, Chapter 2
χ_{lit}	Representative χ value from general literature
χ_s	Entropy component of χ
$'$	(Prime) indicates cohesion parameter used to describe a surface
1	(Subscript) indicates a solvent
2	(Subscript) indicates a polymer (or second material in contact with a solvent)
c	Concentration in Chapter 6, Equation 6.4 (Note: There is also a "c" as entry 2 above)
D	(Subscript) dispersion component
P	(Subscript) polar component
H	(Subscript) hydrogen bonding component
d	(Subscript) dispersion component
p	(Subscript) polar component
h	(Subscript) hydrogen bonding component

Contents

1 Solubility Parameters — An Introduction

CONTENTS

ABSTRACT

Solubility parameters have found their greatest use in the coatings industry to aid in the selection of solvents. They are used in other industries, however, to predict compatibility of polymers, chemical resistance, and permeation rates and even to characterize the surfaces of pigments, fibers, and fillers. Liquids with similar solubility parameters will be miscible, and polymers will dissolve in solvents whose solubility parameters are not too different from their own. The basic principle has been "like dissolves like." More recently, this has been modified to "like seeks like," since many surface characterizations have also been made and surfaces do not (usually) dissolve. Solubility parameters help put numbers into this simple qualitative idea. This chapter describes the tools commonly used in Hansen solubility parameter (HSP) studies. These include liquids used as energy probes and computer programs to process data. The goal is to arrive at the HSP for interesting materials either by calculation or, if necessary, by experiment and preferably with agreement between the two.

INTRODUCTION

The solubility parameter has been used for many years to select solvents for coatings materials. A lack of total success has stimulated research. The skill with which solvents can be optimally selected with respect to cost, solvency, workplace environment, external environment, evaporation rate, flash point, etc. has improved over the years as a result of a series of improvements in the solubility parameter concept and widespread use of computer techniques. Most, if not all, commercial suppliers of solvents have computer programs to help with solvent selection. One can now easily predict how to dissolve a given polymer in a mixture of two solvents, neither of which can dissolve the polymer by itself.

Unfortunately, this book cannot include discussion of all of the significant efforts leading to our present state of knowledge of the solubility parameter. An attempt is made to outline developments, provide some background for a basic understanding, and give examples of uses in practice. The key is to determine which affinities the important components in a system have for each other. For many products this means evaluating or estimating the relative affinities of solvents, polymers, additives, pigment surfaces, filler surfaces, fiber surfaces, and substrates.

It is noteworthy that the concepts presented here have developed toward not just predicting solubility, which requires high affinity between solvent and solute, but to predicting affinities between different polymers leading to compatibility, and affinities to surfaces to improve dispersion and adhesion. In these applications the solubility parameter has become a tool, using well-defined liquids as energy probes, to measure the similarity, or lack of the same, of key components. Materials with widely different chemical structure may be very close in affinities. Only those materials which interact differently with different solvents can be characterized in this manner. It can be expected that many inorganic materials, such as fillers, will not interact differently with these energy probes since their energies are very much higher. An adsorbed layer of water on the high energy surface can also play an important role. Regardless of these concerns, it has been possible to characterize pigments, both organic and inorganic, as well as fillers like barium sulfate, zinc oxide, etc. and also inorganic fibers as discussed in Chapter 5. Changing the surface energies by various treatments can lead to a surface which can be characterized more readily and which often interacts more strongly with given organic solvents. When the same solvents that dissolve a polymeric binder are also those which interact most strongly with a surface, one can expect the binder and the surface to have high affinity for each other.

Solubility parameters are sometimes called cohesion energy parameters since they derive from the energy required to convert a liquid to a gas. The energy of vaporization is a direct measure of the total (cohesive) energy holding the liquid's molecules together. All types of bonds holding the liquid together are broken by evaporation, and this has led to the concepts described in more detail later. The term cohesion energy parameter is more appropriately used when referring to surface phenomena.

HILDEBRAND PARAMETERS AND BASIC POLYMER SOLUTION THERMODYNAMICS

The term solubility parameter was first used by Hildebrand and Scott.[1,2] The earlier work of Scatchard and others was contributory to this development. The Hildebrand solubility parameter is defined as the square root of the cohesive energy density:

$$\delta = (E/V)^{1/2} \tag{1.1}$$

V is the molar volume of the pure solvent, and E is its (measurable) energy of vaporization (see Equation 1.15). The numerical value of the solubility parameter in $MPa^{1/2}$ is 2.0455 times larger than that in $(cal/cm^3)^{1/2}$. The solubility parameter is an important quantity for predicting solubility relations, as can be seen from the following brief introduction.

Thermodynamics requires that the free energy of mixing must be zero or negative for the solution process to occur spontaneously. The free energy change for the solution process is given by the relation

$$\Delta G^M = \Delta H^M - \Delta TS^M \tag{1.2}$$

where ΔG^M is the free energy of mixing, ΔH^M is the heat of mixing, T is the absolute temperature, and ΔS^M is the entropy change in the mixing process.

Equation 1.3 gives the heat of mixing as proposed by Hildebrand and Scott:

$$\Delta H^M = \varphi_1 \varphi_2 V_M (\delta_1 - \delta_2)^2 \quad (1.3)$$

The φs are volume fractions of solvent and polymer, and V_M is the volume of the mixture. Equation 1.3 is not correct. This equation has often been cited as a shortcoming of this theory in that only positive heats of mixing are allowed. It has been shown by Patterson, Delmas, and coworkers that $\Delta G^M_{noncomb}$ is given by the right-hand side of Equation 1.3 and not ΔG^M. This is discussed more in Chapter 2. The correct relation is[3-8]

$$\Delta G^M_{noncomb} = \varphi_1 \varphi_2 V_M (\delta_1 - \delta_2)^2 \quad (1.4)$$

The noncombinatorial free energy of solution, $\Delta G^M_{noncomb}$, includes all free energy effects other than the combinatorial entropy of solution occurring because of simply mixing the components. Equation 1.4 is consistent with the Prigogine corresponding states theory of polymer solutions (see Chapter 2) and can be differentiated to give expressions[3,4] predicting both positive and negative heats of mixing. Therefore, both positive and negative heats of mixing can be expected from theoretical considerations and have been measured accordingly. It has been clearly shown that solubility parameters can be used to predict both positive and negative heats of mixing. Previous objections to the effect that only positive values are allowed in this theory are not correct.

This discussion clearly demonstrates that one should actually consider the solubility parameter as a free energy parameter. This is also more in agreement with the use of the solubility parameter plots to follow, since these use solubility parameters as axes and have the experimentally determined boundaries of solubility defined by the condition that the free energy of mixing is zero. The combinatorial entropy enters as a constant factor in the plots of solubility in different solvents, for example, since the concentrations are usually constant for a given study.

It is important to note that the solubility parameter, or rather the difference in solubility parameters for the solvent–solute combination, is important in determining the solubility of the system. It is clear that a match in solubility parameters leads to a zero change in noncombinatorial free energy, and the positive entropy change (the combinatorial entropy change) found on simple mixing to arrive at the disordered mixture compared to the pure components will ensure solution is possible from a thermodynamic point of view. The maximum difference in solubility parameters, which can be tolerated where solution still occurs, is found by setting the noncombinatorial free energy change equal to the combinatorial entropy change.

$$\Delta G^M_{noncomb} = T \Delta S^M_{comb} \quad (1.5)$$

This equation clearly shows that an alternate view of the solubility situation at the limit of solubility is that it is the entropy change which dictates how closely the solubility parameters must match each other for solution to (just) occur.

It will be seen in Chapter 2 that solvents with smaller molecular volumes will be thermodynamically better than larger ones having identical solubility parameters. A practical aspect of this effect is that solvents with relatively low molecular volumes, such as methanol and acetone, can dissolve a polymer at larger solubility parameter differences than expected from comparisons with other solvents with larger molecular volumes. An average solvent molecular volume is usually taken as about 100 cc/mol. The converse is also true. Larger molecular species may not dissolve, even though solubility parameter considerations might predict this. This can be a difficulty in predicting the behavior of plasticizers based on data for lower molecular weight solvents only. These effects are also discussed elsewhere in this book, particularly Chapters 2, 7, and 8.

A shortcoming of the earlier solubility parameter work is that the approach was limited to regular solutions as defined by Hildebrand and Scott,[2] and does not account for association between molecules, such as those which polar and hydrogen bonding interactions would require. The latter problem seems to have been largely solved with the use of multicomponent solubility parameters. However, the lack of accuracy with which the solubility parameters can be assigned will always remain a problem. Using the difference between two large numbers to calculate a relatively small heat of mixing, for example, will always be problematic.

A more detailed description of the theory presented by Hildebrand and the succession of research reports which have attempted to improve on it can be found in Barton's extensive handbook.[9] The slightly older, excellent contribution of Gardon and Teas[10] is also a good source of related information, particularly for coatings and adhesion phenomena. The approach of Burrell,[11] who divided solvents into hydrogen bonding classes, has found numerous practical applications; the approach of Blanks and Prausnitz[12] divided the solubility parameter into two components, nonpolar and "polar." Both are worthy of mention, however, in that these have found wide use and greatly influenced the author's earlier activities, respectively. The latter article, in particular, was farsighted in that a corresponding states procedure was introduced to calculate the dispersion energy contribution to the cohesive energy. This is discussed in more detail in Chapter 2.

It can be seen from Equation 1.2 that the entropy change is beneficial to mixing. When multiplied by the temperature, this will work in the direction of promoting a more negative free energy of mixing. This is the usual case, although there are exceptions. Increasing temperature does not always lead to improved solubility relations, however. Indeed, this was the basis of the pioneering work of Patterson and co-workers,[3-8] to show that increases in temperature can predictably lead to insolubility. Their work was done in essentially nonpolar systems. Increasing temperature can also lead to a nonsolvent becoming a solvent and, subsequently, a nonsolvent again with still further increase in temperature. Polymer solubility parameters do not change much with temperature, but those of a liquid frequently decrease rapidly with temperature. This situation allows a nonsolvent with a solubility parameter which is initially too high to pass through a soluble condition to once more become a nonsolvent as the temperature increases. These are usually "boundary" solvents on solubility parameter plots.

The entropy changes associated with polymer solutions will be smaller than those associated with liquid–liquid miscibility, for example, since the "monomers" are already bound into the configuration dictated by the polymer they make up. They are no longer free in the sense of a liquid solvent and cannot mix freely to contribute to a larger entropy change. This is one reason polymer–polymer miscibility is difficult to achieve. The free energy criterion dictates that polymer solubility parameters match extremely well for mutual compatibility, since there is little help to be gained from the entropy contribution when progressively larger molecules are involved. However, polymer–polymer miscibility can be promoted by the introduction of suitable copolymers or comonomers which interact specifically within the system. Further discussion of these phenomena is beyond the scope of the present discussion; however, see Chapter 3.

HANSEN SOLUBILITY PARAMETERS

A widely used solubility parameter approach to predicting polymer solubility is that proposed by the author. The basis of these so-called Hansen solubility parameters (HSP) is that the total energy of vaporization of a liquid consists of several individual parts.[13-17] These arise from (atomic) dispersion forces, (molecular) permanent dipole–permanent dipole forces, and (molecular) hydrogen bonding (electron exchange). Needless to say, without the work of Hildebrand and Scott[1,2] and others not specifically referenced here such as Scatchard, this postulate could never have been made. The total cohesive energy, E, can be measured by evaporating the liquid, i.e., breaking all the cohesive bonds. It should also be noted that these cohesive energies arise from interactions of a given solvent molecule with another of its own kind. The basis of the approach is, therefore, very simple, and it is surprising that so many different applications have been possible since 1967 when the idea was first published. A rather large number of applications are discussed in this book. Others are found in Barton.[9] A lucid discussion by Barton[18] enumerates typical situations where problems occur when using solubility parameters. These occur most often where the environment causes the solvent molecules to interact with or within themselves differently than when they make up their own environment, i.e., as pure liquids. Several cases are discussed where appropriate in the following chapters.

Materials having similar HSP have high affinity for each other. The extent of the similarity in a given situation determines the extent of the interaction. The same cannot be said of the total or Hildebrand solubility parameter.[1,2] Ethanol and nitromethane, for example, have similar total solubility parameters (26.1 vs. 25.1 $MPa^{1/2}$, respectively), but their affinities are quite different. Ethanol is water soluble, while nitromethane is not. Indeed, mixtures of nitroparaffins and alcohols were demonstrated in many cases to provide synergistic mixtures of two nonsolvents which dissolved polymers.[13] This could never have been predicted by Hildebrand parameters, whereas the HSP concept readily confirms the reason for this effect.

There are three major types of interaction in common organic materials. The most general are the "non-polar" interactions. These derive from atomic forces. These have also been called dispersion interactions in the literature. Since molecules are built up from atoms, all molecules will contain this type of attractive force. For the saturated aliphatic hydrocarbons, for example, these are essentially the only cohesive interactions, and the energy of vaporization is assumed to be the same as the dispersion cohesive energy, E_D. Finding the dispersion cohesive energy as the cohesion energy of the homomorph, or hydrocarbon counterpart, is the starting point for the calculation of the three Hansen parameters for a given liquid. As discussed in more detail below, this is based on a corresponding states calculation.

The permanent dipole–permanent dipole interactions cause a second type of cohesion energy, the polar cohesive energy, E_P. These are inherently molecular interactions and are found in most molecules to one extent or another. The dipole moment is the primary parameter used to calculate these interactions. A molecule can be mainly polar in character without being water soluble, so there is misuse of the term "polar" in the general literature. The polar solubility parameters referred to here are well-defined, experimentally verified, and can be estimated from molecular parameters as described later. As noted previously, the most polar of the solvents include those with relatively high total solubility parameters which are not particularly water soluble, such as nitroparaffins, propylene carbonate, tri-*n*-butyl phosphate, and the like. Induced dipoles have not been treated specifically in this approach, but are recognized as a potentially important factor, particularly for solvents with zero dipole moments (see the Calculation of the Polar Solubility Parameter section).

The third major cohesive energy source is hydrogen bonding, E_H. This can be called more generally an electron exchange parameter. Hydrogen bonding is a molecular interaction and resembles the polar interactions in this respect. The basis of this type of cohesive energy is attraction among molecules because of the hydrogen bonds. In this perhaps oversimplified approach, the hydrogen bonding parameter has been used to more or less collect the energies from interactions not included in the other two parameters. Alcohols, glycols, carboxylic acids, and other hydrophilic materials have high hydrogen bonding parameters. Other researchers have divided this parameter into separate parts, for example, acid and base cohesion parameters, to allow both positive and negative heats of mixing. These approaches will not be dealt with here, but can be found described in Barton's handbook[9] and elsewhere.[19-21] The most extensive division of the cohesive energy has been done by Karger et al.[22] who developed a system with five parameters — dispersion, orientation, induction, proton donor, and proton acceptor. As a single parameter, the Hansen hydrogen bonding parameter has accounted remarkably well for the experience of the author and keeps the number of parameters to a level which allows ready practical usage.

It is clear that there are other sources of cohesion energy in various types of molecules arising, for example, from induced dipoles, metallic bonds, electrostatic interactions, or whatever type of separate energy can be defined. The author stopped with the three major types found in organic molecules. It was and is recognized that additional parameters could be assigned to separate energy types. The description of organometallic compounds could be an intriguing study, for example. This would presumably parallel similar characterizations of surface active materials, where each separate part of the molecule requires separate characterization for completeness. The Hansen parameters have mainly been used in connection with solubility relations mostly, but not exclusively, in the coatings industry.

Solubility and swelling have been used to confirm the solubility parameter assignments of many of the liquids. These have then been used to derive group contribution methods and suitable equations based on molecular properties to arrive at estimates of the three parameters for additional liquids. The goal of a prediction is to determine the similarity or difference of the cohesion energy parameters. The strength of a particular type of hydrogen bond or other bond, for example, is important only to the extent that it influences the cohesive energy density.

HSP do have direct application in other scientific disciplines such as surface science, where they have been used to characterize the wettability of various surfaces, the adsorption properties of pigment surfaces,[10,14,16,23-26] and have even led to systematic surface treatment of inorganic fibers so they could be readily incorporated into polymers of low solubility parameters such as polypropylene[27] (see also Chapter 5). Many other applications of widely different character have been discussed by Barton[9] and Gardon.[28] Surface characterizations have not been given the attention deserved in terms of a unified similarity-of-energy approach. The author can certify that thinking in terms of similarity of energy, whether surface energy or cohesive energy, can lead to rapid decisions and plans of action in critical situations where data are lacking. In other words, the everyday industrial crisis situation often can be reduced in scope by appropriate systematic approaches based on similarity of energy. The successes using the HSP for surface applications are not surprising in view of the similarity of predictions offered by these and the Prigogine corresponding states theory of polymer solutions discussed in Chapter 2. Flory also emphasized that it is the surfaces of molecules which interact to produce solutions,[29] so the interactions of molecules residing in surfaces should clearly be included in any general approach to interactions among molecules.

The basic equation which governs the assignment of Hansen parameters is that the total cohesion energy, E, must be the sum of the individual energies which make it up.

$$E = E_D + E_P + E_H \tag{1.6}$$

Dividing this by the molar volume gives the square of the total (or Hildebrand) solubility parameter as the sum of the squares of the Hansen D, P, and H components.

$$E/V = E_D/V + E_P/V + E_H/V \tag{1.7}$$

$$\delta^2 = \delta_D^2 + \delta_P^2 + \delta_H^2 \tag{1.8}$$

To sum up this section, it is emphasized that HSP quantitatively account for the cohesion energy (density). An experimental latent heat of vaporization has been considered much more reliable as a method to arrive at a cohesion energy than using molecular orbital calculations or other calculations based on potential functions. Indeed, the goal of such extensive calculations for polar and hydrogen bonding molecules should be to accurately arrive at the energy of vaporization.

METHODS AND PROBLEMS IN THE DETERMINATION OF PARTIAL SOLUBILITY PARAMETERS

The best method to calculate the individual HSP depends to a great extent on what data are available. The author originally adopted an essentially experimental procedure and established values for 90 liquids based on solubility data for 32 polymers.[13] This procedure involved calculation of the nonpolar parameter according to the procedure outlined by Blanks and Prausnitz.[12] This calculational procedure is still in use and is considered the most reliable and consistent for this parameter. It is outlined below. The division of the remaining cohesive energy between the polar and hydrogen bonding interactions was initially done by trial and error to fit experimental polymer solubility data. A key to parameter assignments in this initial trial and error approach was that mixtures of two nonsolvents could be systematically found to synergistically (but predictably) dissolve given polymers. This meant that these had parameters placing them on opposite sides of the solubility region, a spheroid, from each other. By having a large number of such predictably synergistic systems as a basis, reasonably accurate divisions into the three energy types were possible.

Using the experimentally established, approximate, δ_P and δ_H parameters, Hansen and Skaarup[15] found that the Böttcher equation could be used to calculate the polar parameter quite well, and this led to a revision of the earlier values to those now accepted for these same liquids. These values were also consistent with the experimental solubility data for 32 polymers available at that time and with Equation 1.6. Furthermore, Skaarup developed the equation for the solubility parameter "distance," Ra, between two materials based on their respective partial solubility parameter components:

$$(Ra)^2 = 4(\delta_{D2} - \delta_{D1})^2 + (\delta_{P2} - \delta_{P1})^2 + (\delta_{H2} - \delta_{H1})^2 \tag{1.9}$$

This equation was developed from plots of experimental data where the constant "4" was found convenient and correctly represented the solubility data as a sphere encompassing the good solvents (see Chapter 3). When the scale for the dispersion parameter is doubled compared with the other two parameters, essentially spherical rather than spheroidal, regions of solubility are found. This greatly aids two-dimensional plotting and visualization. There are, of course, boundary regions where deviations can occur. These are most frequently found to involve the larger molecular species being less effective solvents compared with the smaller counterparts which define the solubility sphere. Likewise, smaller molecular species such as acetone, methanol, nitromethane, and others often appear as outliers in that they dissolve a polymer even though they have solubility parameters placing them at a distance greater than the experimentally determined radius of the solubility sphere, Ro. This dependence on molar volume is inherent in the theory developed by Hildebrand, Scott, and Scatchard discussed above. Smaller molar volume favors lower ΔG^M, as discussed in Chapter 2. This in turn promotes solubility. Such smaller molecular volume species which dissolve "better" than predicted by comparisons based on solubility parameters alone should not necessarily be considered outliers.

The molar volume is frequently used successfully as a fourth parameter to describe molecular size effects. These are especially important in correlating diffusional phenomena with HSP, for example (see Chapters 7 and 8). The author has preferred to retain the three, well-defined, partial solubility parameters with a separate, fourth, molar volume parameter, rather than to multiply the solubility parameters by the molar volume raised to some power to redefine them.

The reason for the experimentally determined constant 4 in Equation 1.9 will be discussed in more detail in Chapter 2. It will be noted here, however, that the constant 4 is theoretically predicted by the Prigogine corresponding states theory of polymer solutions when the geometric mean is used to estimate the interaction in mixtures of dissimilar molecules.[30] This is exceptionally strong evidence that dispersion, permanent dipole–permanent dipole, and hydrogen bonding interactions all follow the geometric mean rule. Patterson and co-workers have been especially instrumental in relating the Prigogine theory to solubility parameters and to the Flory-Huggins theory of polymer solutions.[3-8] The HSP approach of dividing the cohesive energy into parts derived from different types of cohesive forces has been confirmed both by experimental studies as well as by the Prigogine theory. The use of the geometric mean is basic to this agreement between the HSP approach and that of Prigogine (see Chapter 2).

The approach of optimizing solubility data to spheres is still very much is use. Plotting regions of solubility based on experimental solubility data or computer optimizing boundaries of solubility by locating the maximum difference in solubility parameters allowed by Equation 1.9 are both used. The total free energy of mixing, ΔG^M, is equal to zero on the boundary. It should be recognized that using solubility parameters, which relate to $\Delta G^M_{noncomb}$ in Equation 1.4, differs from this by the combinatorial entropy of mixing.

Another promising approach to arrive at the HSP for materials based on experimental data is to use multivariable analysis of one type or another as discussed in Chapter 3. This type of approach has not been attempted by the author, but it clearly has advantages in some cases. The author's preferred approach of locating the polymer HSP as the center of a sphere has a problem in that it is in reality the poor solvents or nonsolvents located near the boundary of the sphere which fix the boundary (and center) rather than the best solvents in the middle. This may present problems for smaller sets of data, but it is an advantage when extrapolating into regions of HSP higher than those of any liquid which can be used in testing. This is discussed in Chapter 3 in more detail and is based on Equation 1.9 to define the limited segment of the boundary of the HSP sphere derivable from such correlations.

Equation 1.9 is readily used on a computer (or on a hand calculator) and supplementary relations allow easier scanning of large sets for data. It is obvious that solubility, or high affinity, requires that Ra be less than Ro. The ratio Ra/Ro has been called the RED number, reflecting the Relative Energy Difference.

$$RED = Ra/Ro \tag{1.10}$$

An RED number of 0 is found for no energy difference. RED numbers less than 1.0 indicate high affinity; RED equal to or close to 1.0 is a boundary condition; and progressively higher RED numbers indicate progressively lower affinities. Scanning a computer output for RED numbers less than 1.0, for example, rapidly allows location of the most interesting liquids for a given application.

It should be noted parenthetically here that the ratio of Ra to Ro is really a ratio of quantities having the same units as the solubility parameter. The ratio $(Ra/Ro)^2 = (RED)^2$ is a ratio of cohesion energies. This latter quantity is important for relating the HSP approach to that of Huggins and Flory as discussed in Chapter 2.

The revised set of parameters for the 90 original solvents was the basis for group contribution procedures developed by (most notably) van Krevelen,[31] Beerbower,[32] and Hansen and Beerbower,[17] who also used Fedors' work.[33] These various developments have been summarized by Barton,[9] although Beerbower's latest values have only appeared in the National Aeronautics and Space Administration (NASA) document.[32] Table 1.1 is an expanded table of Beerbower group contributions which was distributed among those who were in contact with Beerbower in the late 1970s. The majority of the data in this table, as well as Table 1.2, have also appeared in Reference 34. Beerbower also developed a simple equation for the polar parameter,[17] which involved only the dipole moment and the square root of the molar volume. This is also given later (Equation 1.13) and has been found quite reliable by Koenhen and Smolders.[35] This equation has been found reliable by the author as well, giving results generally consistent with Equations 1.6 to 1.8, which, again, is the basis of the whole approach. Koenhen and Smolders also give correlation coefficients for other calculational procedures to arrive at the individual Hansen parameters.

TABLE 1.1
Group Contributions to Partial Solubility Parameters

Functional Group	Molar Volume,[a] ΔV (cm³/mol)		London Parameter, $\Delta V \delta_D^2$ (cal/mol)			Polar Parameter, $\Delta V \delta_P^2$ (cal/mol)			Electron Transfer Parameter, $\Delta V \delta_H^2$ (cal/mol)		Total Parameter,[a] $\Delta V \delta^2$ (cal/mol)	
	Aliphatic	Aromatic[b]	Alkane	Cyclo	Aromatic	Alkane	Cyclo	Aromatic	Aliphatic	Aromatic	Aliphatic	Aromatic
CH_3–	33.5	Same	1,125	Same	Same	0	0	0	0	0	1,125	Same
CH_2<	16.1	Same	1,180	Same	Same	0	0	0	0	0	1,180	Same
–CH<	–1.0	Same	820	Same	Same	0	0	0	0	0	820	Same
>C<	–19.2	Same	350	Same	Same	0	0	0	0	0	350	Same
CH_2 = olefin	28.5	Same	850 ± 100	?	?	25 ± 10	?	?	180 ± 75	?	1,030	Same
–CH = olefin	13.5	Same	875 ± 100	?	?	18 ± 5	?	?	180 ± 75	?	1,030	Same
>C = olefin	–5.5	Same	800 ± 100	?	?	60 ± 10	?	?	180 ± 75	?	1,030	Same
Phenyl-	—	71.4	—	—	7,530	—	—	50 ± 25	—	50 ± 50[c]	—	7630
C-5 ring (saturated)	16	—	—	250	—	0	0	—	0	—	250	—
C-6 ring	16	Same	—	250	250	0	0	0	0	0	250	250
–F	18.0	22.0	0	0	0	1,000 ± 150	?	700 ± 100	0	0	1,000	800[b]
=F_2 twin[f]	40.0	48.0	0	0	0	700 ± 250[c]	?	500 ± 250[c]	0	0	1,700	1,360[b]
≡F_3 triplet[f]	66.0	78.0	0	0	0	?	?	?	0	0	1,650	1,315[b]
–Cl	24.0	28.0	1,400 ± 100	?	1,300 ± 100	1,250 ± 100	1,450 ± 100	800 ± 100	100 ± 20[c]	Same	2,760	2,200[b]
=Cl_2 twin[f]	52.0	60.0	3,650 ± 160	?	3,100 ± 175[c]	800 ± 150	?	400 ± 150[c]	165 ± 10[c]	180 ± 10[c]	4,600	3,670[b]
≡Cl_3 triplet[f]	81.9	73.9	4,750 ± 300[c]	?	?	300 ± 100	?	?	350 ± 250[c]	?	5,400	4,300[b]
–Br	30.0	34.0	1,950 ± 300[c]	1,500 ± 175	1,650 ± 140	1,250 ± 100	1,700 ± 150	800 ± 100	500 ± 100	500 ± 100	3,700	2,960[b]
=Br_2 twin[f]	62.0	70.0	4,300 ± 300[c]	?	3,500 ± 300[c]	800 ± 250[c]	?	400 ± 150[c]	825 ± 200[c]	800 ± 250[c]	5,900	4,700[b]
≡Br_3 triplet[f]	97.2	109.2	5,800 ± 400[c]	?	?	350 ± 150[c]	?	?	1,500 ± 300[c]	?	7,650	6,100[b]
–I	31.5	35.5	2,350 ± 250[c]	2,200 ± 250[c]	2,000 ± 250[c]	1,250 ± 100	1,350 ± 100	575 ± 100	1,000 ± 200[c]	1,000 ± 200[c]	4,550	3,600[b]
=I_2 twin[e]	66.6	74.6	5,500 ± 300[c]	?	4,200 ± 300[c]	800 ± 250[c]	?	400 ± 150[c]	1,650 ± 250[c]	1,800 ± 250[c]	8,000	6,400[b]
≡I_3 triplet[e]	111.0	123.0	?	?	?	?	?	?	?	?	11,700	9,350[b]
–O– ether	3.8	Same	0	0	0	500 ± 150	600 ± 150	450 ± 150	450 ± 25	1,200 ± 100	800	(1,650 ± 150)
>CO ketone	10.8	Same	—[e]	2,350 ± 400	2,800 ± 325	(15, 000 ± 7%)/V	1,000 ± 300	950 ± 300	800 ± 250[d]	400 ± 125[c]	4,150	Same
–CHO	(23.2)	(31.4)	950 ± 300	?	550 ± 275	2,100 ± 200	3,000 ± 500	2,750 ± 200	1,000 ± 200	750 ± 150	(4,050)	Same
–COO-ester	18.0	Same	—[f]	?	—[f]	(56,000 ± 12%)/V	?	(338,000 ± 10%)/V	1,250 ± 150	475 ± 100[c]	4,300	Same
–COOH	28.5	Same	3,350 ± 300	3,550 ± 250	3,600 ± 400	500 ± 150	300 ± 50	750 ± 350	2,750 ± 250	2,250 ± 250[c]	6,600	Same

TABLE 1.1 (continued)
Group Contributions to Partial Solubility Parameters

Functional Group	Molar Volume,[a] ΔV (cm³/mol)		London Parameter, $\Delta V \delta_D^2$ (cal/mol)			Polar Parameter, $\Delta V \delta_P^2$ (cal/mol)			Electron Transfer Parameter, $\Delta V \delta_H^2$ (cal/mol)		Total Parameter,[a] $\Delta V \delta^2$ (cal/mol)	
	Aliphatic	Aromatic[b]	Alkane	Cyclo	Aromatic	Alkane	Cyclo	Aromatic	Aliphatic	Aromatic	Aliphatic	Aromatic
–OH	10.0	Same	1,770 ± 450	1,370 ± 500	1,870 ± 600	700 ± 200	1,100 ± 300	800 ± 150	4,650 ± 400	4,650 ± 500	7,120	Same
=$(OH)_2$ twin or adjacent	26.0	Same	0	?	?	1,500 ± 100	?	?	9,000 ± 600	9,300 ± 600	10,440	Same
–CN	24.0	Same	1,600 ± 850[c]	?	0	4,000 ± 800[c]	?	3,750 ± 300[c]	500 ± 200[d]	400 ± 125[c]	4,150	Same
$-NO_2$	24.0	32.0	3,000 ± 600	?	2,550 ± 125	3,600 ± 600	?	1,750 ± 100	400 ± 50[d]	350 ± 50[c]	7,000	(4,400)
$-NH_2$ amine	19.2	Same	1,050 ± 300	1,050 ± 450[c]	150 ± 150[c]	600 ± 200	600 ± 350[c]	800 ± 200	1,350 ± 200	2,250 ± 200[d]	3,000	Same
$>NH_2$ amine	4.5	Same	1,150 ± 225	?	?	100 ± 50	?	?	750 ± 200	?	2,000	Same
$-NH_2$ amide	(6.7)	Same	?	?	?	?	?	?	2,700 ± 550[c]	?	(5,850)	Same
$\rightarrow PO_4$ ester	28.0	Same	—[e]	?	?	(81,000 ± 10%)/V	?	?	3,000 ± 500	?	(7,000)	Same

[a] Data from Fedors.[33]

[b] These values apply to halogens attached directly to the ring and also to halogens attached to aliphatic double-bonded C atoms.

[c] Based on very limited data. Limits shown are roughly 95% confidence; in many cases, values are for information only and not to be used for computation.

[d] Includes unpublished infrared data.

[e] Use formula in $\Delta V \delta_P^2$ column to calculate, with V for total compound.

[f] Twin and triplet values apply to halogens on the same C atom, except that $\Delta V \delta_P^2$ also includes those on adjacent C atoms.

From Hansen, C. M., *Paint Testing Manual,* Manual 17, Koleske, J. V., Ed., American Society for Testing and Materials, Philadelphia, 1995, 388. Copyright ASTM. Reprinted with permission.

TABLE 1.2
Lydersen Group Constants

Group	Aliphatic, D*T*	Cyclic, D*T*	D_T^P	Aliphatic, D*P*	Cyclic, D*P*
CH_3	0.020	—	0.0226	0.227	—
CH_2	0.020	0.013	0.0200	0.227	0.184
>CH–	0.012	0.012	0.0131	0.210	0.192
>C<	0.000	–0.007	0.0040	0.210	0.154
$=CH_2$	0.018	—	0.0192	0.198	—
=CH–	0.018	0.011	0.0184	0.198	0.154
=C<	0.000	0.011	0.0129	0.198	0.154
=CH aromatic	—	—	0.0178	—	—
=CH aromatic	—	—	0.0149	—	—
–O–	0.021	0.014	0.0175	0.16	0.12
>O epoxide	—	—	0.0267	—	—
–COO–	0.047	—	0.0497	0.47	—
>C=O	0.040	0.033	0.0400	0.29	0.02
–CHO	0.048	—	0.0445	0.33	—
$–CO_2O$	—	—	0.0863	—	—
–OH→	—	—	0.0343	0.06	—
–H→	—	—	–0.0077	—	—
–OH primary	0.082	—	0.0493	—	—
–OH secondary	—	—	0.0440	—	—
–OH tertiary	—	—	0.0593	—	—
–OH phenolic	0.035	—	0.0060	–0.02	
$–NH_2$	0.031	—	0.0345	0.095	—
–NH–	0.031	0.024	0.0274	0.135	0.09
>N–	0.014	0.007	0.0093	0.17	0.13
–C≡N	0.060	—	0.0539	0.36	—
–NCO	—	—	0.0539	—	—
HCON<	—	—	0.0546	—	—
–CONH–	—	—	0.0843	—	—
–CON<	—	—	0.0729	—	—
$–CONH_2$	—	—	0.0897	—	—
–OCONH–	—	—	0.0938	—	—
–S–	0.015	0.008	0.0318	0.27	0.24
–SH	0.015	—	—	—	—
–Cl 1°	0.017	—	0.0311	0.320	—
–Cl 2°	—	—	0.0317	—	—
Cl_2 twin	—	—	0.0521	—	—
Cl aromatic	—	—	0.0245	—	—
–Br	0.010	—	0.0392	0.50	—
–Br aromatic	—	—	0.0313	—	—
–F	0.018	—	0.006	0.224	—
–I	0.012	—	—	0.83	—

TABLE 1.2 (continued)
Lydersen Group Constants

Group	Aliphatic, D*T*	Cyclic, D*T*	D_T^P	Aliphatic, D*P*	Cyclic, D*P*
Conjugation	—	—	0.0035	—	—
cis double bond	—	—	–0.0010	—	—
trans double bond	—	—	–0.0020	—	—
4 Member ring	—	—	0.0118	—	—
5 Member ring	—	—	0.003	—	—
6 Member ring	—	—	–0.0035	—	—
7 Member ring	—	—	0.0069	—	—
Ortho	—	—	0.0015	—	—
Meta	—	—	0.0010	—	—
Para	—	—	0.0060	—	—
Bicycloheptyl	—	—	0.0034	—	—
Tricyclodecane	—	—	0.0095	—	—

From Hansen, C. M., *Paint Testing Manual,* Manual 17, Koleske, J. V., Ed., American Society for Testing and Materials, Philadelphia, 1995, 391. Copyright ASTM. Reprinted with permission.

A sizable number of materials have been assigned Hansen parameters using the procedures described here. Many of these have not been published. Exxon Chemical Corporation[36,37] has indicated a computer program with data for over 500 solvents and plasticizers, 450 resins and polymers, and 500 pesticides. The author's files contain the three parameters for about 850 liquids, although several of them appear with two or three sets of possible values awaiting experimental confirmation. In some cases, this is due to questionable physical data, for example, for latent heats of vaporization, or wide variations in reported dipole moments. Another reason for this is that some liquids are chameleonic,[38] as defined by Hoy in that they adopt configurations depending on their environment. Hoy[38] cites the formation of cyclic structures for glycol ethers with (nominally) linear structure. The formation of hydrogen bonded tetramers of alcohols in a fluoropolymer has also been pointed out.[39] The term "compound formation" can be found in the older literature, particularly where mixtures with water were involved, and structured species were postulated to explain phenomena based on specific interactions among the components of the mixtures. Barton has discussed some of the situations where cohesion parameters need more careful use and points out that Hildebrand or Hansen parameters must be used with particular caution where the extent of donor–acceptor interactions, and in particular hydrogen bonding within a compound, is very different from that between compounds.[18] Amines, for example, are known to associate with each other. Pure component data cannot be expected to predict the behavior in such cases.

Still another reason for difficulties is the large variation of dipole moments reported for the same liquid. The dipole moment for some liquids depends on their environment, as discussed later. A given solvent can be listed with different values in files to keep these phenomena in mind.

Large data sources greatly enhance searching for similar materials and locating new solvents for a polymer based on limited data, for example. Unfortunately, different authors have used different group contribution techniques, and there is a proliferation of different "Hansen" parameters for the same chemicals in the literature. This would seem to be an unfortunate situation, but may ultimately provide benefits. In particular, partial solubility parameter values found in Hoy's extensive tables[9,40] are not compatible with the customary Hansen parameters reported here. Hoy has provided an excellent source of total solubility parameters. He independently arrived at the same type division of cohesion energies as Hansen, although the methods of calculation are quite different.

Many solvent suppliers have also presented tables of solvent properties and/or use computer techniques using these in their technical service. Partial solubility parameters not taken directly from earlier well-documented sources should be used with caution. In particular, it can be noted that the Hoy dispersion parameter is consistently lower than that found by Hansen. Hoy subtracts estimated values of the polar and hydrogen bonding energies from the total energy to find the dispersion energy. This allows for more calculational error and underestimates the dispersion energy, since the Hoy procedure does not appear to fully separate the polar and hydrogen bonding energies. The van Krevelen dispersion parameters appear to be too low. The author has not attempted these calculations, being completely dedicated to the full procedure based on corresponding states described here, but values estimated independently using the van Krevelen dispersion parameters are clearly low. A comparison with related compounds, or similarity principle, gives better results than those found from the van Krevelen dispersion group contributions.

In the following, calculational procedures and experience are presented according to the procedures found most reliable for the experimental and/or physical data available for a given liquid.

CALCULATION OF THE DISPERSION SOLUBILITY PARAMETER, d_D

The δ_D parameter is calculated according to the procedures outlined by Blanks and Prausnitz.[12] Figures 1.1, 1.2, or 1.3 can be used to find this parameter, depending on whether the molecule of interest is aliphatic, cycloaliphatic, or aromatic. These figures have been inspired by Barton,[9] who converted earlier data to Standard International (SI) units. All three of these figures have been straight-line extrapolated into a higher range of molar volumes than that reported by Barton. Energies found with these extrapolations have also provided consistent results. As noted earlier, the solubility parameters in SI units, $MPa^{½}$, are 2.0455 times larger than those in the older cgs (centimeter gram second) system, $(cal/cc)^{½}$, which still finds extensive use in the U.S., for example.

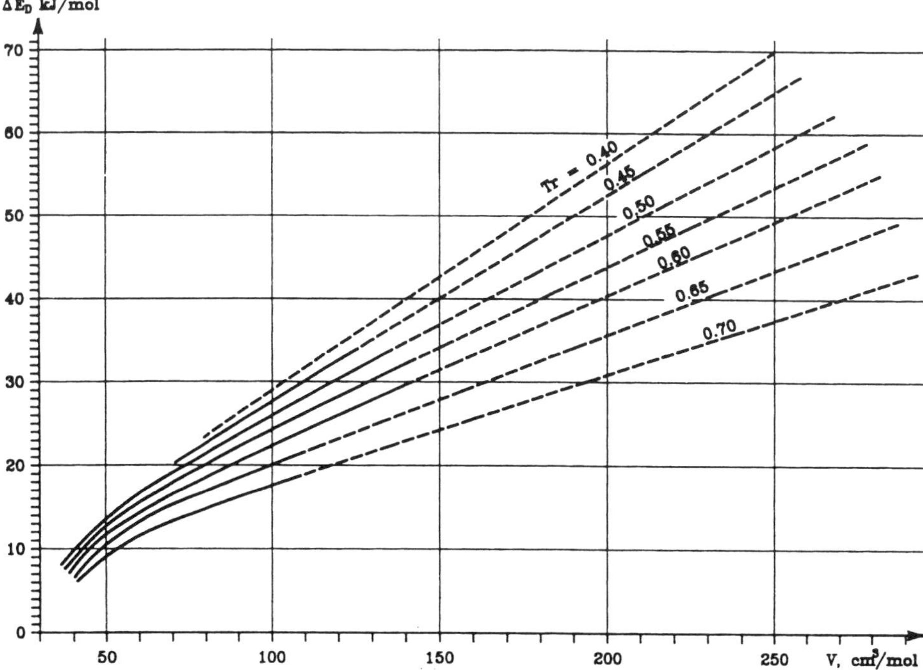

FIGURE 1.1 Energy of vaporization for straight chain hydrocarbons as a function of molar volume and reduced temperature.[34] (From Hansen, C. M., *Paint Testing Manual,* Manual 17, Koleske, J. V., Ed., American Society for Testing and Materials, Philadelphia, 1995, 389. Copyright ASTM. Reprinted with permission.)

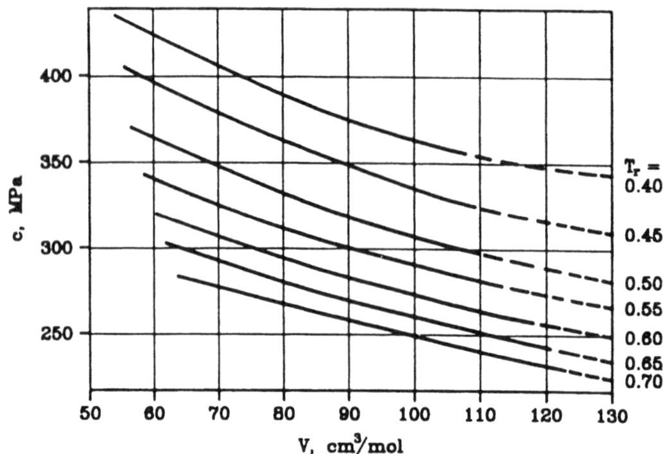

FIGURE 1.2 Cohesive energy density for cycloalkanes as a function of molar volume and reduced temperature.[34] (From Hansen, C. M., *Paint Testing Manual,* Manual 17, Koleske, J. V., Ed., American Society for Testing and Materials, Philadelphia, 1995, 389. Copyright ASTM. Reprinted with permission.)

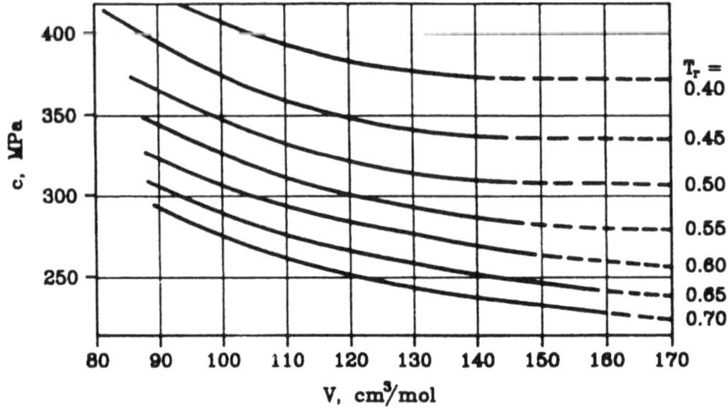

FIGURE 1.3 Cohesive energy density for aromatic hydrocarbons as a function of molar volume and reduced temperature.[34] (From Hansen, C. M., *Paint Testing Manual,* Manual 17, Koleske, J. V., Ed., American Society for Testing and Materials, Philadelphia, 1995, 389. Copyright ASTM. Reprinted with permission.)

The figure for the aliphatic liquids gives the dispersion cohesive energy, E_D, whereas the other two figures directly report the dispersion cohesive energy density, c. The latter is much simpler to use since one need only take the square root of the value found from the figure to find the respective partial solubility parameter. Barton also presented a similar figure for the aliphatic solvents, but it is inconsistent with the energy figure and in error. Its use is not recommended. When substituted cycloaliphatics or substituted aromatics are considered, simultaneous consideration of the two separate parts of the molecules is required. The dispersion energies are evaluated for each of the types of molecules involved, and a weighted average for the molecule of interest based on numbers of significant atoms is taken. For example, hexyl benzene would be the arithmetic average of the dispersion energies for an aliphatic and an aromatic liquid, each with the given molar volume of hexyl benzene. Liquids such as chlorobenzene, toluene, and ring compounds with alkyl substitutions with only two or three carbon atoms have been considered as cyclic compounds only. Such weighting has been found necessary to satisfy Equation 1.6.

The critical temperature, T_c, is required to use the dispersion energy figures. If the critical temperature cannot be found, it must be estimated. A table of the Lydersen group contributions,[41]

Δ_T, as given by Hoy[40] for calculation of the critical temperature is included as Table 1.2. In some cases, the desired groups may not be in the table, which means some educated guessing is required. The end result does not appear too sensitive to these situations. The normal boiling temperature, T_b, is also required in this calculation. This is not always available either and must be estimated by similarity, group contribution, or some other technique. The Lydersen group contribution method involves the use of Equations 1.11 and 1.12.

$$T_b/T_c = 0.567 + \Sigma\Delta_T - (\Sigma\Delta_T)^2 \tag{1.11}$$

and

$$T_r = T/T_c \tag{1.12}$$

where T has been taken as 298.15 K.

The dispersion parameter is based on atomic forces. The size of the atom is important. It has been found that corrections are required for atoms significantly larger than carbon, such as chlorine, sulfur, bromine, etc., but not for oxygen or nitrogen which have a similar size. The carbon atom in hydrocarbons is the basis of the dispersion parameter in its present form. These corrections are applied by first finding the dispersion cohesive energy from the appropriate figure. This requires multiplication by the molar volume for the cyclic compounds using data from the figures here, since these figures give the cohesive energy densities. The dispersion cohesive energy is then increased by adding on the correction factor. This correction factor for chlorine, bromine, and sulfur has been taken as 1650 J/mol for each of these atoms in the molecule. Dividing by the molar volume and then taking the square root gives the (large atom corrected) dispersion solubility parameter.

The need for these corrections has been confirmed many times, both for interpretation of experimental data and to allow Equations 1.6 to 1.8 to balance. Research is definitely needed in this area. The impact of these corrections is, of course, larger for the smaller molecular species. The taking of square roots of the larger numbers involved with the larger molecular species reduces the errors involved in these cases, since the corrections themselves are relatively small.

It can be seen from the dispersion parameters of the cyclic compounds that the ring also has an effect similar to increasing the effective size of the interacting species. The dispersion energies are larger for cycloaliphatic compounds than for their aliphatic counterparts, and they are higher for aromatic compounds than for the corresponding cycloaliphatics. Similar effects also appear with the ester group. This group appears to act as if it were, in effect, an entity which is larger than the corresponding compound containing only carbon (i.e., its homomorph), and it has a higher dispersion solubility parameter without any special need for corrections.

The careful evaluation of the dispersion cohesive energy may not have a major impact on the value of the dispersion solubility parameter itself because of the taking of square roots of rather large numbers. Larger problems arise because of Equation 1.6. Energy assigned to the dispersion portion cannot be reused when finding the other partial parameters using Equation 1.6 (or Equation 1.8). This is one reason group contributions are recommended in some cases as discussed below.

CALCULATION OF THE POLAR SOLUBILITY PARAMETER, d_P

The earliest assignments of a "polar" solubility parameter were given by Blanks and Prausnitz.[12] These parameters were, in fact, the combined polar and hydrogen bonding parameters as used by Hansen, and they cannot be considered polar in the current context. The first Hansen polar parameters[13] were reassigned new values by Hansen and Skaarup according to the Böttcher equation.[15] This equation requires the molar volume, the dipole moment (DM), the refractive index, and the dielectric constant. These are not available for many compounds, and the calculation is somewhat more difficult than using the much simpler equation developed by Hansen and Beerbower:[17]

$$\delta_P = 37.4(DM)/V^{1/2} \tag{1.13}$$

The constant 37.4 gives this parameter in SI units.

Equation 1.13 has been consistently used by the author over the past years, particularly in view of its reported reliability.[35] This reported reliability appears to be correct. The molar volume must be known or estimated in one way or another. This leaves only the dipole moment to be found or estimated. Standard reference works have tables of dipole moments, with the most extensive listing still being McClellan.[42] Other data sources also have this parameter as well as other relevant parameters, and data such as latent heats and critical temperatures. The so-called DIPPR[43] database has been found useful for many compounds of reasonably common usage, but many interesting compounds are not included in the DIPPR. When no dipole moment is available, similarity with other compounds, group contributions, or experimental data can be used to estimate the polar solubility parameter.

It must be noted that the fact of zero dipole moment in symmetrical molecules is not basis enough to assign a zero polar solubility parameter. An outstanding example of variations of this kind can be found with carbon disulfide. The reported dipole moments are mostly 0 for gas phase measurements, supplemented by 0.08 in hexane, 0.4 in carbon tetrachloride, 0.49 in chlorobenzene, and 1.21 in nitrobenzene. There is a clear increase with increasing solubility parameter of the media. The latter and highest value has been found experimentally most fitting for correlating permeation through a fluoropolymer film used for chemical protective clothing.[44] Many fluoropolymers have considerable polarity. The lower dipole moments seem to fit in other instances. Diethyl ether has also presented problems as an outlier in terms of dissolving or not, and rapid permeation or not. Here, the reported dipole moments[42] vary from 0.74 to 2.0 with a preferred value of 1.17, and with 1.79 in chloroform. Choosing a given value seems rather arbitrary. The chameleonic cyclic forms of the linear glycol ethers would also seem to provide for a basis of altered dipole moments in various media.[38]

When Equation 1.13 cannot be used, the polar solubility parameter has been found using the Beerbower table of group contributions, by similarity to related compounds and/or by subtraction of the dispersion and hydrogen bonding cohesive energies from the total cohesive energy. The question in each case is, "Which data are available and judged most reliable?" New group contributions can also be developed from related compounds where their dipole moments are available. These new polar group contributions then become supplementary to the Beerbower table.

For large molecules, and especially those with long hydrocarbon chains, the accurate calculation of the relatively small polar (and hydrogen bonding) contributions present special difficulties. The latent heats are not generally available with sufficient accuracy to allow subtraction of two large numbers from each other to find a very small one. In such cases, the similarity and group contribution methods are thought best. Unfortunately, latent heats found in a widely used handbook[45] are not clearly reported as to the reference temperature. There is an indication that these are 25°C data, but checking indicated many of the data were identical with boiling point data reported elsewhere in the literature. Subsequent editions of this handbook[46] have a completely different section for the latent heat of evaporation. Again, even moderate variations in reported heats of vaporization can cause severe problems in calculating the polar (or hydrogen bonding) parameter when Equations 1.6 or 1.8 are strictly adhered to.

CALCULATION OF THE HYDROGEN BONDING SOLUBILITY PARAMETER, d_H

In the earliest work, the hydrogen bonding parameter was almost always found from the subtraction of the polar and dispersion energies of vaporization from the total energy of vaporization. This is still widely used where the required data are available and reliable. At this stage, however, the group contribution techniques are considered reasonably reliable for most of the required calculations and, in fact, more reliable than estimating several of the other parameters to ultimately arrive at the subtraction step just mentioned. Therefore, in the absence of reliable latent heat and dipole moment data, group contributions are judged to be the best alternative. Similarity to related compounds can also be used, of course, and the result of such a procedure should be essentially the same as for using group contributions.

SUPPLEMENTARY CALCULATIONS AND PROCEDURES

The procedures listed previously are those most frequently used by the author in calculating the three partial solubility parameters for liquids where some data are available. There are a number of other calculations and procedures which are also helpful. Latent heat data at 25°C have been found consistently from latent heats at another temperature using the relation given by Fishtine.[47]

$$\Delta H_v(T_1)/\Delta H_v(T_2) = \left[(1-T_{r1})/(1-T_{r2})\right]^{0.38} \tag{1.14}$$

This is done even if the melting point of the compound being considered is higher than 25°C. The result is consistent with all the other parameters, and to date no problems with particularly faulty predictions have been noted in this respect, i.e., it appears as if the predictions are not significantly in error when experimental data have been available for checking. When the latent heat at the boiling point is given in cal/mol, Equation 1.14 is used to estimate the latent heat at 25°C. RT equal to 592 cal/mol is then subtracted from this according to Equation 1.15 to find the total cohesion energy, E, in cgs units at this temperature:

$$E = \Delta E_v = \Delta H_v - RT \tag{1.15}$$

A computer program has been developed by the author to assign HSP to solvents based on experimental data alone. This has been used in several cases where the parameters for the given liquids were desired with a high degree of accuracy. The procedure is to enter solvent quality, good or bad, into the program for a reasonably large number of polymers where the solubility parameters and appropriate radius of interaction for the polymers are known. The program then locates that set of δ_D, δ_P, and δ_H parameters for the solvent which best satisfies the requirements of a location within the spheres of the appropriate polymers where solvent quality is good and outside of the appropriate spheres where it is bad.

An additional aid in estimating HSP for many compounds is that these parameters can be found by interpolation or extrapolation, especially for homologous series. The first member may not necessarily be a straight-line extrapolation, but comparisons with related compounds should always be made where possible to confirm assignments. Plotting the parameters for homologous series among the esters, nitroparaffins, ketones, alcohols, and glycol ethers has aided in finding the parameters for related compounds.

TEMPERATURE DEPENDENCE

Only very limited attempts have been made to calculate solubility parameters at a higher temperature. Solubility parameter correlations of phenomena at higher temperatures have generally been found satisfactory when the established 25°C parameters have been used. Recalculation to higher temperatures is possible, but has not been found necessary. In this direct but approximate approach, it is assumed that the parameters all demonstrate the same temperature dependence, which, of course, is not the case. It might be noted in this connection that the hydrogen bonding parameter, in particular, is the most sensitive to temperature. As the temperature is increased, more and more hydrogen bonds are progressively broken, and this parameter will decrease more rapidly than the others.

The gas phase dipole moment is not temperature dependent, although the volume of a fluid does change with temperature, which will change its cohesive energy density. The change of the δ_D, δ_P, and δ_H parameters for liquids with temperature, T, can be estimated by the following equations where α is the coefficient of thermal expansion:[17]

$$d\delta_D/dT = -1.25\alpha\delta_D \tag{1.16}$$

$$d\delta_P/dT = -0.5\alpha\delta_P \tag{1.17}$$

$$d\delta_H/dT = -\delta_H\left(1.22\times10^{-3} + 0.5_\alpha\right) \tag{1.18}$$

Higher temperature means a general increase in rate of solubility/diffusion/permeation, as well as larger solubility parameter spheres. δ_D, δ_P, and δ_H decrease with increased temperature, as can be seen by a comparison of Equations 1.16, 1.17, and 1.18. This means that alcohols, phenols, glycols, and glycol ethers become better solvents for polymers of lower solubility parameters as the temperature increases. Thus, increasing the temperature can cause a nonsolvent to become a good solvent, a fact which is often noted in practice. As mentioned earlier, it is possible that a boundary solvent can be a good solvent at a given temperature, but become bad with either an increase in temperature or with a decrease in temperature. These phenomena are discussed in great detail by Patterson and co-workers.[3,4] They can be explained either by the change in solubility parameter with temperature or more completely by the Prigogine corresponding states theory (CST). The effects of temperature changes on solubility relations discussed here are most obvious with systems having high hydrogen bonding character. Examples are given in the next section for some special situations involving water and methanol.

Some Special Effects Temperature Changes

Water (and methanol) uptake in most polymers increases with increasing temperature. This is because the solubility parameters of the water and polymer are closer at higher temperatures. The δ_H parameter of water (and methanol) falls with increasing temperature, while that of most polymers remains reasonably constant. Water is also well known as an exceptionally good plasticizer because of its small molecular size. The presence of dissolved water not only softens (reduces the glass transition temperature) a polymer as such, but it also means diffusion rates of other species will be increased. The presence of water in a film can also influence the uptake of other materials, such as during solubility parameter studies or resistance testing, with hydrophilic materials being more prone to enter the film than when the extra water is not present.

This can cause blistering on rapid cooling as discussed in Chapter 7 and in Reference 48 (see Chapters 6 and 7: Figure 6.3 shows how rapid cooling from a water-saturated state at higher temperature can lead to blistering; Figures 7.3 and 7.4 show how this effect can be measured experimentally as an increase in water content above the equilibrium value when temperature cycling is encountered). This leads to premature failure of polymeric products used in such environments.

A related problem has been encountered with methanol. It was intended to follow the rate of uptake of methanol in an epoxy coating at room temperature by weighing coated metal panels periodically on an analytical balance. Blistering was encountered in the coating near the air surface shortly after the experiment was started. The methanol which had absorbed into the coating near the surface became insoluble as the temperature of the coating near the surface was lowered by the evaporation of excess methanol during the handling and weighing of the panels. This is a rather extreme case, and, as mentioned earlier, use of the HSP determined at 25°C at elevated temperatures can most often be done without too much trouble from a practical point of view. One should be aware that the changes in the δ_H parameter will be larger than those in the other parameters, and this effect will be most significant for those liquids with larger δ_H values.

Effects of Solvent Molecular Size

The size of both solvent and solute molecules is important for solubility, permeation, diffusion, and chemical resistance phenomena. Smaller molecules tend to be more readily soluble than larger ones. As stated previously, the Hildebrand solubility parameter theory also points to smaller molar volume solvents as being better than those with larger molar volumes, even though they may have identical solubility parameters.[1,2] This fact of expected improved solvency for smaller molecules is also known from the Flory-Huggins theory of polymer solutions.[29] Smaller molecular solvents have also been regularly noted as being superior to those with larger molecular size when highly crystalline polymers or solids are being tested for solubility. So it is not surprising that solvent molecular size can be an important fourth parameter in solubility and, in some cases, in chemical resistance. Specific examples are given in Chapters 3 and 7.

The size and shape of the solvent molecule are also very important for kinetic phenomena such as diffusion, permeation, and attainment of equilibrium. Smaller and more linear molecules diffuse more rapidly than larger and more bulky ones. The diffusion coefficient may be so low that equilibrium is not attained for hundreds of years at room temperature in common solvent exposures of rigid polymers like polyphenylene sulfide (PPS) with thicknesses of several millimeters.[49] Likewise, the second stage in the two-stage drying process in polymer film formation by solvent evaporation can last for many years.[16,50] Polymer samples used for solubility parameter or other testing may well retain solvent or monomer for many years, and this may affect the evaluations.

Attempts to include the molecular volume into a new composite solubility parameter and size parameter have not been particularly successful.[20,21] This may be because the size effect is most often not caused through the thermodynamic considerations on which the solubility parameters are based, but rather through a kinetic effect of diffusion rates or other free volume consideration. The similarities in the HSP approach and the Prigogine theory discussed in Chapter 2 indicate a remarkably close, if not identical, relation between the Prigogine ρ (segment size parameter) and the δ_D parameter, suggesting that molecular size differences are at least partially accounted for in the δ_D parameter. The Prigogine theory also has a parameter to describe "structural effects," including size of polymer molecules, but this has not been explored in relation to the present discussion. The increase of δ_D with increasing molecular size among the aliphatic hydrocarbons, the higher δ_D values for the larger "units" represented by cycloaliphatic and aromatic rings, and the need for corrections for larger atoms discussed earlier all tend to support this.

Sorting output data according to the molecular volume of the test solvents in a computer analysis helps to discover whether solvent molecular size is indeed an additional significant factor in a given correlation or testing program.

Computer Programs

The author has used two computer programs extensively in his own studies and in collecting material for this book. These are called SPHERE and SPHERE1. They are very similar, the only difference being that SPHERE optimizes the polymer (or other material, of course) parameters based on all the data, whereas SPHERE1 considers data for those solvents considered as "good" only. It neglects the nonsolvent data. SPHERE1 has been most useful in correlations with pigments, fillers, and fibers, as described in Chapter 5.

The data input is by solvent number followed by an indication of the quality of interaction with that solvent. A "1" indicates a "good" solvent, while a "0" is used for a "bad" solvent. What is considered good or bad varies according to the level of interaction being studied. This can be solution or not, a given percentage of swelling or uptake, breakthrough time being less than a given interval, permeation coefficients higher than a given value, long-time suspension of a pigment, etc.

The program systematically evaluates the input data using a quality of fit function called the "Desirability Function."[51] This suggestion was made by a reputed statistician many years ago as the most appropriate statistical treatment for this type of problem. It has been in use since the late 1960s. The function has the form:

$$\text{DATA FIT} = \left(A_1 * A_2 * \ldots A_n\right)^{1/n} \tag{1.19}$$

where n is the number of solvents for which there is experimental data in the correlation. The DATA FIT approaches 1.0 as the fit improves during an optimization and reaches 1.0 when all the good solvents are included within the sphere and all the bad ones are outside of it.

$$A_i = e^{-(\text{ERROR DISTANCE})} \tag{1.20}$$

The A_i for a given good solvent within the sphere and for a given bad solvent outside the sphere will be 1.0. The error distance is the distance to the sphere boundary for the solvent in error either as being good and outside the sphere or being bad and inside the sphere.

Ro is the radius of the sphere, and Ra is the distance from a given solvent point to the center of the sphere. For a good solvent outside the sphere, an error enters the DATA FIT according to

$$A_i = e^{+(Ro-Ra)} \tag{1.21}$$

Such errors are often found for solvents having low molecular volumes.

For a bad solvent inside the sphere, the contribution to the DATA FIT is

$$A_i = e^{+(Ra-Ro)} \tag{1.22}$$

Such errors can sometimes be found for larger molecular species such as plasticizers. This is not unexpected for the reasons mentioned earlier.

The solvents with large and/or small molecules which give the "errors" can sometimes be (temporarily) disregarded by generating a new correlation giving an excellent DATA FIT for an abbreviated range of molecular volumes. There is a special printout with the solvents arranged in order of molecular volume which helps to analyze such situations. The computer printouts all include a column for the RED number.

The program assumes a starting point based on an average for each of the HSP for the good solvents only. The program then evaluates eight points at the corners of a cube, with the current best values as center. Different radii are evaluated at each of these points in the optimization process. When better fits are found among the eight points, the point with the best fit is taken as a new center and eight points around it are evaluated in a similar manner. This continues until the DATA FIT cannot be improved upon. The length of the edge of the cube is then reduced in size to fine tune the fit. The initial length of the cube is 1 unit, which is reduced to 0.3 units, and finally to 0.1 units in the final optimization step.

Experimental data for the solvents are entered with solvent number (comma) and a 1 for a good solvent or a 0 for a bad one.

Errors in the correlations are indicated with an "*" in the SOLUB column where the experimental input data are indicated. As stated above, systematic errors can sometimes be seen in the molar volume

TABLE 1.3
HSP Correlations Related to Water

Material	δ_D	δ_P	δ_H	Ro	FIT	G/T
Water — Single molecule	15.5	16.0	42.3	—	—	—
Water — >1% soluble in	15.1	20.4	16.5	18.1	0.856	88/167
Water — Total miscibility	18.1	17.1	16.9	13.0	0.880	47/166
Water — Total miscibility "1"	18.1	12.9	15.5	13.9	1.000	47

printout. This may suggest a new analysis of the data. Nonsystematic errors may be real, such as for reactions or some extraneous effect not predictable by the solubility parameter. They may also be bad data, and rechecking data indicated with an "*" in the output has become a routine practice. The output of this program is for the least radius allowing the maximum DATA FIT.

Results from the SPHERE program reported in this book generally include the HSP, given as D, P, and H, respectively, and Ro for the correlation in question, as well as the DATA FIT, the number of good solvents (G), and the total solvents (T) in the correlation. This latter information has not always been recorded and may be lacking for some correlations, especially the older ones.

HANSEN SOLUBILITY PARAMETERS FOR WATER

Water is such an important material that a special section is dedicated to its HSP at this point. The behavior of water often depends on its local environment, which makes general predictions very difficult. Water is still so unpredictable that its use as a test solvent in solubility parameter studies is not recommended. This is true of water as a pure liquid or water in mixtures. Table 1.3 includes data from various HSP analyses of the behavior of water. The first set of data is derived from the energy of vaporization of water at 25°C. The second set of data is based on a correlation of the solubility of various solvents in water, where good solvents are soluble to more than 1% in water and bad ones dissolve to a lesser extent. The third set of data is for a correlation of total miscibility of the given solvents with water. The second and third entries in Table 1.3 are based on the SPHERE program where both good and bad solvents affect the DATA FIT and hence the result of the optimization. The last entry in Table 1.3 is for an analysis using the SPHERE1 program. The HSP data are for the minimum sphere which encompasses the good solvents and only the good solvents. The bad solvents are simply not considered in the data processing. This type of comparison usually results in some of the parameters being lower than when all the data are included. One also has the problem that a considerable portion of the sphere found by the correlation covers such high energies that no liquids have such high solubility parameters. The cohesion energy is so high as to require solids. The constant "4" in the correlations (Equation 1.9) is still used for these correlations, primarily based on successes at lower levels of cohesion energies, but this is also supported by the comparison with the Prigogine corresponding states theory of polymer solutions discussed at some length in Chapter 2.

The HSP for water as a single molecule based on the latent heat at 25°C are sometimes used in connection with mixtures with water to estimate average HSP. More recently, it has been found in a study involving water, ethanol, and 1,2-propanediol that the HSP for water indicated by the total water solubility correlation could be used to explain the behavior of the mixtures involved. The averaged values are very questionable since water can associate and water has a very small molar volume as a single molecule. It almost appears to have a dual character. The data for the 1% correlation[52] as well as for the total water miscibility suggest that about six water molecules associate into units.

CONCLUSION

This chapter has been dedicated to describing the tools with which different HSP characterizations can be made and some of the pitfalls which may be encountered in the process. The justification for the tools is further confirmed in Chapter 2, and their use is demonstrated in all the subsequent chapters. Figure 1.4 is included to show where many common solvents are located on a δ_P vs. δ_H plot.

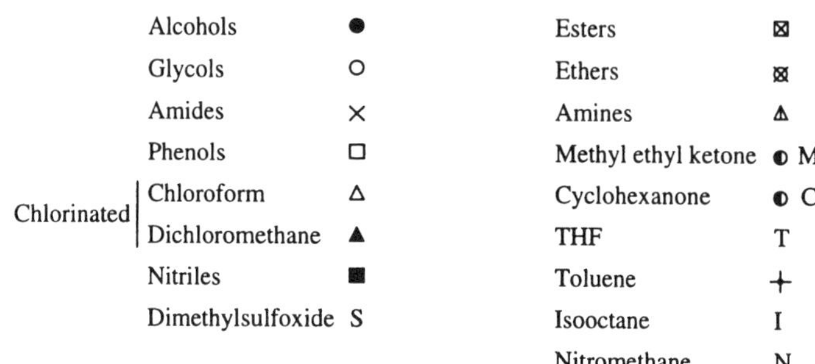

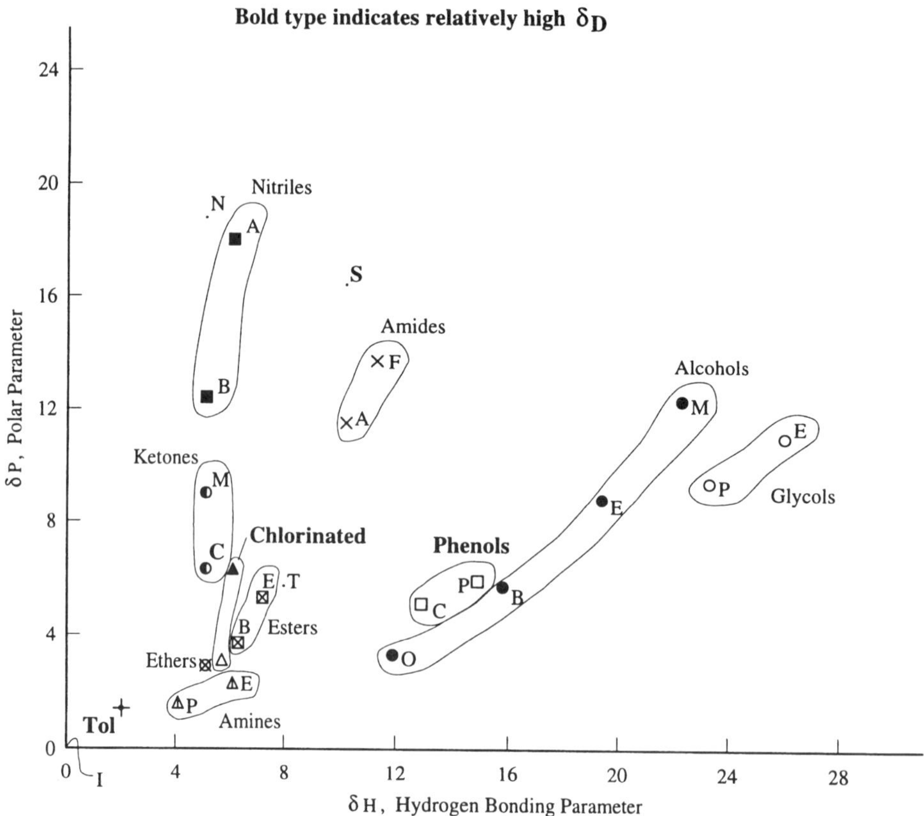

FIGURE 1.4 δ_P vs. δ_H plot showing the location of various common solvents. The glycols are ethylene glycol and propylene glycol. The alcohols include methanol (M), ethanol (E), 1-butanol (B), and 1-octanol (O). The amides include dimethyl formamide (F) and dimethyl acetamide (D). The nitriles are acetonitrile (A) and butyronitrile (B). The esters are ethyl acetate (E) and *n*-butyl acetate (B). The amines are ethyl amine (E) and propyl amine (P). The phenols are phenol (P) and *m*-cresol (C). The ethers are symbolized by diethyl ether.

REFERENCES

1. Hildebrand, J. and Scott, R. L., *The Solubility of Nonelectrolytes,* 3rd Ed., Reinhold, New York, 1950.
2. Hildebrand, J. and Scott, R. L., *Regular Solutions,* Prentice-Hall Inc., Englewood Cliffs, NJ, 1962.
3. Patterson, D., and Delmas, G., New Aspects of Polymer Solution Thermodynamics, *Off. Dig. Fed. Soc. Paint Technol.,* 34(450), 677–692, 1962.
4. Delmas, D., Patterson, D., and Somcynsky, T., Thermodynamics of Polyisobutylene-n-Alkane Systems, *J. Polym. Sci.,* 57, 79–98, 1962.
5. Bhattacharyya, S. N., Patterson, D., and Somcynsky, T., The Principle of Corresponding States and the Excess Functions of n-Alkane Mixtures, *Physica,* 30, 1276–1292, 1964.
6. Patterson, D., Role of Free Volume Changes in Polymer Solution Thermodynamics, *J. Polym. Sci. Part C,* 16, 3379–3389, 1968.
7. Patterson, D. D., Introduction to Thermodynamics of Polymer Solubility, *J. Paint Technol.,* 41(536), 489–493, 1969.
8. Biros, J., Zeman, L., and Patterson, D., Prediction of the X Parameter by the Solubility Parameter and Corresponding States Theories, *Macromolecules,* 4(1), 30–35, 1971.
9. Barton, A. F. M., *Handbook of Solubility Parameters and Other Cohesion Parameters*, CRC Press, Boca Raton, FL, 1983.
10. Gardon, J. L. and Teas, J. P., Solubility Parameters, in *Treatise on Coatings, Vol. 2, Characterization of Coatings: Physical Techniques, Part II,* Myers, R. R. and Long, J. S., Eds., Marcel Dekker, New York, 1976, chap. 8.
11. Burrell, H., Solubility Parameters for Film Formers, *Off. Dig. Fed. Soc. Paint Technol.,* 27(369), 726–758, 1972; Burrell, H., A Solvent Formulating Chart, *Off. Dig. Fed. Soc. Paint Technol.,* 29(394), 1159–1173, 1957; Burrell, H., The Use of the Solubility Parameter Concept in the United States, *VI Federation d'Associations de Techniciens des Industries des Peintures, Vernis, Emaux et Encres d'Imprimerie de l'Europe Continentale, Congress Book*, 21–30, 1962..
12. Blanks, R. F. and Prausnitz, J. M., Thermodynamics of Polymer Solubility in Polar and Nonpolar Systems, *Ind. Eng. Chem. Fundam.,* 3(1), 1–8, 1964.
13. Hansen, C. M., The Three Dimensional Solubility Parameter — Key to Paint Component Affinities I, *J. Paint Technol.,* 39(505), 104–117, 1967.
14. Hansen, C. M., The Three Dimensional Solubility Parameter — Key to Paint Component Affinities II, *J. Paint Technol.,* 39(511), 505–510, 1967.
15. Hansen, C. M. and Skaarup, K., The Three Dimensional Solubility Parameter — Key to Paint Component Affinities III, *J. Paint Technol.,* 39(511), 511–514, 1967.
16. Hansen, C. M., The Three Dimensional Solubility Parameter and Solvent Diffusion Coefficient, Doctoral Dissertation, Danish Technical Press, Copenhagen, 1967.
17. Hansen, C. M. and Beerbower, A., Solubility Parameters, in *Kirk-Othmer Encyclopedia of Chemical Technology,* Suppl. Vol., 2nd ed., Standen, A., Ed., Interscience, New York, 1971, 889–910.
18. Barton, A. F. M., Applications of Solubility Parameters and Other Cohesion Energy Parameters, *Polym. Sci. Technol. Pure Appl. Chem.,* 57(7), 905–912, 1985.
19. Sørensen, P., Application of the Acid/Base Concept Describing the Interaction Between Pigments, Binders, and Solvents, *J. Paint Technol.,* 47(602), 31–39, 1975.
20. Van Dyk, J. W., Paper presented at the Fourth Chemical Congress of America, New York, August 25–30, 1991.
21. Anonymous (Note: This was, in fact, Van Dyk, J. W., but this does not appear on the bulletin), Using Dimethyl Sulfoxide (DMSO) in Industrial Formulations, Bulletin #102, Gaylord Chemical Corp., Slidell, LA, 1992.
22. Karger, B. L., Snyder, L. R., and Eon, C., Expanded Solubility Parameter Treatment for Classification and Use of Chromatographic Solvents and Adsorbents, *Anal. Chem.,* 50(14), 2126–2136, 1978.
23. Hansen, C. M. and Wallström, E., On the Use of Cohesion Parameters to Characterize Surfaces, *J. Adhes.,* 15, 275–286, 1983.
24. Hansen, C. M., Characterization of Surfaces by Spreading Liquids, *J. Paint Technol.,* 42(550), 660–664, 1970.
25. Hansen, C. M., Surface Dewetting and Coatings Performance, *J. Paint Technol.,* 44(570), 57–60, 1972.
26. Hansen, C. M. and Pierce, P. E., Surface Effects in Coatings Processes, *Ind. Eng. Chem. Prod. Res. Dev.,* 13(4), 218–225, 1974.

27. Hennissen, L., Systematic Modification of Filler/Fibre Surfaces to Achieve Maximum Compatibility with Matrix Polymers, Lecture for the Danish Society for Polymer Technology, Copenhagen, February 10, 1996.
28. Gardon, J. L., Critical Review of Concepts Common to Cohesive Energy Density, Surface Tension, Tensile Strength, Heat of Mixing, Interfacial Tension and Butt Joint Strength, *J. Colloid Interface Sci.,* 59(3), 582–596, 1977.
29. Flory, P. J., *Principles of Polymer Chemistry,* Cornell University Press, New York, 1953.
30. Prigogine, I. (with the collaboration of A. Bellemans and A. Mathot), *The Molecular Theory of Solutions*, North-Holland, Amsterdam, 1957, chaps. 16, 17.
31. van Krevelen, D. W. and Hoftyzer, P. J., *Properties of Polymers: Their Estimation and Correlation with Chemical Structure,* 2nd ed., Elsevier, Amsterdam, 1976.
32. Beerbower, A., Environmental Capability of Liquids, in *Interdisciplinary Approach to Liquid Lubricant Technology,* NASA Publication SP-318, 1973, 365–431.
33. Fedors, R. F., A Method for Estimating both the Solubility Parameters and Molar Volumes of Liquids, *Polym. Eng. Sci.,* 14(2), 147–154, 472, 1974.
34. Hansen, C. M., Solubility Parameters, in *Paint Testing Manual,* Manual 17, Koleske, J. V., Ed., American Society for Testing and Materials, Philadelphia, 1995, 383–404.
35. Koenhen, D. N. and Smolders, C. A., The Determination of Solubility Parameters of Solvents and Polymers by Means of Correlation with Other Physical Quantities, *J. Appl. Polym. Sci.,* 19, 1163–1179, 1975.
36. Anonymous, Brochure: Co-Act — A Dynamic Program for Solvent Selection, Exxon Chemical International Inc., 1989.
37. Dante, M. F., Bittar, A. D., and Caillault, J. J., Program Calculates Solvent Properties and Solubility Parameters, *Mod. Paint Coat.,* 79(9), 46–51, 1989.
38. Hoy, K. L., New Values of the Solubility Parameters from Vapor Pressure Data, *J. Paint Technol.,* 42(541), 76–118, 1970.
39. Myers, M. M. and Abu-Isa, I. A., Elastomer Solvent Interactions III — Effects of Methanol Mixtures on Fluorocarbon Elastomers, *J. Appl. Polym. Sci.,* 32, 3515–3539, 1986.
40. Hoy, K. L., *Tables of Solubility Parameters,* Union Carbide Corp., Research and Development Dept., South Charleston, WV, 1985 (1st ed. 1969).
41. Reid, R. C. and Sherwood, T. K., *Properties of Gases and Liquids,* McGraw-Hill, New York, 1958. (Lydersen Method — see also Reference 31).
42. McClellan, A. L., *Tables of Experimental Dipole Moments,* W. H. Freeman, San Francisco, 1963.
43. Tables of Physical and Thermodynamic Properties of Pure Compounds, American Institute of Chemical Engineers Design Institute for Physical Property Research, Project 801, Data Compilation, R. P. Danner and T. E. Daubert, Project Supervisors, DIPPR Data Compilation Project, Department of Chemical Engineering, Pennsylvania State University, University Park.
44. Hansen, C. M., Selection of Chemicals for Permeation Testing Based on New Solubility Parameter Models for Challenge 5100 and Challenge 5200, under contract DTCG50-89-P-0333 for the U.S. Coast Guard, June 1989, Danish Isotope Centre, Copenhagen.
45. Weast, R. C., Editor-in-Chief, *CRC Handbook of Chemistry and Physics,* 65 ed., CRC Press, Boca Raton, FL, 1988–1989, C-672–C-683.
46. Majer, V., Enthalpy of Vaporization of Organic Compounds, in *Handbook of Chemistry and Physics,* 72nd ed., Lide, D. R., Editor-in-Chief, CRC Press, Boca Raton, 1991–1992, 6-100–6-107.
47. Fishtine, S. H., Reliable Latent Heats of Vaporization, *Ind. Eng. Chem.,* 55(4), 20–28, 1963; ibid. (5), 55–60; ibid. (6), 47–56.
48. Hansen, C. M., New Developments in Corrosion and Blister Formation in Coatings, *Prog. Org. Coat.,* 26, 113–120, 1995.
49. Hansen, C. M., *Solvent Resistance of Polymer Composites — Glass Fibre Reinforced Polyphenylene Sulfide,* Centre for Polymer Composites (Denmark), Danish Technological Institute, Taastrup, 1993, 1–62, ISBN 87-7756-286-0.
50. Hansen, C. M., A Mathematical Description of Film Drying by Solvent Evaporation, *J. Oil Colour Chem. Assoc.,* 51(1), 27–43, 1968.
51. Harrington, E. C., Jr., The Desirability Function, *Ind. Qual. Control,* 21(10), 494–498, April 1965.
52. Hansen, C. M. and Andersen B. H., The Affinities of Organic Solvents in Biological Systems, *Am. Ind. Hyg. Assoc. J.,* 49(6), 301–308, 1988.

2 Theory — The Prigogine Corresponding States Theory, the χ_{12} Interaction Parameter, and the Hansen Solubility Parameters

CONTENTS

ABSTRACT

Patterson has shown that the χ_{12} interaction parameter can be estimated from the corresponding states theory of Prigogine. Correlations using Hansen solubility parameters (HSP) confirm the treatment of the cohesive energy difference term proposed in the Prigogine corresponding states theory. Therefore, the HSP approach can be expected to be useful to predict the Flory χ_{12} interaction coefficient. Equations for this purpose are presented and discussed based on comparisons of calculated and experimental values for five polymers. There is agreement in many cases, especially for essentially nonpolar systems, but full understanding of the interrelationship has not yet been achieved. The lack of accounting for permanent dipole–permanent dipole and hydrogen bonding (electron interchange) in the "New Flory" theory leading to χ_{12} is thought to be largely responsible for this.

It does appear, however, that the "4" (or ¼) in the HSP correlations and the ¼ in the leading term of the Prigogine theory have identical functions. They modify the specific interactions described by the Prigogine **d** and also the polar and hydrogen bonding HSP (δ_P and δ_H). These lead to the specific interactions. This could imply that the Prigogine ρ attempts to describe what the δ_D parameter describes, that is, the nondirectional (nonpolar) interactions.

Neither the Flory nor the Prigogine approaches can lead to the type of predictions possible with the HSP approach. The many correlations and other predictions contained in this book would not be possible with these theories since they do not separate the polar and hydrogen bonding effects independently. The Prigogine theory must be used with the geometric mean to estimate the interaction between different species. The Hildebrand and HSP approaches inherently use the geometric mean. This implies that the geometric mean is capable of describing not only dispersion interactions, but also those due to permanent dipoles and hydrogen bonding.

INTRODUCTION

The Flory-Huggins chi parameter, χ, has been used for many years in connection with polymer solution behavior,[1,2] with the χ_{12} parameter derived from the "New Flory Theory" being currently accepted for general use instead of the older χ. It would be desirable to relate the widely used HSP[3-10] more directly to χ_{12}. This would allow estimates of χ_{12} for systems where the HSP are known, but χ_{12} is not. The reverse is not possible since a single χ_{12} parameter cannot be used to divide the cohesion energy into contributions from dispersion (nonpolar) forces, permanent dipole–permanent dipole forces, and hydrogen bonding (electron interchange), which is the basis of the HSP. Reliable χ_{12} values for numerous solvents and the same polymer can be used to determine the HSP for the polymer in the same manner as solvency or swelling data are used for this purpose, however. In principle, the weighting schemes described in Chapter 3 to average solvent parameters to arrive at polymer HSP can also be used with the χ_{12} parameter, just as they are with weight gain or intrinsic viscosity.

Patterson[15] and co-workers[17] have shown how to predict the χ_{12} parameter using corresponding states theories (CST),[2,11-17] as well as using the Hildebrand solubility parameter in (strictly) nonpolar systems. They use the symbol ν^2 instead of χ_{12} for the same quantity. The Hildebrand solubility parameter is the square root of the cohesive energy density (ced).[18,19] Also, it has been shown recently that HSP and the Prigogine/Patterson CST are mutually confirming and give similar predictions.[20,21] This is discussed in more detail below.

The customary equation to calculate χ_{12} from the Hildebrand solubility parameters for a nonpolar solvent and a nonpolar polymer is

$$\chi_{12} = \left[V(\delta_1 - \delta_2)^2\right] / RT + \beta \tag{2.1}$$

where V is the molar volume of the solvent, δ is the Hildebrand solubility parameter for the solvent (1) and polymer (2), R is the gas constant, and T is the absolute temperature.

The empirical constant β has been discussed as being necessary for polymer systems[22] as a correction to the Flory combinatorial entropy. β, although combinatorial in origin, was attached to χ_{12} in order to preserve the Flory form of the chemical potential expression. β has a generally accepted average value near 0.34. Biros et al.[17] state that this value of β presents difficulties as an explanation of an error in the Flory combinatorial entropy approximation. These authors state that β should be interpreted as bringing the χ_{12} values found from the solubility parameters in line with those found from CST. The CST predict a χ about 0.3 units larger than that found when using the Hildebrand solubility parameter. β is not required for essentially nonpolar systems when HSP are used in a relation similar to Equation 2.1, as shown below.

The Hildebrand parameters are applicable to regular solutions, which in the current context implies strictly nonpolar systems. Hildebrand solubility parameters have been shown to reflect the noncombinatorial free energy directly via the first term in Equation 2.1[12,13] (see also Chapter 1, Equations 1.3 and 1.4). It was previously thought that the heat of mixing was given by the Hildebrand theory by $\phi_1\phi_2V_M(\delta_1 - \delta_2)^2$, where V_M is the volume of the mixture and the φs are volume fractions of the solvent and polymer. This is not true. The heat of mixing must be found by differentiating this relation as shown by Delmas and co-workers.[12,13] The work of Delmas, Patterson, and co-workers has shown that predictions with the nonpolar Hildebrand solubility parameter and the Prigogine CST are in excellent agreement with each other with regard to heats of mixing in essentially nonpolar systems. Both positive and negative heats of mixing are allowed, predicted, and found. The argument that solubility parameters are inadequate since they do not allow for negative heats of mixing is not valid. These studies also show that increases in temperature lead to improved solvency when a solvent has higher solubility parameters than the polymer. When the solvent has lower solubility parameters than the polymer, an increase in temperature leads to poorer solvent quality. Precipitation can even occur with increasing temperature at what is called the lower critical solution temperature. See also the discussion in Chapter 1.

HANSEN SOLUBILITY PARAMETERS (HSP)

It has been shown that the total energy of vaporization can be divided into at least three separate parts.[6] These parts come from the nonpolar/dispersion (atomic) forces, E_D; the permanent dipole–permanent dipole (molecular) forces, E_P; and hydrogen bonding (molecular) forces, E_H. The latter is more generally called the electron exchange energy.

$$E = E_D + E_P + E_H \tag{2.2}$$

$$E/V = E_D/V + E_P/V + E_H/V \tag{2.3}$$

$$\delta^2 = \delta_D^2 + \delta_P^2 + \delta_H^2 \tag{2.4}$$

δ_D, δ_P, and δ_H are the HSP for the dispersion, polar, and hydrogen bonding interactions, respectively. δ is the Hildebrand solubility parameter, $(E/V)^{1/2}$. It might be noted that the value of a solubility parameter in $MPa^{1/2}$ is 2.0455 times larger than in the often used $(cal/cm^3)^{1/2}$ units.

As described in Chapter 1, a corresponding states calculation using hydrocarbons as reference is used to find that part of the cohesive energy of a liquid which is attributable to dispersion (nonpolar) forces. Subtracting this nonpolar contribution from the total cohesion energy then gives the sum of the permanent dipole–permanent dipole and hydrogen bonding (electron interchange) contributions to the total cohesion energy. These can then be separated by calculation and/or experiment into the polar and hydrogen bonding parameters. HSP also inherently include volume effects, since they are based on cohesion energy density. Volume effects are also basic to the Prigogine CST.

HSP have been applied to the study of polymer solubility and swelling, biological materials, barrier properties of polymers, surfaces,[4,20,23-26] etc. and have been described in greater detail elsewhere[7,8,10] (see also the following chapters). The three parameters described in Equation 2.4 are fundamental energy parameters which can be calculated from the mutual interactions of identical molecules in a pure liquid. The quantities required are E, V, the dipole moment (and perhaps the refractive index and the dielectric constant), and generalized, corresponding states correlations for E for hydrocarbons, (E_D). Group contribution methods and simpler calculational procedures have also been established.[10] These procedures are described in Chapter 1. The calculated values for a large number of the liquids have been confirmed experimentally by solubility tests.

The usual equation used in HSP correlations is

$$(\mathrm{Ra})^2 = 4(\delta_{D2} - \delta_{D1})^2 + (\delta_{P2} - \delta_{P1})^2 + (\delta_{H2} - \delta_{H1})^2 \tag{2.5}$$

Ra in this equation is a modified difference between the HSP for a solvent (1) and polymer (2). Ra must not exceed Ro, the radius of interaction of an HSP solubility sphere, if solubility is to be maintained. Both Ra and Ro have the same units as solubility parameters. These correlations have been very convenient for practical use in solvent selection, for example. The constant "4" has been found empirically useful to convert spheroidal plots of solubility using δ_D and either of the other parameters into spherical ones (see Chapter 3). It has been used with success in well over 1000 HSP correlations with a computer program which optimizes a solubility sphere according to Equation 2.5 where all the "good" solvents are within the sphere and the "bad" ones are outside. This program is described in Chapter 1. This experimental procedure is still thought to be the best way to arrive at the HSP for polymers; the polymer HSP are given by the coordinates for the center of the sphere. The reliability of the spherical characterizations and the need to divide the total cohesion energy into at least three parts has been confirmed by systematically locating nondissolving solvents which are synergistic and dissolve a given polymer when mixed.[3] They only need to be located on opposite sides of the sphere of solubility for the given polymer.

For present purposes of comparison, Equation 2.5 must be normalized by $4R_M^2$ to make its predictions consistent with the quantities commonly used in the literature in connection with the corresponding states theories and χ_{12}:

$$H = (\mathrm{RA})^2/R_M^2 = \left[(\delta_{D2} - \delta_{D1})^2 + (\delta_{P2} - \delta_{P1})^2/4 + (\delta_{H2} - \delta_{H1})^2/4\right]/R_M^2 \tag{2.6}$$

$$\mathrm{RA} = \mathrm{Ra}/2;\ R_M = \mathrm{Ro}/2 \tag{2.7}$$

R_M is the maximum solubility parameter difference which still allows the polymer to dissolve based on Equation 2.6. R_M is the radius of an HSP sphere (spheroid) based on Equation 2.6. The HSP difference between solvent and polymer, RA, must be less than R_M for solution to occur. It can be seen that the quantities RA/R_M and $H = (\mathrm{RA}/R_M)^2$ will be 1.0 on the boundary surface of a sphere describing polymer solubility. RA/R_M is a ratio of solubility parameters, while H is a ratio of cohesion energy densities. H is zero when the solubility parameters for the solvent and polymer match and continually increases to larger values as the differences between solvent and polymer increase.

RESEMBLANCE BETWEEN PREDICTIONS OF HANSEN SOLUBILITY PARAMETERS AND CORRESPONDING STATES THEORIES

Patterson and co-workers[15,17] have explained the Prigogine theory in concise form and simplified some of the most important aspects. A key parameter is the Prigogine **d**. This describes normalized cohesive energy differences between polymer segments and solvents.

The Prigogine **d** parameter is defined as

$$\boldsymbol{\delta} = (\varepsilon_2 - \varepsilon_1)/\varepsilon_1 \tag{2.8}$$

ε is the cohesive energy for a polymer segment (2) or for a solvent (1).

For the present discussion, it is advantageous to define the Prigogine **d** using cohesive energy densities as follows:

$$\delta = \left[\left(ced_2\right)^{1/2} - \left(ced_1\right)^{1/2} \right]^2 / \left(ced_1\right) = \left(\delta_1 - \delta_2\right)^2 / \delta_1^2 \tag{2.9}$$

The numerator is the difference in cohesion energy densities between solvent and polymer, and this is normalized by the ced of the solvent. As indicated above, cohesion energies (HSP) for solvents can be calculated, while those for polymers currently require experimental data on solubility or other relevant testing procedures.

The Prigogine ρ accounts for differences in the size of the solvent and polymer segments. The segmental distance parameter is σ. ρ is defined as

$$\rho = \left(\sigma_2 - \sigma_1\right) / \sigma_1 \tag{2.10}$$

Another key parameter in the Prigogine/Patterson CST is ν^2, which is in fact equal to χ_{12}.[15] "ν^2" is approximated by

$$\nu^2 \text{ app} = \left(\delta^2/4 + 9\rho^2\right) \text{ app} = \left(\delta/2 - 3\rho/2\right)^2 \tag{2.11}$$

"ν^2" includes effects from differences in segmental energy (the **d** effect) and in segmental size (the ρ effect). The geometric mean rule $[\varepsilon_{12} = (\varepsilon_1\varepsilon_2)^{1/2}]$ was used to arrive at this result, just as it was used to arrive at the equations having differences in solubility parameters. Patterson states that the coefficient in front of ρ is uncertain[15] and, furthermore,[27] that "ρ was a (misguided) attempt to take into account segment size-differences."

Patterson[27] has recently helped the author to clarify some points about the relations among the theories. I feel a few quotes from these communications in addition to the one just cited are in order at this point.

> sic In my opinion the Flory theory was a very usable and successful particular case of the Prigogine theory (which was in fact difficult to use). An additional thing which Flory made a lot of, and which was only touched by Prigogine, is the point that surface/volume fraction of the polymers are very different from those of solvents, i.e., the polymer is very bulky. These are interesting differences between the Flory theory and that of Prigogine. But, again in my opinion, Flory always presented his theory as something absolutely different from that of Prigogine's, using different symbols, different names for concepts, etc. In reaction against this I have always liked to call the whole thing the Prigogine-Flory theory. However, since about 1970, I have done very little with polymer solutions, and hence when using the term "Prigogine-Flory theory" I have used it with respect to mixtures of small molecules and not polymer solutions. I think that the work by a lot of people has established the utility of the Prigogine-Flory theory, or if you like, the Flory theory, for small-molecule mixtures. sic

Also,

> sic Very specifically, the Prigogine parameters delta and rho are now out of fashion, and they got lumped together in the χ_{12} parameter. Particularly, the rho parameter does not have nearly as much importance as Prigogine thought and Flory completely discarded it.
>
> I think too the origin of the parameter beta in the solubility parameter approach would in the Prigogine-Flory approach be ascribed to free volume differences which must inevitably exist between any polymer and any solvent and which gives a contribution to the chi parameter… sic

This demonstrates that there is not complete agreement among those who have concerned themselves with these theories. The following is an attempt to unify all of these thoughts. The ideas have not been fully tested as of yet, but the implications appear very clear to the author, at least. The discussion concentrates on the ν^2 parameter, being loyal to the Patterson article[15] (sometimes referred to as the Prigogine-Patterson theory) where this part of this book got its start.

More specifically, ν^2 accounts for segmental energy differences and differences in size of solvent and polymer segments for breakage of solvent–solvent (1-1) bonds and polymer–polymer (2-2) bonds to allow formation of solvent–polymer (1-2) bonds.

In nonpolar systems, the Prigogine **d** is small (perhaps essentially zero in this context), and the quantity ν^2 depends on segmental size differences only. The Prigogine **d** parameter becomes important in systems with specific interactions, i.e., those with polar and hydrogen bonding. Differences in cohesive energy (density) arising from these sources in such systems are modified by a factor of ¼ according to Equation 2.11.

If we now consider Equation 2.6, it can be seen that each of the three terms in this equation is in the form of a Prigogine **d** as given by Equation 2.9. These terms describe normalized differences in the respective types of cohesive energy in corresponding states terminology. The cohesive energy differences in Equation 2.6 are normalized by $R_M{}^2$, the ced of the worst possible "good" solvent, i.e., a solvent located on the boundary of a Hansen solubility sphere, rather than by the ced of a given solvent under consideration.

In strictly nonpolar systems, the polar and hydrogen bonding terms in Equation 2.6 are zero, and the interaction is described by the difference in δ_D. One is led to the conclusion that the first term in Equation 2.6 relates directly to the second term (the ρ effect) in Equation 2.11. This relationship could be explored in more detail in the future with the hope of experimental verification of the coefficient in front of ρ.

If we now consider a system where δ_{D1} is equal to δ_{D2}, the polymer–solvent interaction will be either polar or hydrogen bonding (or both) in nature, i.e., there will be specific interactions and only specific interactions. Such differences in δ_P and δ_H will be modified by ¼ in Equation 2.6. It is noteworthy that the same factor, ¼, modifies the Prigogine **d** term in Equation 2.11, i.e., when there are specific interactions between solvent and polymer.

The same ¼ is present for the same purpose both in Equation 2.6 (HSP) where the geometric mean is used, and in Equation 2.11 (CST) where the geometric mean was also used. The geometric mean appears to be applicable to all the types of energies discussed here.

THE χ_{12} PARAMETER AND HANSEN SOLUBILITY PARAMETERS

Patterson and co-workers have shown that χ_{12} can be calculated, using ν^2 as a symbol for the same quantity.[17] Therefore, according to the previous discussion, it is expected that Equation 2.6 can be used in a similar way to predict χ_{12}.

There is still a general belief that χ_{12} can be calculated by Equation 2.1 using Hildebrand solubility parameters and a value of 0.34 for β. The change required to progress from the nonpolar Hildebrand solubility parameter to include polar and hydrogen bonding effects with the HSP in calculating χ_{12} is to replace the Hildebrand solubility parameter difference of Equation 2.1 by a corresponding HSP term, i.e., $A_{1,2}$

$$A_{1,2} = \left[(\delta_{D2} - \delta_{D1})^2 + 0.25(\delta_{P2} - \delta_{P1})^2 + 0.25(\delta_{H2} - \delta_{H1})^2\right] \qquad (2.12)$$

and χ_{12} is estimated from

$$\chi_{12} = VA_{1,2}/RT \qquad (2.13)$$

The empirical factor β (0.34) in Equation 2.1 was found from studies on almost nonpolar systems using the Hildebrand solubility parameters. This is an average correction to these calculations because of the neglect of some relatively small but significant values of $(\delta_{P2} - \delta_{P1})$ and/or $(\delta_{H2} - \delta_{H1})$. β is not required in Equation 2.13. This same assumption was made by Zellers and coworkers in their approach to correlate the swelling and permeation of elastomers used in chemical protective clothing.[28-31]

An estimate for χ_{12} can also be found by noting that the total χ_{12} parameter in common solutions of polymers having high molecular weight is required to be close to 0.5 at the point of marginal solution/precipitation.[1] This boundary value is called the critical chi parameter, χ_c. In HSP terminology, this is a boundary solvent with a placement directly on the sphere of solubility, and the quality is indicated by H in Equation 2.6 being equal to 1.0. This allows a simple estimate for χ_{12} for higher molecular weight polymers by the relation

$$\chi_{12} = \chi_c (\text{RED})^2 = \text{H}/2 \tag{2.14}$$

This last equation assumes an average V, just like the HSP correlations have done up to this point. It has been noted many times that liquids with lower V are often better solvents than liquids having essentially identical HSP but with larger V. This is seen with liquids like methanol (V = 40.7 cc/mol) and acetone (V = 74.0 cc/mol) which are sometimes good solvents, even when H is greater than 1.0, and for liquids like the phthalate and other plasticizers (V > 150 cc/mol) which are not good solvents in spite of H being less than 1.0. An explanation for this is found by comparing Equation 2.13 with Equation 2.14. Equation 2.15 can be derived from this comparison. The dependency of χ_c on polymer molecular size is also included in Equation 2.15, since this is a partial explanation for some of the results discussed later.[1]

$$\text{R}_\text{M}^2 = (\text{Ro}/2)^2 = \left\{0.5\left(1 + 1/\text{r}^{1/2}\right) \text{RT/V}\right\} \tag{2.15}$$

"r" is the ratio of the polymer size to that of the solvent. "r" is usually taken as being approximately the degree of polymerization, assuming the size of the solvent molecule is close to that of the polymer segment size. For a solvent with V = 100 cc/mol this added term leads to a correction of 1.1 for a polymer molecular weight of 10,000 and to a correction of 1.03 for a polymer molecular weight of 100,000.

The correction term is expressed more generally by the relation in Equation 2.16.

$$\text{Correction} = \left\{0.5\left(\text{r}_1^{-1/2} + \text{r}_2^{-1/2}\right)^2\right\} \tag{2.16}$$

Here, the r_1 and r_2 are the number of statistical segments in the molecules in question. For mixtures of low molecular weight, where r_1 is approximately equal to r_2 and both are approximately equal to 1, the correction amounts to a factor of about 2. Where one of the molecules is a high polymer the correction amounts to 0.5, as discussed earlier. When both of the molecules have very high molecular weights, the correction approaches 0, meaning compatible mixtures of polymers are very difficult to achieve. A study of the kind discussed in the following for smaller molecules would seem appropriate to help clarify some of the questions raised.

Table 2.1 gives the expected R_M and Ro values based on Equation 2.15 for polymers of molecular weight high enough to neglect the effect of r.

It is usually assumed that V for the average solvent is near 100 cc/mole. Table 2.1 indicates that all polymers of reasonably high molecular weight will be insoluble in solvents with V greater than about 100 cc/mol for RA greater than an R_M of 3.5 MPa½ (Ra greater than 7.0 MPa½). This

TABLE 2.1
Expected Solubility Parameter Differences for Marginal Solubility as a Function of the Molecular Volume of the Solvent

V (cc/mol)	R_M (MPa$^{1/2}$)	Ro (MPa$^{1/2}$)
50	5.0	10.0
100	3.5	7.0
200	2.5	5.0

is not generally the case since many values of R_M have been reported which are much higher than this[10] (see also Appendix, Table A.2), meaning they are more easily dissolved than Equation 2.15 indicates. Values for R_M greater than 5 MPa$^{1/2}$ are common, with some polymer R_M values being considerably larger, although these are generally for lower molecular weight materials. This immediately points to potential problems in directly calculating χ_{12} from HSP data when R_M is significantly larger than about 3.5 MPa$^{1/2}$.

Some improvement in the estimates of χ_{12} using Equation 2.14 is possible by including V in a correction term. Equation 2.14 has been retained for present purposes of comparison, however, because of its simplicity. The column for χ_{12} estimated from Equation 2.14 in Tables 2.3 to 2.7 is placed adjacent to that of the solvent molar volume to allow an easy mental multiplication by V/100 if desired.

COMPARISON OF CALCULATED AND EXPERIMENTAL c_{12} PARAMETERS

The predictions of χ_{12} using Equations 2.13 and 2.14 have been compared with χ_{12} parameter data in standard references.[32,33] The polymers used for the comparisons are listed in Table 2.2 with their HSP data. The calculated and indicative experimental values for χ_{12} for the given solvent–polymer systems are reported in Tables 2.3 to 2.7. The polymers were chosen because of similarities/differences in the HSP data as well as the availability of sufficient data on the χ_{12} parameter. This is not a complete evaluation of Equations 2.13 and 2.14, but it points out clearly that there are factors which are not completely understood. The solvents discussed are found with their HSP in Table 2.8.

TABLE 2.2
Hansen Solubility Parameter Data for Polymers Selected for the Comparisons Given in Tables 2.2 to 2.6.

Polymers	d_D	d_P	d_H	Ro
Polybutadiene	17.5	2.3	3.4	6.5
Polyisobutylene	16.9	2.5	4.0	7.2
Polystyrene	21.3	5.8	4.3	12.7
Polyvinylacetate	20.9	11.3	9.7	13.7
Polyacrylonitrile	21.7	14.1	9.1	10.9

Note: Units are MPa$^{1/2}$. Values in these units are 2.0455 times larger than in (cal/cm^3)$^{1/2}$.

TABLE 2.3
Comparison of Experimental, Indicative Chi Parameter Data, χ_{lit}, with Calculations Based on HSP for Polybutadiene, Buna Hüls CB 10 *cis*-Polybutadiene Raw Elastomer, Chemische Werke Hüls

Solvent	V	c_{12} (Equation 2.14)	c_{12} (Equation 2.13)	c_{lit}
Benzene	89.4	0.12	0.10	0.4
Toluene	106.8	0.04	0.04	0.3
Xylene	123.3	0.02	0.02	0.3
Styrene	115.6	0.08	0.08	0.4
Pentane	116.2	0.62	0.62	0.7
n-Hexane	131.6	0.52	0.58	0.6
n-Heptane	147.4	0.43	0.54	0.5
n-Octane	163.5	0.39	0.54	0.6
Chloroform	80.7	0.07	0.05	0.15
Carbon tetrachloride	97.1	0.16	0.13	0.3
Methanol	40.7	5.68	1.97	3.3
Water	18.0	20.3	3.1	3.5

Note: Solubility data from Hansen.[3,6]

TABLE 2.4
Comparison of Experimental, Indicative Chi Parameter Data, χ_{lit}, with Calculations Based on HSP for Polyisobutylene, Lutonal® I60, BASF

Solvent	V	c_{12} (Equation 2.14)	c_{12} (Equation 2.13)	c_{lit}
Benzene	89.4	0.19	0.17	0.5
Toluene	106.8	0.10	0.11	0.5
Decalin	156.9	0.35	0.43	0.4
Cyclohexane	108.7	0.20	0.23	0.45
Pentane	116.2	0.44	0.53	0.5
n-Hexane	131.6	0.37	0.51	0.5
n-Heptane	147.4	0.31	0.49	0.5
n-Octane	163.5	0.29	0.50	0.5
n-Nonane	179.7	0.27	0.51	0.3+
Chloroform	80.7	0.06	0.05	1.0
Carbon tetrachloride	97.1	0.20	0.21	0.5
Methylene dichloride	63.9	0.25	0.17	0.6

Note: Solubility parameter data from Hansen.[10]

A casual inspection of the measured and calculated χ_{12} values in Tables 2.3 to 2.7 quickly gives the impression that there are significant discrepancies between these which require further explanation. The calculated and literature values for χ_{12} agree in some cases and differ significantly in others. Some possible reasons for this are discussed in the following.

TABLE 2.5
Comparison of Experimental, Indicative Chi Parameter Data, χ_{lit}, with Calculations Based on HSP for Polystyrene, Polystyrene LG, BASF

Solvent	V	χ_{12} (Equation 2.14)	χ_{12} (Equation 2.13)	χ_{lit}
Benzene	89.4	0.23	0.66	0.40–0.44
Toluene	106.8	0.21	0.73	0.40–0.44
Xylene	123.3	0.25	0.99	0.4
Ethyl benzene	123.1	0.26	1.05	0.45
Styrene	115.6	0.16	0.61	0.35
Tetralin	136.0	0.09	0.38	0.4
Decalin (*cis*)	156.9	0.24	1.22	0.5
Cyclohexane	108.7	0.41	1.44	0.50–1.0
Methyl cyclohexane	128.3	0.49	2.03	0.5
n-Hexane	131.6	0.67	2.87	0.8
n-Heptane	147.1	0.61	2.92	0.8
n-Octane	163.5	0.58	3.08	0.9
Acetone	74.0	0.51	1.22	0.6
Methyl ethyl ketone	90.1	0.38	1.12	0.49
Methyl isobutyl ketone	125.8	0.45	1.83	0.5
Cyclohexanone	104.0	0.15	0.52	0.5
Ethyl acetate	98.5	0.40	1.29	0.5
n-Butyl acetate	132.5	0.40	1.72	0.5
sec-Butyl acetate	133.6	0.54	2.35	0.4

Note: Solubility data from Hansen.[3,6]

Polybutadiene

The calculated and experimental χ_{12} values for polybutadiene are given in Table 2.3. The first four entries are for aromatic solvents (including styrene). The solubility parameter predictions indicate that these are exceptionally good solvents, while the χ_{12} values indicate that they are moderately good. Before one adds on a constant value of about 0.3 to bring agreement, it should be noted that the calculated and experimental χ_{12} values for the aliphatic solvents are in good agreement. The solubility parameters for the higher molecular weight homologs are closer to those of the polymer, but the size effect reduces solvent quality. Agreement for the aliphatic solvents is considered excellent. It should be noted that Ro is very near the ideal value for such calculations according to Table 2.1.

Chloroform and carbon tetrachloride are predicted by HSP to be very good solvents, and the χ_{lit} for chloroform especially confirms this. HSP considerations indicate that chloroform and the aromatic solvents are near neighbors with similar HSP and that their quality should be similar. This is not borne out by the χ_{lit} values for the aromatics, which are suspected as being too high for presently unknown reasons.

The calculated and literature values for methanol and water are different enough to warrant a comment. HSP considerations indicate the difference in behavior between these two liquids should be sizable, which the χ_{lit} values do not indicate. A problem of some significance in any study of solvents at low concentrations in polymers is that the smaller amounts of solvent relative to the polymer can lead to preferential association of the solvent with those local regions/segments/groups in the polymer which have energies (HSP) most similar to their own. These local regions may not necessarily reflect the same affinities as the polymer as a whole, such as are reflected by the soluble-or-not approach commonly used in HSP evaluations. These local association effects can influence

TABLE 2.6
Comparison of Experimental, Indicative Chi Parameter Data, χ_{lit}, with Calculations Based on HSP for Polyvinylacetate, Mowilith® 50, Farbwerke Hoechst

Solvent	V	χ_{12} (Equation 2.14)	χ_{12} (Equation 2.13)	χ_{lit}
Benzene	89.4	0.56	1.91	0.3–0.5
Toluene	106.8	0.51	2.05	0.5
Decalin (*cis*)	156.9	0.64	3.80	2.7
Tetralin	136.0	0.37	1.92	1.3
Cyclohexane	108.7	0.76	3.13	2.4
Methyl cyclohexane	128.3	0.49	2.03	0.5
n-Nonane	179.7	0.88	6.00	3.3
n-Decane	195.9	0.88	6.54	3.4
Acetone	74.0	0.33	0.92	0.3–0.46
Methyl ethyl ketone	90.1	0.33	1.11	0.4–0.44
Methyl isobutyl ketone	125.8	0.45	1.83	0.5
Ethyl acetate	98.5	0.39	1.46	0.4
Dimethyl phthalate	163.0	0.12	0.58	0.4
Dioxane	85.7	0.29	0.95	0.4
Chloroform	80.7	0.32	0.99	0.4
Chlorobenzene	102.1	0.33	1.27	0.5
n-Propanol	75.2	0.47	1.34	1.2–1.6

Note: Solubility data from Hansen.[3,6]

TABLE 2.7
Comparison of Experimental, Indicative Chi Parameter Data, χ_{lit}, with Calculations Based on HSP for Polyacrylonitrile

Solvent	V	χ_{12} (Equation 2.14)	χ_{12} (Equation 2.13)	χ_{lit}
Ethylene carbonate	66.0	0.40	0.63	0.4
gamma-Butyrolactone	76.8	0.16	0.30	0.36–0.40
Ethanol	58.5	1.15	1.61	4.0
Water	18.0	5.3	2.3	2.0
N,N-Dimethyl formamide	77.0	0.33	0.61	0.2–0.3
N,N-Dimethyl acetamide	92.5	0.44	0.97	0.4
Dimethyl sulfoxide	71.3	0.21	0.36	0.3–0.4
Tetramethylene sulfoxide	90.0	0.25	0.53	0.3

Note: Solubility data from Brandrup and Immergut.[33]

results on swelling studies in both good and bad solvents, for example. Other types of studies carried out at low solvent concentrations can also be influenced by this segregation/association phenomena. An extension of this type of situation can be cited in the tendencies of water to associate with itself as well as with local regions within polymers. This has made simple predictions of its behavior impossible. A detailed discussion of this is beyond the scope of this chapter. It is suggested, however, that the potential differences observed here between HSP predictions and observed χ_{lit} may derive from such phenomena.

TABLE 2.8
Hansen Solubility Parameters for the Solvent Included in Tables 2.3 to 2.7

Solvent	d_D	d_P	d_H	Solvent	d_D	d_P	d_H
Benzene	18.4	0.0	2.0	*n*-Butyl acetate	15.8	3.7	6.3
Toluene	18.0	1.4	2.0	sec-Butyl acetate	15.0	3.7	7.6
Xylene	17.6	1.0	3.1	Dimethyl phthalate	18.6	10.8	4.9
Ethyl benzene	17.8	0.6	1.4	1,4-Dioxane	19.0	1.8	7.4
Styrene	18.6	1.0	4.1	Chloroform	17.8	3.1	5.7
Decalin (*cis*)	18.0	0.0	0.0	Chlorobenzene	19.0	4.3	2.0
Tetralin	19.6	2.0	2.9	Carbon tetrachloride	17.8	0.0[b]	0.6
Cyclohexane	16.8	0.0	0.2	Methylene dichloride	18.2	6.3	6.1
Methyl cyclohexane	16.0	0.0	1.0	Methanol	15.1	12.3	22.3
n-Pentane	15.6	0.0	0.0	Ethanol	15.8	8.8	19.4
n-Hexane	14.9	0.0	0.0	*n*-Propanol	16.0	6.8	17.4
n-Heptane	15.3	0.0	0.0	Ethylene carbonate	19.4	21.7	5.1
n-Octane	15.5	0.0	0.0	gamma-Butyrolactone	19.0	16.6	7.4
n-Nonane	15.7	0.0	0.0	*N,N*-Dimethyl formamide	17.4	13.7	11.3
Acetone	15.5	10.4	7.0	*N,N*-Dimethyl acetamide	16.8	11.5	10.2
Methyl ethyl ketone	16.0	9.0	5.1	Dimethyl sulfoxide	18.4	16.4	10.2
Methyl isobutyl ketone	15.3	6.1	4.1	Tetramethylene sulfoxide	18.2	11.0	9.1
Cyclohexanone	17.8	6.3	5.1	Water	15.5	16.0	42.3
Ethyl acetate	15.8	5.3	7.2				

[a] Units are $MPa^{½}$.

[b] The value 0.0 is valid in nonpolar media and derives from a zero dipole moment; a progressively higher value in increasingly polar media is required because of induced dipoles[10] (see also Chapter 1).

Polyisobutylene

The calculated and experimental χ_{12} values for polyisobutylene are given in Table 2.4. There are some similarities with polybutadiene both chemically and in the Ro value of 7.2 $MPa^{½}$ being near the ideal for a polymer of very high molecular weight. The cyclic and aromatic solvents are again better as judged by HSP than the χ_{lit} values indicate, while the estimates for the aliphatic solvents are in excellent agreement with Equation 2.13, in particular. Again, HSP finds chloroform, methylene dichloride, and carbon tetrachloride as being very good, in agreement with solubility-or-not experiments, while the χ_{lit} values indicate these are not good or at best marginal in quality.

The results of Equation 2.13 for the aliphatic hydrocarbons are in particularly good agreement with χ_{lit}.

Polystyrene

The calculated and experimental χ_{12} values for polystyrene are given in Table 2.5. The Ro value of 12.7 $MPa^{½}$ is now much higher than the ideal one indicated in Table 2.1. The polymer molecular weight is thought to be reasonably high, but is unknown. As a consequence of the Ro value, practically all the χ_{12} values calculated by Equation 2.13 are too high. One is tempted to divide by a factor of 2 or 3, but there is no consistent pattern. Equation 2.14 includes the boundary value of χ_c equal to 0.5, so the results are more in agreement with χ_{lit}. HSP predicts that the aromatic and cyclic solvents are somewhat better that expected from χ_{lit}. The agreement would be better if the χ_{12} values found from Equation 2.14 for these were increased by a factor 2. The values found by Equation 2.14 for the aliphatic hydrocarbons are also lower than χ_{lit}, but are qualitatively in agreement. The χ_{12} values found for ketones and esters using Equation 2.14 are in generally good agreement with the literature values. The one exception of some note is the well-known good solvent cyclohexanone, which is predicted as a much better solvent by HSP than the χ_{lit} value would indicate. There is considerably more differentiation in predictions of solvent quality found by Equation 2.14 than the χ_{lit} values indicate.

Polyvinylacetate

The calculated and experimental χ_{12} values for polyvinylacetate are given in Table 2.6. The molecular weight for this polymer is reported as being 260,000. It can initially be noted that Ro is 13.7 MPa^{½}, which again means χ_{12} values found from Equation 2.13 will be higher than those found in the literature. This difference varies considerably, but a factor of 2 to 4 is generally required to give reasonable agreement. Equation 2.13 is certainly not generally acceptable as an instrument to predict χ_{lit}. Equation 2.14 gives reasonably good approximations to χ_{lit} as long as the solvents are good enough to dissolve the polymer. However, there are some major disagreements. Tetralin dissolves the polymer, but has χ_{lit} equal to 1.3. *n*-Propanol is an error from the HSP prediction of it being a marginal solvent, whereas it is a nonsolvent. Alcohols of higher and lower molecular weight do have a significant effect on this polymer, however, and the azeotropic mixture of ethanol and water actually dissolves it.[3,6] When dealing with nonsolvents, the HSP predictions of χ_{12} are generally lower than the data found in the literature. Once again, a factor of 3 to 4 is required to bring the values into agreement.

Polyacrylonitrile

The calculated and experimental χ_{12} values for polyacrylonitrile are given in Table 2.7. This polymer has high polar and hydrogen bonding parameters and Ro equal to 10.9 MPa^{½} which, once more, is somewhat above the ideal. The agreement with Equation 2.14 is reasonably good for the good solvents. The nonsolvents are not in good agreement. Equation 2.13 agrees surprisingly well with the best solvents, gamma-butyrolactone and dimethyl sulfoxide, but the agreement is not uniform when all the solvents are considered.

GENERAL DISCUSSION

It should be noted in general that χ_{12} can either increase or decrease with concentration of the polymer. Barton[32] presents data to examine the potential magnitude of this effect. The correlations given in Table 2.2 are based on whether or not the polymer dissolves at a concentration of 10%, with the exception of the data for polyacrylonitrile where no polymer concentration is indicated in the original solubility data.[33] The HSP data for correlations of the type given in Table 2.2 can also be expected to change for higher polymer concentration and higher polymer molecular weight. R_M is expected to decrease only slightly for marginally higher polymer molecular weight, once a reasonably high molecular weight is being considered, and R_M is expected to decrease somewhat for higher polymer concentration, although this can vary as well, especially for lower molecular weight species. An interesting fact to keep in mind is that a polymer with molecular weights in the millions will only swell in even the best solvents. The present evaluations are at the same polymer concentration unless otherwise noted. No significant corrections of the type included in Equation 2.16 are required since the polymer molecular weight is very high in all cases. Corrections of this type are therefore not responsible for the differences in the calculated and observed χ_{12} parameters.

A point of some concern is that negative values for χ_{12} are found in the literature, but these are not allowed in either the CST or HSP approaches. There is no obvious general explanation for this situation. A negative χ_{12} implies a solvent of a quality superior to anything normal polymer–liquid interactions could provide. Normal here also includes what have been called specific interactions attributable to permanent dipole–permanent dipole and hydrogen bonding interactions as discussed earlier. A closer review of this situation is desired. No systems with negative χ_{12} are included in Tables 2.2 to 2.7.

An additional problem of some concern is that, in general, there is considerable scatter in the χ_{12} parameter data from different sources. Clearing up this situation is far beyond the scope of this book. However, one cannot help but wonder why, and the seeming discrepancies do not contribute to blind confidence in the any of the reported χ_{12} values. Indicative χ_{12} values are used here. One can also find variations in HSP values for polymers from different sources,[32] so there are also problems in determining which values are best in this approach. The χ_{12} parameter does not specifically account for permanent dipole–permanent dipole or hydrogen bonding interactions, which must be considered a major source of potential differences.

There has been some discussion as to whether the coefficient "4" in Equation 2.5 (corresponding to a coefficient of 0.25 in Equation 2.12) should be a different number. Barton cites a case where a coefficient of 0.2 (rather than 0.25) in Equation 2.12 was determined.[34] Skaarup has mentioned a case of 5 (rather than 4) as a value for the coefficient in Equation 2.5, which, of course, gives 0.2 in Equation 2.12.[35] The author has also explored situations where water was involved where the DATA FIT was equal for either a 4 or 5 in Equation 2.5. Zellers and co-workers use this coefficient as an adjustable parameter for individual solvents in their studies.[28-31] One significant factor in this discussion is that the solvents with higher solubility parameters generally have lower molecular volumes. This means they will be better than expected by average comparisons of behavior. This fact tends to lead one to stretch the spheres a little more in the δ_P and δ_H directions to encompass these good solvents which would otherwise lie outside of the spheres (spheroids). This would lead to a number slightly higher than the 4 in Equation 2.5, and it is the author's feeling that this will be shown to be the case whenever a complete understanding of the effect of solvent molecular volume and other size effects is accomplished.

The Prigogine theory contains structural parameters which have not been explored in this context. There are also structural parameters in the New Flory theory. It is possible that the use of structural parameters will allow better understanding, and perhaps the possibility of improved calculations, and reduce the need for experimental studies. The experimentally determined radius for the solubility spheres automatically takes these factors into account, but reliable calculation of the radius of interaction has not been possible as yet.

POSTSCRIPT

The author has always experienced consistency in the quality of the predictions using the HSP. Care is required to generate the necessary data, and there should always be a reevaluation of experimental data based on an initial correlation. The solvent parameters have been used with success for many years in industrial practice to predict solvent quality using computer techniques by most, if not all, major solvent suppliers. Mixing rules have been established for even complicated solvent blends. These are usually based on summing up simple volume fraction times solubility parameter values. (An evaluation of the quality of the χ_{12} values in the literature could be made with precipitation experiments for mixtures to see whether a mixing rule gives consistent results for these as well.)

The solvent δ_D, δ_P, and δ_H values which were established with extensive calculations have been supported by many tens of thousands of experimental data points based on solubility, permeation, surface wetting, etc.[10] It has become quite clear that the HSP for the solvents are not precise enough for sophisticated calculations, but they certainly represent a good average satisfactory for practical applications. The HSP for the solvents relative to each other are correct for the majority of the common solvents. The "nearest neighbors" to a good solvent are clearly expected to be of nearly comparable quality unless they are in a boundary region of the HSP solubility sphere. The solvent quality indicated by the ratio Ra/Ro (the RED number) has been particularly satisfactory. This ratio was defined years ago as a ratio of solubility parameters, since plotting and interpretation of data used solubility parameters. Use of the ratio of cohesion energy densities is also possible, of course, since this is indeed closer to an energy difference number and would agree more with the Flory approach as seen in Equations 2.6 and 2.14, since H is really nothing other than $(RED)^2$.

The result of having written all of this is that the author senses that the HSP approach is a practical extension in complete agreement with the Prigogine-Flory theory when the geometric mean is used, at least as far as the major factors discussed earlier are concerned. The comparisons presented previously confirm some relation, but the single χ_{12} parameter may have been oversimplified such that the more complete HSP approach cannot be immediately recognized. The ability of HSP to describe molecular affinities among so many different materials listed in this book speaks for the general application of both the Prigogine and the HSP treatments. The Prigogine treatment is acknowledged as difficult to use in practice. This is not true of the HSP approach.

CONCLUSION

The Prigogine/Patterson CST and the HSP approach (which also involves a corresponding states calculation) are shown to have very close resemblance. Both can be used to estimate the Flory χ_{12}. Two equations involving HSP are given for this purpose. Reasonably good predictions are possible under favorable circumstances. Favorable circumstances involve a system with an essentially nonpolar polymer whose Ro value is not too different from 7.0 $MPa^{1/2}$. χ_{12} values for the better solvents are calculated by HSP at lower levels than those found in the literature. χ_{12} values for nonsolvents are also generally calculated by HSP as being significantly lower than the reported literature values. The most favorable circumstances are, of course, not always present, and some problems still exist and need to be solved before these calculations can be used with confidence to estimate χ_{12} values for any solvent–polymer system. The HSP values for the polymers used for the present comparisons are based on solubility-or-not type experiments which reflect the properties of the polymer as a whole. These may not completely correspond to the type evaluation often used to find χ_{12}, since less-than-dissolving amounts of solvent may be used, and the solvent may associate with given segments/groups in the polymer and not reflect the behavior of the polymer as a whole (see also the discussion in Chapter 3).

An empirical factor, β, equal to about 0.34 appears in many sources in the literature in connection with calculation of χ_{12} using Hildebrand solubility parameters. β disappears when HSP are used for this purpose, but the resulting equation has not been studied enough yet to allow general use of HSP to calculate χ_{12} parameters.

Studies on the effect of molecular size, segmental size, and polymer size are still required. It is suggested that the structural factors discussed by Prigogine be tried in this respect.[11] Use of the geometric mean in conjunction with the Prigogine theory brings the HSP and Prigogine approaches into agreement. The massive amount of experimental data presented in this book strongly supports the use of the geometric mean. As a curiosity, it might be noted that the use of the geometric mean (Lorenz-Berthelot mixtures) was used to generate an ellipsoidal miscibility plot essentially identical to those given in Chapter 3, Figures 3.1 and 3.2.[36] This approach was not continued because "The boundary of this ellipse is of little practical importance as there are no cases known of immiscibility in mixtures known to conform to the Lorenz-Berthelot equations."

As stated in the Preface to this book, it has not been its purpose to recite the developments of polymer solution thermodynamics in a historical manner with full explanations of each theory or modifications thereof. The references cited in the Preface do this already. This chapter has attempted to show relations between the classical theories and the HSP approach, which includes a quantitative accounting of both permanent dipole–permanent dipole and hydrogen bonding interactions as an integral part. The relation between the Prigogine-Patterson theory and HSP was the most obvious.

REFERENCES

1. Flory, P. J., *Principles of Polymer Chemistry*, Cornell University Press, New York, 1953.
2. Eichinger, B. E. and Flory, P. J., Thermodynamics of Polymer Solutions, *Trans. Faraday Soc.*, 64(1), 2035–2052, 1968; ibid. (2), 2053–2060; ibid. (3), 2061–2065; ibid. (4), 2066–2072.
3. Hansen, C. M., The Three Dimensional Solubility Parameter — Key to Paint Component Affinities I. Solvents, Plasticizers, Polymers, and Resins, *J. Paint Technol.*, 39(505), 104–117, 1967.
4. Hansen, C. M., The Three Dimensional Solubility Parameter — Key to Paint Component Affinities II. Dyes, Emulsifiers, Mutual Solubility and Compatibility, and Pigments, *J. Paint Technol.*, 39(511), 505–510, 1967.
5. Hansen, C. M. and Skaarup, K., The Three Dimensional Solubility Parameter — Key to Paint Component Affinities III. Independent Calculation of the Parameter Components, *J. Paint Technol.*, 39(511), 511–514, 1967.
6. Hansen, C. M., The Three Dimensional Solubility Parameter and Solvent Diffusion Coefficient, Their Importance in Surface Coating Formulation, Doctoral Dissertation, Danish Technical Press, Copenhagen, 1967.
7. Hansen, C. M., The Universality of the Solubility Parameter, *Ind. Eng. Chem. Prod. Res. Dev.*, 8(1), 2–11, 1969.
8. Hansen, C. M. and Beerbower, A., Solubility Parameters, in *Kirk-Othmer Encyclopedia of Chemical Technology*, Suppl. Vol., 2nd ed., Standen, A., Ed., Interscience, New York, 1971, 889–910.
9. Hansen, C. M., 25 Years with Solubility Parameters (in Danish: 25 År med Opløselighedsparametrene), *Dan. Kemi*, 73(8), 18–22, 1992.
10. Hansen, C. M., Solubility Parameters, in *Paint Testing Manual*, Manual 17, Koleske, J. V., Ed., American Society for Testing and Materials, Philadelphia, 1995, 383–404.
11. Prigogine, I. (with the collaboration of A. Bellemans and A. Mathot), *The Molecular Theory of Solutions*, North-Holland, Amsterdam, 1957, chaps. 16,17.
12. Patterson, D. and Delmas, G., New Aspects of Polymer Solution Thermodynamics, *Off. Dig. Fed. Soc. Paint Technol.*, 34(450), 677–692, 1962.
13. Delmas, D., Patterson, D., and Somcynsky, T., Thermodynamics of Polyisobutylene-n-Alkane Systems, *J. Polym. Sci.*, 57, 79–98, 1962.
14. Bhattacharyya, S. N., Patterson, D., and Somcynsky, T., The Principle of Corresponding States and the Excess Functions of n-Alkane Mixtures, *Physica*, 30, 1276–1292, 1964.
15. Patterson, D., Role of Free Volume Changes in Polymer Solution Thermodynamics, *J. Polym. Sci. Part C*, 16, 3379–3389, 1968.
16. Patterson, D. D., Introduction to Thermodynamics of Polymer Solubility, *J. Paint Technol.*, 41(536), 489–493, 1969.
17. Biros, J., Zeman, L., and Patterson, D., Prediction of the χ Parameter by the Solubility Parameter and Corresponding States Theories, *Macromolecules*, 4(1), 30–35, 1971.
18. Hildebrand, J. and Scott, R. L., *The Solubility of Nonelectrolytes,* 3rd ed., Reinhold, New York, 1950.
19. Hildebrand, J. and Scott, R. L., *Regular Solutions,* Prentice-Hall Inc., Englewood Cliffs, NJ, 1962.
20. Hansen, C. M., Cohesion Parameters for Surfaces, Pigments, and Fillers (in Danish: Kohæsionsparametre for Overflader, Pigmenter, og Fyldstoffer), *Färg och Lack Scand.*, 43(1), 5–10, 1997.
21. Hansen, C. M., Polymer Solubility — Prigogine Theory and Hansen Solubility Parameter Theory Mutually Confirmed (In Danish: Polymeropløselighed — Prigogine Teori og Hansen Opløselighedsparameterteori Gensidigt Bekræftet), *Dan. Kemi,* 78(9), 4–6, 1997.
22. Hildebrand, J. and Scott, R. L., *The Solubility of Nonelectrolytes,* 3rd ed., Reinhold, New York, 1950, chap. 20.
23. Hansen, C. M., Characterization of Surfaces by Spreading Liquids, *J. Paint Technol.*, 42(550), 660–664, 1970.
24. Hansen, C. M., Surface Dewetting and Coatings Performance, *J. Paint Technol.*, 44(570), 57–60, 1972.
25. Hansen, C. M. and Pierce, P. E., Surface Effects in Coatings Processes, *XII Federation d'Associations de Techniciens des Industries des Peintures, Vernis, Emaux et Encres d'Imprimerie de l'Europe Continentale, Congress Book*, Verlag Chemie, Weinheim/Bergstrasse, 1974, 91–99; *Ind. Eng. Chem., Prod. Res. Dev.*, 13(4), 218–225, 1974.

26. Hansen, C. M. and Wallström, E., On the Use of Cohesion Parameters to Characterize Surfaces, *J. Adhes.*, 15(3/4), 275–286, 1983.
27. Patterson, D., Personal communication, 1997.
28. Zellers, E. T., Three-Dimensional Solubility Parameters and Chemical Protective Clothing Permeation. I. Modeling the Solubility of Organic Solvents in Viton® Gloves, *J. Appl. Polym. Sci.*, 50, 513–530, 1993.
29. Zellers, E. T. and Zhang G.-Z., Three-Dimensional Solubility Parameters and Chemical Protective Clothing Permeation. II. Modeling Diffusion Coefficients, Breakthrough Times, and Steady-State Permeation Rates of Organic Solvents in Viton® Gloves, *J. Appl. Polym. Sci.*, 50, 531–540, 1993.
30. Zellers, E. T., Anna, D. H., Sulewski, R., and Wei, X., Critical Analysis of the Graphical Determination of Hansen's Solubility Parameters for Lightly Crosslinked Polymers, *J. Appl. Polym. Sci.*, 62, 2069–2080, 1996.
31. Zellers, E. T., Anna, D. H., Sulewski, R., and Wei, X., Improved Methods for the Determination of Hansen's Solubility Parameters and the Estimation of Solvent Uptake for Lightly Crosslinked Polymers, *J. Appl. Polym. Sci.*, 62, 2081–2096, 1996.
32. Barton, A. F. M., *Handbook of Polymer-Liquid Interaction Parameters and Solubility Parameters*, CRC Press, Boca Raton, FL, 1990.
33. Brandrup, J. and Immergut, E. H., Eds., *Polymer Handbook*, 3rd ed., Wiley-Interscience, New York, 1989. (a) Gundert, F. and Wolf, B. A., Polymer-Solvent Interaction Parameters, VII/173–182; (b) Fuchs, O., Solvents and Non-Solvents for Polymers, VII/379–407.
34. Barton, A. F. M., Applications of Solubility Parameters and other Cohesion Energy Parameters, *Polym. Sci. Technol. Pure Appl. Chem.*, 57(7), 905–912, 1985.
35. Skaarup, K., private communication, 1997.
36. Rowlinson, J. S., *Liquids and Liquid Mixtures*, Butterworths Scientific Publications, London, 1959, chap. 9.

3 Methods of Characterization — Polymers

CONTENTS

ABSTRACT

The simplest experimental method to determine the Hansen solubility parameters (HSP) for a polymer is to evaluate whether or not it dissolves in selected solvents. Those solvents dissolving the polymer will have HSP closer to those of the polymer than those which do not. A computer program or graphical method can then be used to find the HSP for the polymer. Other types of evaluations can also lead to polymer HSP. These include swelling, melting point reduction, surface attack, chemical resistance, barrier properties, viscosity measurements, and any other measurement reflecting differences in polymer affinities among the solvents.

Polymer HSP can be higher than the HSP of any of the test solvents. This means that some of the methods suggested in the literature to interpret data, i.e., those which use averages of solvent HSP to arrive at the polymer HSP, must be used with care.

INTRODUCTION

Experience has shown that if it is at all possible, an experimental evaluation of the behavior of a polymer in contact with a series of selected liquids is the best way to arrive at its HSP. Experimental data can be generated and treated in various ways to arrive at the values of interest. Examples are included in the following.

The author's usual approach to generate data in solubility parameter studies is to contact a polymer of interest with 40 to 45 well-chosen liquids. One may then observe or measure a number of different phenomena including full solution at a given concentration, degree of swelling by visual observation or by measurement of weight change, volume change, clarity, surface attack, etc. The object of the studies is to determine differences in affinity of the polymer for the different solvents. These differences are then traditionally used to divide the solvents into two groups, one which is considered "good" and the other which is considered "bad." Such data can be entered into the

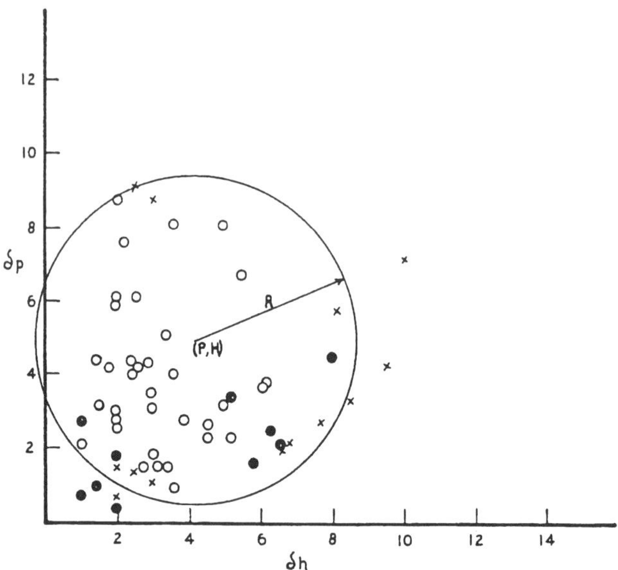

FIGURE 3.1 Two-dimensional plot of δ_P vs. δ_H for the solubility of polymethyl methacrylate (Polymer B in Table 3.2). The circle is the projection of a sphere on the given coordinates. (From Hansen, C. M., *Färg och Lack,* 17(4), 71, 1971. With permission.)

SPHERE program as discussed in Chapter 1. Whenever possible, the author uses a set of solvents as described below, often supplemented by selected solvents depending on the purpose of the investigation.

The goal of the experimental work is to arrive at a set of data showing differences in behavior among the test solvents. These data are then processed to arrive at the four parameters characteristic of HSP correlations, the three describing the nonpolar, polar, and hydrogen bonding interactions as for the liquids and the fourth, Ro, a radius of interaction for the type of interaction described.

The author has most often used computer techniques to evaluate the data to find the polymer HSP. In earlier work simple plots were used. A simple plot of δ_P vs. δ_H is also helpful in many practical situations to get guidance as discussed in Chapter 6. The approximate determination of polymer HSP can be done using three plots of experimental data using the HSP parameters pairwise. Figures 3.1 to 3.3 demonstrate how this was attempted initially.[1] The spheroids in the figures including the δ_D parameter gave problems. Hansen and Skaarup[2] simply used a scaling factor of 2 (the coefficient "4" in Chapter 1, Equation 1.9) to produce spheres in all three plots. Since Ro must be the same in all of these plots, a single compass setting is tried for a set of $\delta_D;\delta_P;\delta_H$ to see how well the separation into good and bad solvents is accomplished. Calculations for points in doubt can be made using Chapter 1, Equation 1.9. Plots with the modified δ_D axis are given for the solubility of polystyrene[3] shown in Figures 3.4 to 3.6. These are the original figures from this thesis, and the numbers refer to a table of solvents found there. An idea of the accuracy of the graphical approach can be found in Table 3.1, where comparisons are made with the SPHERE program approach. Table 3.2 contains a listing of the polymers included in Table 3.1.

Teas[4] has developed a triangular plotting technique which helps visualization of three parameters on a plane sheet of paper. Examples are found in References 5 to 7 and in Chapter 6. The triangular plotting technique uses parameters for the solvents, which, in fact, are modified HSP parameters. The individual Hansen parameters are normalized by the sum of the three parameters. This gives three fractional parameters defined by Equations 3.1 to 3.3.

$$f_d = 100\,\delta_D/\left(\delta_D + \delta_P + \delta_H\right) \tag{3.1}$$

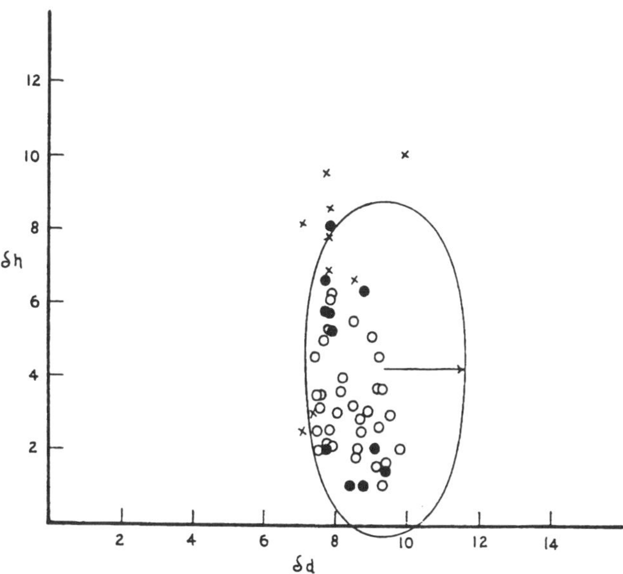

FIGURE 3.2 Two-dimensional plot of δ_H vs. δ_D for the solubility of polymethyl methacrylate (Polymer B in Table 3.2). Expansion of the δ_D scale by a factor of 2 would yield a circle (a sphere in projection). (From Hansen, C. M., *Färg och Lack,* 17(4), 71, 1971. With permission.)

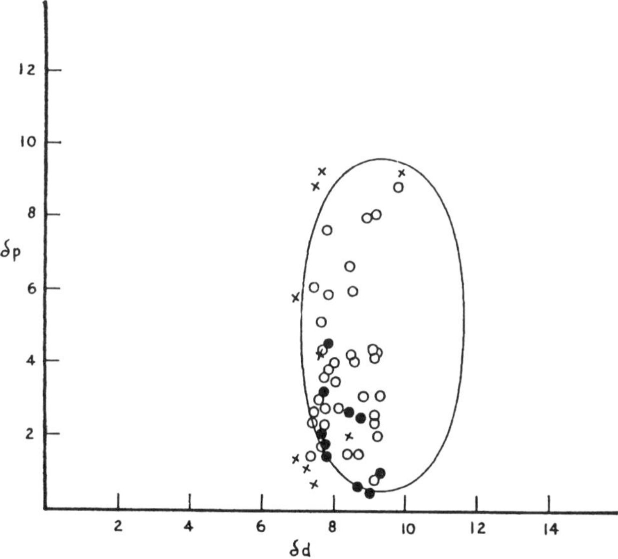

FIGURE 3.3 Two-dimensional plot of δ_P vs. δ_D for the solubility of polymethyl methacrylate (Polymer B in Table 3.2). Expansion of the δ_D scale by a factor of 2 would yield a circle (a sphere in projection). (From Hansen, C. M., *Färg och Lack,* 17(4), 72, 1971. With permission.)

$$f_p = 100\,\delta_P/(\delta_D + \delta_P + \delta_H) \quad (3.2)$$

$$f_h = 100\,\delta_H/(\delta_D + \delta_P + \delta_H) \quad (3.3)$$

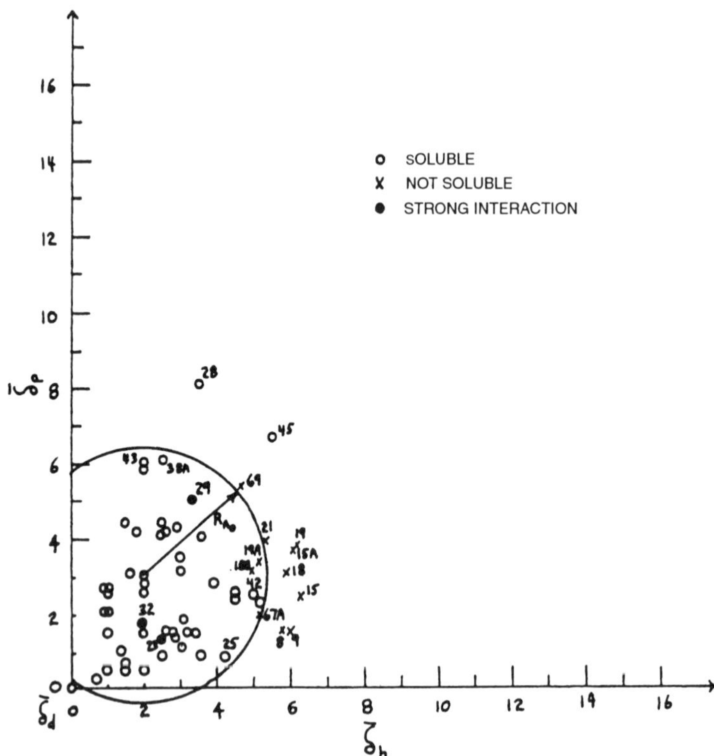

FIGURE 3.4 Two-dimensional plot δ_P vs. δ_H of solubility data for polystyrene (Polymer G in Table 3.2).

The sum of these three fractional parameters is 1.0. This allows the use of the special triangular technique. Some accuracy is lost, and there is no theoretical justification for this plotting technique, but one does get all three parameters onto a two-dimensional plot. This plotting technique is often used by those who conserve old paintings, because it was described in a standard reference book very shortly after it was developed.[7] Chapter 6, Figure 6.4 shows how such a plot can be used in finding a suitable solvent when dealing with such an older oil painting.

HSP for the polymers and film formers discussed in the following examples are given in Table 3.3. These data are based on solubility determinations unless otherwise noted. Barton[6,8] has also provided solubility parameters for many polymers. Values for a number of acrylic, epoxy, and other polymers potentially useful in self-stratifying coatings have been reported by Benjamin et al.[9] (see Chapter 6). Rasmussen and Wahlström[10] provide additional HSP data in relation to the use of replenishable natural products (oils) in connection with solvents. The data processing techniques and data accumulated by Zellers and co-workers[11-14] on elastomers used in chemical protective clothing are also useful. Zellers et al. also point out many of the problems encountered with these characterizations. Such problems are also discussed below. There are other sources of HSP for polymers in the literature, but a full review of these and their uses is beyond the scope of this book.

CALCULATION OF POLYMER HSP

Calculation of the HSP for polymers is also possible. The results are not yet fully satisfactory, but there is hope for the future. One of the more significant efforts in this has been made by Utracki and co-workers.[15,16] They assumed the δ_D parameter for polymers did not differ too much between polymers and interpreted evaluations of polymer–polymer compatibility using calculated values for δ_P and δ_H. Group calculations were used. This is probably the best calculation approach currently

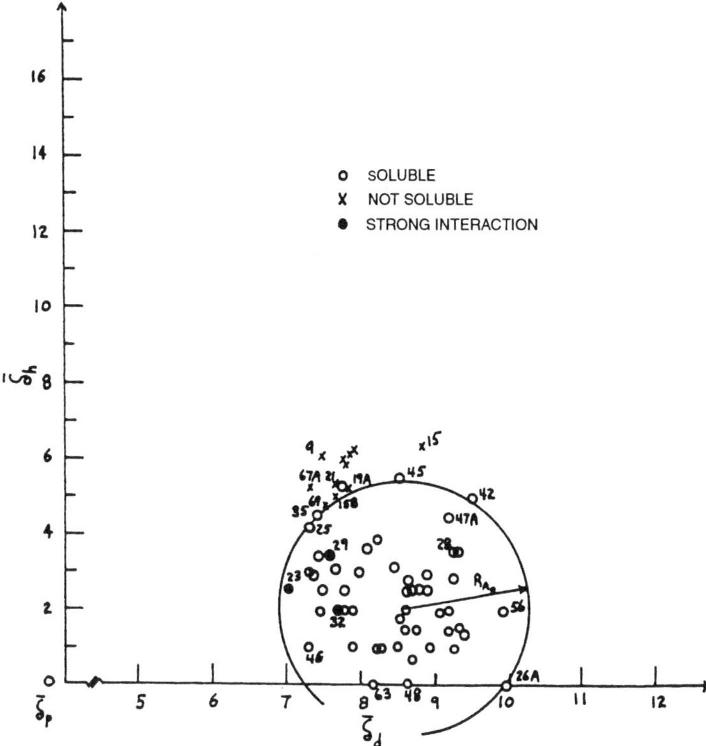

FIGURE 3.5 Two-dimensional plot δ_H vs. δ_D of solubility data for polystyrene (Polymer G in Table 3.2). Expansion of the δ_D scale by a factor of 2 has given a spherical representation according to Chapter 1, Equation 1.9.

available, but improvements are thought possible. The group contributions given in Chapter 1 can be used for this purpose. It is suggested that HSP for polymers determined by these calculations not be mixed with experimentally determined HSP until confirmation of agreement is found. It can be presumed that the errors involved in either process will cancel internally, but these may not necessarily be the same for the calculated results as for the experimental ones.

The author has never been particularly successful in calculating the same values as were found experimentally, although a serious effort to use weighting factors and the like as discussed in the following has never been tried.

SOLUBILITY — EXAMPLES

The most direct method to determine the three HSP for polymers or other soluble materials is to evaluate their solubility or degree of swelling/uptake in a series of well-defined solvents. The solvents should have different HSP chosen for systematic exploration of the three parameters at all levels. As indicated earlier, a starting point could be the series of liquids used by the author for many years. These are essentially those included in Table 3.4. Sometimes boundaries are defined better by inclusion of additional test solvents. A computer analysis quickly gives a choice of many of these, since solvents with RED numbers (Chapter 1, Equation 1.10) near 1.0 are located near the sphere boundary. It is actually the boundary which is used to define the center point of the sphere using Chapter 1, Equation 1.9. Some changes are also possible to remove or replace solvents which are now considered too hazardous, although good laboratory practice should allow use of the ones indicated. The HSP generally in use for liquids have all been evaluated/calculated at 25°C.

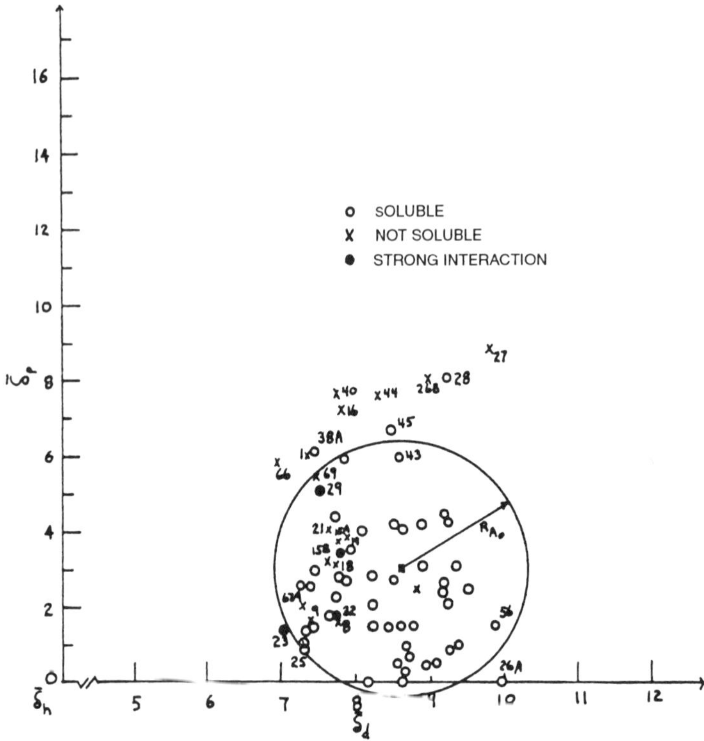

FIGURE 3.6 Two-dimensional plot δ_P vs. δ_D of solubility data for polystyrene (Polymer G in Table 3.2). Expansion of the δ_D scale by a factor of 2 has given a spherical representation according to Chapter 1, Equation 1.9.

These same values can also be used to correlate physical phenomena related to solubility at other test temperatures with some care as noted below.

Several examples of HSP correlations based on solubility are found in Table 3.3. The entry for polyethersulfone (PES) found in Table 3.3 was determined from data included in the computer output reported in Table 3.4. The solubility of PES was evaluated in 41 different solvents. It was found that five of them actually dissolved the polymer. The input data to the SPHERE program described in Chapter 1 are included in Table 3.4 in the SOLUB column. A "1" means a good solvent and a "0" means a bad solvent. A 1* means that a good solvent lies outside the sphere, where it should not, and a 0* means a bad solvent lies inside the sphere, which means it is an outlier. Each of these error situations reduces the data fit. D, P, H, and R for the solubility of PES are given at the top. In addition, there is an indication of the data fit, which is 0.999 here. A perfect fit is 1.000. A data fit slightly less than 1.0 is actually preferred, since the computer program has then optimized the data to a single set of values which are so close to being correct as they can be within experimental error. An unknown number of sets of the parameters can give a data fit of 1.0 whenever this result is found. Perfect fits are rather easily obtained with small sets of data, and the boundaries are rather poorly defined, which means the center is also poorly defined. One can continue testing with additional solvents located in the boundary regions of the established sphere as stated previously. These can be found easily by listing the solvents in order of their RED numbers and choosing those with values not too different from 1.0. The RED number is given for each solvent in the RED column.

Finally, there is a column in Table 3.4 indicating the molar volume, V, of the solvents in cc/mol. There was no need to analyze the influence of this parameter in the present case.

TABLE 3.1
Calculated and Trial and Error Solubility Parameter Data for Various Polymers[a]

Computed Handtrials and Error	d_D	d_P	d_H	R	Max D	Computed Handtrials and Error	d_D	d_P	d_H	R	Max D
A	8.60	4.72	1.94	5.20	0.960	R	9.04	4.50	2.40	5.20	0.985
	9.2	5.3	2.1	5.3	0.923		9.2	4.5	2.6	5.0	0.972
B	9.11	5.14	3.67	4.20	0.945	S	10.53	7.30	6.00	8.20	0.910
	9.2	5.0	4.2	4.0	0.923		8.8	7.0	5.5	6.0	0.879
C	9.95	5.88	5.61	6.20	0.853	T	8.58	1.64	1.32	3.20	0.974
	8.5	5.5	5.5	4.7	0.829		8.7	1.8	1.8	3.5	0.965
D	9.98	1.68	2.23	6.70	0.974	U	9.10	4.29	2.04	4.70	0.969
	8.5	2.5	3.0	5.3	0.957		9.3	4.5	2.0	4.7	0.950
E	9.79	2.84	5.34	5.70	0.930	V	8.10	0.69	–0.40	4.70	0.974
	9.4	3.2	5.1	5.0	0.929		8.5	1.5	1.5	3.4	0.964
F	9.09	2.13	6.37	5.20	0.948	X	7.10	1.23	2.28	6.20	0.921
	8.5	4.3	5.5	4.8	0.871		7.8	1.0	3.6	4.0	0.881
G	10.40	2.81	2.10	6.20	0.955	Y	8.57	1.10	1.67	3.20	0.950
	8.6	3.0	2.0	3.5	0.915		8.8	2.5	1.2	3.8	0.914
H	10.23	5.51	4.72	6.70	0.891	Z	8.52	–0.94	7.28	4.70	0.971
	9.3	5.0	4.0	4.9	0.855		8.2	0.8	5.7	2.9	0.954
I	10.17	4.05	7.31	6.20	0.924	$\overline{A}$	9.60	2.31	3.80	5.20	0.942
	9.5	4.0	6.4	4.7	0.909		8.7	2.5	3.5	4.2	0.951
J	7.53	7.20	4.32	5.60	0.933	$\overline{B}$	9.95	4.17	5.20	7.20	0.980
	7.0	7.0	4.3	5.5	0.918		9.5	4.0	5.5	7.0	0.976
K	9.90	3.09	2.64	5.20	0.949	$\overline{C}$	8.05	0.18	1.39	4.20	0.966
	9.3	3.7	2.1	4.2	0.933		8.5	1.0	2.0	3.4	0.960
L	9.08	6.22	5.38	3.70	0.921	$\overline{D}$	10.34	6.63	6.26	6.70	0.964
	9.5	6.0	6.0	4.5	0.896		9.2	5.8	4.2	5.0	0.868
M	11.37	3.20	4.08	9.70	0.978	$\overline{E}$	8.58	0.58	1.76	3.20	0.968
	9.0	4.0	5.5	6.4	0.923		8.5	1.5	1.8	2.6	0.956
N	9.65	5.68	7.13	6.20	0.897	$\overline{F}$	8.91	3.68	4.08	1.70	0.992
	9.4	5.3	7.4	5.5	0.867		9.4	4.5	3.5	3.2	0.895
O	10.62	0.46	4.17	7.70	1.000	$\overline{G}$	9.49	2.68	2.82	4.70	0.961
	8.9	3.0	3.8	4.5	0.952		8.8	2.7	2.7	4.0	0.963
P	8.58	4.58	7.00	5.20	0.942	$\overline{L}$	9.86	7.14	7.35	5.70	0.970
	8.5	4.7	6.5	5.0	0.940		10.8	7.0	8.8	7.1	0.936
Q	9.87	6.43	6.39	5.70	0.942						
	9.3	6.2	4.7	4.2	0.892						

[a] See Table 3.2 for polymer types.

From Hansen, C. M., *Färg och Lack,* 17(4), 73, 1971. With permission.

A second example of this type of approach is given in Table 3.3. Data on good and bad solvents[17] for polyacrylonitrile (PAN) have been used as input to the computer program. Thirteen solvents are indicated as good, and 23 are indicated as bad. These test solvents do not differ as widely from each other as the test series suggested earlier, but the data are still useful in finding the HSP for this polymer. These are reported in Table 3.3. The data fit of 0.931 is good. Having found the HSP for a polymer in this manner, one can then search a database for additional solvents for the polymer in question. This was done for a database with over 800 solvent entries. A significantly large number of the 123 additional solvents found to have RED numbers less than 1.0 can be expected to dissolve this polymer, but such an extensive experimental study was not undertaken to confirm the predictions.

TABLE 3.2
List of Polymers and Resins Studied

A	Lucite® 2042-poly (ethyl methacrylate), E. I. du Pont de Nemours & Co., Inc.
B	Poly (methyl methacrylate), Rohm and Haas Co.
C	Epikote® 1001-epoxy, Shell Chemical Co.
D	Plexal P65-66% oil length alkyd, Polyplex.
E	Pentalyn® 830-alcohol soluble rosin resin, Hercules Incorporated.
F	Butvar® B76-poly (vinyl butyral), Shawinigan Resins Co.
G	Polystyrene LG, Badische Anilin- und Soda Fabrik.
H	Mowilith® 50-poly (vinyl acetate), Farbwerke Hoechst.
I	Plastopal H-urea formaldehyde resin, Badische Anilin- und Soda Fabrik.
J	½ Sec. Nitrocellulose-H 23, A. Hagedorn and Co., Osnabrück, W. Germany.
K	Parlon® P10-chlorinated poly (propylene), Hercules Incorporated.
L	Cellulose acetate, Cellidora A-Bayer AG.
M	Super Beckacite® 1001-Pure Phenolic Resin, Reichhold Chemicals Co., Hamburg.
N	Phenodur 373U-phenol-resol resin, Chemische Werke Albert-Wiesbaden.
O	Cellolyn 102-modified pentaerythritol ester of rosin, Hercules Incorporated.
P	Pentalyn 255-alcohol soluble resin, Hercules Incorporated.
Q	Suprasec F5100-blocked isocyanate (phenol), Imperial Chemical Ind. Ltd.
R	Plexal C34-34% coconut oil-phthalic anhydride alkyd, Polyplex.
S	Desmophen 850, Polyester-Farbenfabriken Bayer AG. Leverkusen.
T	Polysar 5630 — styrene-butadiene (SBR) raw elastomer, Polymer Corp.
U	Hycar® 1052-acrylonitrile-butadiene raw elastomer, B. F. Goodrich Chemical Corp.
V	Cariflex IR 305-isoprene raw elastomer, Shell Chemical Co.
X	Lutanol IC/123-poly (isobutylene), Badische Anilin- und Soda Fabrik.
Y	Buna Huls CB 10-*cis* poly butadiene raw elastomer, Chemische Werke Huels.
Z	Versamid® 930-polyamide, General Mills, Inc.
$\bar{A}$	Ester gum BL, Hercules Incorporated.
$\bar{B}$	Cymel® 300-hexamethoxy melamine, American Cyanamid Co.
$\bar{C}$	Piccolyte® S100-terpene resin, Pennsylvania Industrial Chemical Corp.
$\bar{D}$	Durez® 14383-furfuryl alcohol resin, Hooker Chemical Co.
$\bar{E}$	Piccopale® 110-petroleum hydrocarbon resin, Pennsylvania Industrial Chemical Corp.
$\bar{F}$	Vipla KR-poly (vinyl chloride), K = 50, Montecatini.
$\bar{G}$	Piccoumarone 450L-cumarone-indene resin, Pennsylvania Industrial Chemical Corp.
$\bar{L}$	Milled wood lignin — special sample from Prof. A. Björkman.

TABLE 3.3
Hansen Solubility Parameter Correlations for Selected Materials

Material	d_D	d_P	d_H	Ro	FIT	G/T
PES solubility	19.6	10.8	9.2	6.2	0.999	5/41
PAN solubility	21.7	14.1	9.1	10.9	0.931	13/36
PP swelling	18.0	3.0	3.0	8.0	1.00	13/21
Polyvinyl alcohol (see text)	17.0	9.0	18.0	4.0	1.00	2/56
Hexamethylphosphoramide	18.5	8.6	11.3	—	—	—
PVDC melting temperature 110°C	17.6	9.1	7.8	3.9	0.992	6/24
PVDC melting temperature 130°C	20.4	10.0	10.2	7.6	0.826	13/24
Dextran C solubility	24.3	19.9	22.5	17.4	0.999	5/50

TABLE 3.4
Calculated Solubility SPHERE for PES Solubility

D = 19.6 P = 10.8 H = 9.2 RAD = 6.2 FIT = 0.999 NO = 41

No.	Solvent	D	P	H	SOLUB	RED	V
4	Acetone	15.5	10.4	7.0	0	1.371	74.0
6	Acetophenone	19.6	8.6	3.7	1	0.955	117.4
13	Benzene	18.4	0.0	2.0	0	2.129	89.4
28	1-Butanol	16.0	5.7	15.8	0	1.777	91.5
30	Butyl acetate	15.8	3.7	6.3	0	1.741	132.5
37	gamma-Butyrolactone	19.0	16.6	7.4	1	0.998	76.8
40	Carbon tetrachloride	17.8	0.0	0.6	0	2.301	97.1
41	Chlorobenzene	19.0	4.3	2.0	0	1.576	102.1
44	Chloroform	17.8	3.1	5.7	0	1.483	80.7
48	Cyclohexanol	17.4	4.1	13.5	0	1.467	106.0
56	Diacetone alcohol	15.8	8.2	10.8	0	1.321	124.2
61	*o*-Dichlorobenzene	19.2	6.3	3.3	0	1.204	112.8
75	Diethylene glycol	16.6	12.0	20.7	0	2.101	94.9
82	Diethyl ether	14.5	2.9	5.1	0	2.183	104.8
90	Dimethyl formamide	17.4	13.7	11.3	1	0.915	77.0
94	Dimethyl sulfoxide	18.4	16.4	10.2	0*	0.996	71.3
96	1,4-Dioxane	19.0	1.8	7.4	0	1.493	85.7
104	Ethanol	15.8	8.8	19.4	0	2.077	58.5
105	Ethanolamine	17.0	15.5	21.2	0	2.241	59.8
106	Ethyl acetate	15.8	5.3	7.2	0	1.547	98.5
120	Ethylene dichloride	19.0	7.4	4.1	0	1.007	79.4
121	Ethylene glycol	17.0	11.0	26.0	0	2.837	55.8
123	Ethylene glycol monobutyl ether	16.0	5.1	12.3	0	1.563	131.6
124	Ethylene glycol monoethyl ether	16.2	9.2	14.3	0	1.395	97.8
126	Ethylene glycol monomethyl ether	16.2	9.2	16.4	0	1.618	79.1
131	Formamide	17.2	26.2	19.0	0	3.044	39.8
140	Hexane	14.9	0.0	0.0	0	2.745	131.6
148	Isophorone	16.6	8.2	7.4	0	1.094	150.5
153	Methanol	15.1	12.3	22.3	0	2.575	40.7
162	Methylene dichloride	18.2	6.3	6.1	1	0.990	63.9
164	Methyl ethyl ketone	16.0	9.0	5.1	0	1.368	90.1
167	Methyl isobutyl ketone	15.3	6.1	4.1	0	1.782	125.8
172	Methyl-2-pyrrolidone	18.0	12.3	7.2	1	0.655	96.5
178	Nitroethane	16.0	15.5	4.5	0	1.580	71.5
179	Nitromethane	15.8	18.8	5.1	0	1.899	54.3
181	2-Nitropropane	16.2	12.1	4.1	0	1.387	86.9
204	Propylene carbonate	20.0	18.0	4.1	0	1.429	85.0
205	Propylene glycol	16.8	9.4	23.3	0	2.457	73.6
222	Tetrahydrofuran	16.8	5.7	8.0	0	1.237	81.7
225	Toluene	18.0	1.4	2.0	0	1.978	106.8
229	Trichloroethylene	18.0	3.1	5.3	0	1.485	90.2

A special problem which can be encountered is when only a few solvents with very high solubility parameters dissolve a polymer. An example is polyvinyl alcohol with true solvents being 1-propanol and ethanol in a data set with 56 solvents.[6] The entry in Table 3.3 places a big question mark with the solubility parameters, as well as with the radius 4.0 and the perfect fit of the data.

ALTERNATE TABLE 3.4
Calculated Solubility SPHERE for PES Solubility (Listed in RED Order)

D = 19.6 P = 10.8 H = 9.2 RAD = 6.2 FIT = 0.999 NO = 41

No.	Solvent	D	P	H	SOLUB	RED	V
172	Methyl-2-pyrrolidone	18.0	12.3	7.2	1	0.655	96.5
90	Dimethyl formamide	17.4	13.7	11.3	1	0.915	77.0
6	Acetophenone	19.6	8.6	3.7	1	0.955	117.4
162	Methylene dichloride	18.2	6.3	6.1	1	0.990	63.9
94	Dimethyl sulfoxide	18.4	16.4	10.2	0*	0.996	71.3
37	gamma-Butyrolactone	19.0	16.6	7.4	1	0.998	76.8
120	Ethylene dichloride	19.0	7.4	4.1	0	1.007	79.4
148	Isophorone	16.6	8.2	7.4	0	1.094	150.5
61	*o*-Dichlorobenzene	19.2	6.3	3.3	0	1.204	112.8
222	Tetrahydrofuran	16.8	5.7	8.0	0	1.237	81.7
56	Diacetone alcohol	15.8	8.2	10.8	0	1.321	124.2
164	Methyl ethyl ketone	16.0	9.0	5.1	0	1.368	90.1
4	Acetone	15.5	10.4	7.0	0	1.371	74.0
181	2-Nitropropane	16.2	12.1	4.1	0	1.387	86.9
124	Ethylene glycol monoethyl ether	16.2	9.2	14.3	0	1.395	97.8
204	Propylene carbonate	20.0	18.0	4.1	0	1.429	85.0
48	Cyclohexanol	17.4	4.1	13.5	0	1.467	106.0
44	Chloroform	17.8	3.1	5.7	0	1.483	80.7
229	Trichloroethylene	18.0	3.1	5.3	0	1.485	90.2
96	1,4-Dioxane	19.0	1.8	7.4	0	1.493	85.7
106	Ethyl acetate	15.8	5.3	7.2	0	1.547	98.5
123	Ethylene glycol monobutyl ether	16.0	5.1	12.3	0	1.563	131.6
41	Chlorobenzene	19.0	4.3	2.0	0	1.576	102.1
178	Nitroethane	16.0	15.5	4.5	0	1.580	71.5
126	Ethylene glycol monomethyl ether	16.2	9.2	16.4	0	1.618	79.1
30	Butyl acetate	15.8	3.7	6.3	0	1.741	132.5
28	1-Butanol	16.0	5.7	15.8	0	1.777	91.5
167	Methyl isobutyl ketone	15.3	6.1	4.1	0	1.782	125.8
179	Nitromethane	15.8	18.8	5.1	0	1.899	54.3
225	Toluene	18.0	1.4	2.0	0	1.978	106.8
104	Ethanol	15.8	8.8	19.4	0	2.077	58.5
75	Diethylene glycol	16.6	12.0	20.7	0	2.101	94.9
13	Benzene	18.4	0.0	2.0	0	2.129	89.4
82	Diethyl ether	14.5	2.9	5.1	0	2.183	104.8
105	Ethanolamine	17.0	15.5	21.2	0	2.241	59.8
40	Carbon tetrachloride	17.8	0.0	0.6	0	2.301	97.1
205	Propylene glycol	16.8	9.4	23.3	0	2.457	73.6
153	Methanol	15.1	12.3	22.3	0	2.575	40.7
140	Hexane	14.9	0.0	0.0	0	2.745	131.6
121	Ethylene glycol	17.0	11.0	26.0	0	2.837	55.8
131	Formamide	17.2	26.2	19.0	0	3.044	39.8

The computer analysis quickly encompasses the two good solvents in the data set within a small sphere since they have reasonably similar parameters. Based on reasonable similarity with other solubility correlations for water soluble polymers, one anticipates spheres with a radius much larger than the distance between these solvents. The properties of good solvents alone cannot always lead to a good estimate of the solubility parameters for these polymers, and the radii of spheres using only a few solvents with high solubility parameters will be very uncertain. One can sometimes find better results by correlating degrees of swelling or uptake, rather than correlate on solubility or not. The work of Zellers and co-workers reports extensive studies of this type.[11-14] It should be noted, however, that the HSP-sphere parameters usually vary some from correlation to correlation based on the same data when different criteria are used for "good" and "bad" solvents. This is because the absorbed solvent tends to locate in regions with similar solubility parameter, and there are local variations in HSP within most, if not all, polymers. This is particularly true of polymers which are not homopolymers. This situation relates to *self-assembly*.

Another example of determining HSP for a polymer with very high solubility parameters is Dextran C (British Drug Houses). Only 5 out of 50 solvents were found to dissolve Dextran C.[18] In this case, there was enough spread in the solubility parameters of the test solvents such that the spherical model correlation (Chapter 1, Equation 1.9) forced the program to find a radius of 17.4 MPa$^{1/2}$. This appears to be a reasonable number for this situation. The problem can be made clearer by noting the dissolving solvents with their RED numbers in parenthesis. These were dimethyl sulfoxide (1.000), ethanolamine (0.880), ethylene glycol (0.880), formamide (0.915), and glycerol (0.991). Some dissolving liquids had RED equal to 1.0 or higher and included diethylene glycol (1.000), propylene glycol (1.053), and 1,3-butanediol (1.054). These helped to define the boundary of the Hansen solubility sphere. Note that the HSP for the polymer are in a region higher than that defined by the values of test liquids. Any technique using an average of the HSP for interacting test solvents will inherently underestimate the solute HSP.

SWELLING — EXAMPLES

The correlation for swelling of polypropylene reported in Table 3.3 is based on solvent uptake data reported by Lieberman and Barbe.[19] The limit of 0.5% was arbitrarily set to differentiate "good" solvents from "bad" ones. As mentioned earlier, experience has shown that a different limit usually gives different parameters. It should be noted that swelling data reflect the properties of the regions in the polymer where the solvent has chosen to reside because of energetic similarity (self-assembly). The principle is not necessarily "like dissolves like," but rather "like seeks like." If the solvent is homogeneously distributed in the polymer, the solubility parameters found will reflect the properties of the whole polymer. Crystalline regions will not contain solvent. If the solvent collects locally in regions with chemical groups different from the bulk of the polymer, then the HSP so derived will reflect at least partially the physical nature of these chemical groups. The parameters reported in Table 3.3 seem appropriate for what is expected in terms of low polarity and low hydrogen bonding properties for a polypropylene-type polymer.

Another example of a characterization using swelling data which did not result in a good correlation is that for Viton® (The Du Pont Company, Wilmington, DE). This problem has been discussed by Zellers and Zhang[11,12] and also is discussed in Chapter 8. One reason for the poor correlation of swelling behavior is that Viton is not a homopolymer. The different segments have different affinities. Indeed, there are several qualities of Viton, each of which has significantly differing chemical resistance.

MELTING POINT DETERMINATIONS — EFFECT OF TEMPERATURE

Partly crystalline polymers which are placed in different liquids will have melting points which are lowered to a degree depending on the solvent quality of the individual liquids. The melting points of polyvinylidine chloride (PVDC) have been measured in different solvents.[20] These data have been analyzed by evaluating solubility parameter regions based on those solvents which dissolve the polymer at 110°C and above and also at 130°C and above. As expected, there are more solvents which dissolve the semicrystalline polymer at the higher temperature. The results for these correlations are included in Table 3.3. The main reasons for the somewhat lower data fit at 130°C include two nondissolving solvents within the solubility parameter sphere. These are dimethyl phthalate, where the large molecular size is a factor, and benzyl alcohol, where temperature effects can be larger than expected compared with the other solvents as discussed later and in Chapter 1. The solubility parameters for PVDC at this temperature, based on tabulated solvent values at 25°C, are not affected significantly by this type of situation. A single room temperature solvent for PVDC is reported by Wessling.[20] This is hexamethylphosphoramide, and its solubility parameters are also reported in Table 3.3 for comparison. The change in the values of the individual solubility parameters with temperature is discussed in Chapter 1 (Equations 1.16 to 1.18).

INTRINSIC VISCOSITY MEASUREMENTS

One of the more promising methods to evaluate polymer HSP for limited data is that using the intrinsic viscosity. Van Dyk et al. found a correlation with the intrinsic viscosity of an acrylic polymer (polyethyl methacrylate) in various solvents and the polymer HSP[24] (see the discussion on polymer compatibility in Chapter 6).

Segarceanu and Leca[25] have devised a method to calculate the polymer HSP from data on its intrinsic viscosity in different solvents. The intrinsic viscosities will be higher in the better solvents because of greater interaction and greater polymer chain extension. The intrinsic viscosity gives an indication of the solvent quality. It has been used earlier to calculate the Flory-Huggins chi parameter, for example.[26]

In the new technique, the intrinsic viscosities are normalized by the intrinsic viscosity of that solvent giving the highest value. These normalized data (numbers are 1.0 or less) are then used in a weighted averaging technique to arrive at the center of the Hansen sphere.

$$\delta_{D2} = \Sigma\left(\delta_{Di} \times [\eta]_i\right) / \Sigma[\eta]_i \tag{3.4}$$

$$\delta_{P2} = \Sigma\left(\delta_{Pi} \times [\eta]_i\right) / \Sigma[\eta]_i \tag{3.5}$$

$$\delta_{H2} = \Sigma\left(\delta_{Hi} \times [\eta]_i\right) / \Sigma[\eta]_i \tag{3.6}$$

The subscript 2 is for the polymer, and the respective solvents are indicated by an "i." The intrinsic viscosity in the "ith" solvent is given by $[\eta]_i$.

Those solvents with the greatest weighting factor have higher intrinsic viscosities and are closest to the geometric center of the sphere. Those solvents which do not dissolve the polymer were assumed to have a zero weighting factor. The HSP for a polyesterimide were reported as an example. HSP values were assigned both by the "classical" evaluation and with this newer approach. These data are included as the first entries in Table 3.5. This is a very promising method of arriving at the polymer HSP with limited data.

However, there are several aspects of this work which deserve comment. It was demonstrated earlier that many polymers have higher solubility parameters than any of the solvents which are (can be) used to test them. The present method only allows for polymer HSP within the range attainable by the test solvents. The methods will lead to values which are too low in some cases, including the example with the polyesterimide used as an example in Segarceanu and Leca.[25] It is not surprising that the polymer HSP are often higher than solvent HSP, since they are in a physical state between that of a liquid and a solid. When the cohesion energy becomes too high, a material

TABLE 3.5
HSP Data for the Same Polyesterimide Polymer Based on Data Given in Reference 25

Correlation	d_D	d_P	d_H	Ro	FIT
Classical[a1]	17.4	12.3	8.6	4.1	—
New[a1]	18.0	11.1	8.8	8.6	—
HSP SPHERE[a]	20.0	11.0	10.0	8.3	1.000
HSP SPHERE[b]	19.0	11.0	9.0	7.0	1.000
Classical[a]					0.426
Classical[b]					0.447
New[a]					0.506
New[b]					0.364

Note: Use of the solubility parameters for the solvents given in Reference 25 are indicated with a superscript a, while use of the solvent HSP data in the author's files is indicated with a superscript b.

is a solid rather than a liquid. Low molecular weight solids frequently have HSP somewhat higher than the HSP of liquids. Many examples can be given, including urea, ethylene carbonate, etc.

When the data (as soluble or not) for the 11 solvents were processed by the SPHERE computer program, the parameters found were those given by the third set of HSP in Table 3.5. The agreement with the "New" method is acceptable, even though none of the test solvents have δ_D as high as that of the polymer. Further inspection showed that the solubility parameters used in the study were not in agreement with those published in the latest reference to Hansen listed by Segarceanu and Leca.[25] It also appears that the radius of the HSP sphere for the "classical" determination is in error, being far too low.

To further clarify the situation, several runs with the SPHERE program were done with the parameters listed in this book, as well as with those listed in the article being discussed. In both cases the data fit is not good for the HSP reported by Segarceanu and Leca.[25] In the "Classical" case, the data fit is only 0.426 (1.0 is perfect), and four of the five "good" solvents are located outside of the sphere. Only *N*-methyl-2-pyrrolidone is inside. In the "new" case, the data fit is not much better, being 0.506. Here, four of the five "bad" solvents are inside the sphere with only one being outside. It has been possible to estimate the polymer parameters within acceptable variation, but the radius of the sphere has not been accounted for in a satisfactory manner.

Further inspection of the data suggests that morpholine, the solvent with the highest $[\eta]$ which was used to normalize the data, is not as good as might have been expected from the intrinsic viscosity data. This can be seen in Table 3.6. The reason for this is unknown, but experience has shown that amines often are seen to react with various materials in a manner which does not allow their inclusion in correlations of the type discussed here.

To conclude this section, it is noted that a similar weighting technique was used by Zellers et al.[13,14] where the weighted measurements were solvent uptake by elastomers customarily used to make chemical protective clothing. The same precautions must be taken in analyzing this type of measurement, but since the polymers studied were reasonably "nonpolar" some of the solvents had HSP which were higher than those of the polymers studied. Zellers et al.[14] and Athey[27] also describe multiple variable statistical analysis techniques to find the HSP of a given polymer. Barton's work[6] contains many literature sources of intrinsic viscosity studies using the solubility parameter for interpretation.

TABLE 3.6
Calculated Solubility SPHERE for Polyesterimide (Listed in RED Order)

D = 19.0 P = 11.0 H = 9.0 RAD = 7.0 FIT = 1.000 NO = 11

No.	Solvent	[η][a]	D	P	H	SOLUB	RED	V
172	Methyl-2-pyrrolidone	0.970	18.0	12.3	7.2	1	0.427	96.5
90	Dimethyl formamide	0.947	17.4	13.7	11.3	1	0.682	77.0
94	Dimethyl sulfoxide	0.182	18.4	16.4	10.2	1	0.809	71.3
37	gamma-Butyrolactone	0.689	19.0	16.6	7.4	1	0.832	76.8
174	Morpholine	1.000	18.8	4.9	9.2	1	0.874	87.1
49	Cyclohexanone	0.718	17.8	6.3	5.1	1	0.937	104.0
56	Diacetone alcohol	—	15.8	8.2	10.8	0	1.031	124.2
4	Acetone	—	15.5	10.4	7.0	0	1.044	74.0
80	Diethylene glycol monomethyl ether	—	16.2	7.8	12.6	0	1.055	118.0
124	Ethylene glycol monoethyl ether	—	16.2	9.2	14.3	0	1.131	97.8
125	Ethylene glycol monoethyl ether acetate	—	15.9	4.7	10.6	0	1.283	136.1

[a] Intrinsic viscosity data from Reference 25.

OTHER MEASUREMENT TECHNIQUES

There are many other techniques to differentiate between the behavior of different solvents in contact with a polymer. Many of these are discussed in the following chapters and will not be treated here. These include permeation measurements, chemical resistance determinations of various kinds, surface attack, etc. Some of the techniques can be very useful depending on the polymer involved. Others will present problems because of the probable influence of other factors such as solvent molar volume, length of time before attainment of equilibrium, and the like. Several of these phenomena can be correlated with HSP, but the techniques used in the measurements will present problems in using the data for direct HSP characterization of polymers because other effects are also important.

An example is polymer stress cracking. Polymer stress cracking in connection with solvent contact can be correlated with HSP, but care should be taken if such data are used to determine polymer HSP. One must consider the absorption rate, which is expected to be a function of the molecular size and shape, as well as solubility parameter relations. Several authors have discussed stress cracking in terms of HSP.[21-23] The general picture, including the results of unpublished studies by the author, is that one can correlate stress cracking at a given initial stress with the result that those solvents which truly dissolve the polymer will have low RED numbers. Larger molecules with low RED numbers may also lead to cracking. The liquids producing cracking will generally be closer to the boundary than the true solvents and have RED numbers approaching 1.0. A liquid with small molecular size with a RED number close to 1.0 may not necessarily crack the polymer if it can absorb very rapidly and only to a lower equilibrium concentration than that truly dissolving the polymer.

CONCLUSION

HSP for polymers can be evaluated experimentally by correlations of data where a suitably large number of well-chosen solvents are brought into contact with the polymer. The observed behavior which can be correlated includes true solubility, swelling, weight gain, dimensional change, degree of surface attack, reduction of melting point, permeation rate, breakthrough time, and tensile strength reduction. Correlations for simple evaluations of chemical resistance of the suitable-or-not type are also possible.

In each case, the molecular size of the liquids used can affect the result and should be considered in some way. The use of water as a test liquid is not recommended for these purposes.

REFERENCES

1. Hansen, C. M., Solubility in the Coatings Industry, *Färg och Lack*, 17(4), 69–77, 1971.
2. Hansen, C. M. and Skaarup, K., The Three Dimensional Solubility Parameter — Key to Paint Component Affinities III. Independent Calculation of the Parameter Components, *J. Paint Technol.*, 39(511), 511–514, 1967.
3. Hansen, C. M., The Three Dimensional Solubility Parameter and Solvent Diffusion Coefficient, Their Importance in Surface Coating Formulation, Doctoral Dissertation, Danish Technical Press, Copenhagen, 1967.
4. Teas, J. P., Graphic Analysis of Resin Solubilities, *J. Paint Technol.*, 40(516), 19–25, 1968.
5. Gardon, J. L. and Teas, J. P., Solubility Parameters, in *Treatise on Coatings, Vol. 2, Characterization of Coatings: Physical Techniques, Part II*, Myers, R. R. and Long, J. S., Eds., Marcel Dekker, New York, 1976, chap. 8.
6. Barton, A. F. M., *Handbook of Solubility Parameters and Other Cohesion Parameters*, CRC Press, Boca Raton, FL, 1983.
7. Torraca, G., *Solubility and Solvents for Conservation Problems 2nd ed.*, International Centre for the Study of the Preservation and the Restoration of Cultural Property (ICCROM), Rome, 1978.
8. Barton, A. F. M., *Handbook of Polymer-Liquid Interaction Parameters and Solubility Parameters*, CRC Press, Boca Raton, FL, 1990.
9. Benjamin, S., Carr, C., and Wallbridge, D. J., Self-Stratifying Coatings for Metallic Substrates, *Prog. Org. Coat.*, 28(3), 197–207, 1996.
10. Rasmussen, D. and Wahlström, E., HSP — Solubility Parameters: A Tool for Development of New Products — Modelling of the Solubility of Binders in Pure and Used Solvents, *Surf. Coat. Intl.*, 77(8), 323–333, 1994.
11. Zellers, E. T., Three-Dimensional Solubility Parameters and Chemical Protective Clothing Permeation. I. Modeling the Solubility of Organic Solvents in Viton® Gloves, *J. Appl. Polym. Sci.*, 50, 513–530, 1993.
12. Zellers, E. T. and Zhang, G.-Z., Three-Dimensional Solubility Parameters and Chemical Protective Clothing Permeation. II. Modeling Diffusion Coefficients, Breakthrough Times, and Steady-State Permeation Rates of Organic Solvents in Viton® Gloves, *J. Appl. Polym. Sci.*, 50, 531–540, 1993.
13. Zellers, E. T., Anna, D. H., Sulewski, R., and Wei, X., Critical Analysis of the Graphical Determination of Hansen's Solubility Parameters for Lightly Crosslinked Polymers, *J. Appl. Polym. Sci.*, 62, 2069–2080, 1996.
14. Zellers, E. T., Anna, D. H., Sulewski, R., and Wei, X., Improved Methods for the Determination of Hansen's Solubility Parameters and the Estimation of Solvent Uptake for Lightly Crosslinked Polymers, *J. Appl. Polym. Sci.*, 62, 2081–2096, 1996.
15. Luciani, A., Champagne, M. F., and Utracki, L. A., Interfacial Tension in Polymer Blends. Part 1: Theory, *Polym. Networks Blends,* 6(1), 41–50, 1996.
16. Luciani, A., Champagne, M. F., and Utracki, L. A., Interfacial Tension in Polymer Blends. Part 2: Measurements, *Polym. Networks Blends,* 6(2), 51–62, 1996.
17. Fuchs, O., Solvents and Non-Solvents for Polymers, in *Polymer Handbook*, 3rd ed., Brandrup, J. and Immergut, E. H., Eds., Wiley-Interscience, New York, 1989, VII/385.
18. Hansen, C. M., The Universality of the Solubility Parameter, *Ind. Eng. Chem. Prod. Res. Dev.*, 8(1), 2–11, 1969.
19. Lieberman, R. B. and Barbe, P. C., Polypropylene Polymers, in *Encylopedia of Polymer Science and Engineering,* Vol. 13, 2nd ed., Mark, H. F., Bikales, N. M., Overberger, C. G., Menges, G., and Kroschwitz, J. I., Eds., Wiley-Interscience, New York, 1988, 482–483.
20. Wessling, R. A., The Solubility of Poly(vinylidine Chloride), *J. Appl. Polym. Sci.*, 14, 1531–1545, 1970.
21. Wyzgoski, M. G., The Role of Solubility in Stress Cracking of Nylon 6,6, in *Macromolecular Solutions*, Seymour, R. B. and Stahl, G. A., Eds., Pergamon Press, New York, 1982, 41–60.
22. Mai, Y.-W., Environmental Stress Cracking of Glassy Polymers and Solubility Parameters, *J. Mater. Sci.*, 21, 904–916, 1986.
23. White, S. A., Weissman, S. R., and Kambour, R. P., Resistance of a Polyetherimide to Environmental Stress Crazing and Cracking, *J. Appl. Polym. Sci.*, 27, 2675–2682, 1982.

24. Van Dyk, J. W., Frisch, H. L., and Wu, D. T., Solubility, Solvency, and Solubility Parameters, *Ind. Eng. Chem. Prod. Res. Dev.,* 24(3), 473–478, 1985.
25. Segarceanu, O. and Leca, M., Improved Method to Calculate Hansen Solubility Parameters of a Polymer, *Prog. Org. Coat.,* 31(4), 307–310, 1997.
26. Kok, C. M. and Rudin, A., Prediction of Flory-Huggins Interaction Parameters from Intrinsic Viscosities, *J. Appl. Polym. Sci.,* 27, 353–362, 1982.
27. Athey, R. D., Testing Coatings: 6. Solubility Parameter Determination, *Eur. Coat. J.,* 5, 367–372, 1993.

4 Methods of Characterization — Surfaces

CONTENTS

ABSTRACT
Relations between cohesion parameters and surface energy parameters and their practical significance are discussed. Cohesion parameters (solubility parameters) can be used with full theoretical justification to characterize many surfaces, including substrates, coatings, plastics, pigment and filler surfaces, etc., in addition to the binder or polymer used in a given product. Important molecular relations between a binder in a coating or adhesive and its surroundings then become obvious. Use of cohesion parameters, i.e., Hansen solubility parameters in a total characterization of surface energy, clearly shows how the single point concepts of the (Zisman) critical surface tension and the wetting tension fit into a larger energy concept. A complete match of surface energies of two surfaces requires that exactly the same liquids (in a larger number of well-chosen test liquids) spontaneously spread on both surfaces. Also, the dewetting behavior (wetting tension test) of the liquids must also be the same, in that the same liquids should not retract when applied to the surfaces as films.

INTRODUCTION

Interfacial free energy and adhesion properties result from intermolecular forces. It has been recognized for many years that molecules interact by (molecular) surface to (molecular) surface contacts to enable solutions to be formed.[1] Since molecular surface-to-surface contacts control both solution phenomena and surface phenomena, it is not surprising that various correlations of cohesion parameters and surface phenomena can be found. This idea has been well explored and dealt with elsewhere.[2] The various treatments and correlations in the literature will not be explicitly dealt with here, other than those directly related to Hansen solubility parameters (HSP). In this chapter, solubility parameters are called cohesion (energy) parameters and refer more specifically to HSP. Solubility as such does not necessarily enter into the energetics of interfacial phenomena, but the energy characteristics of surfaces can still be correlated with HSP.

This chapter will emphasize methods of surface characterization using HSP. The orientation of adsorbed molecules is a significant added effect which must also be considered in many cases. The "like dissolves like" concept is extended and applied as "like seeks like"(self-assembly).

HANSEN SOLUBILITY PARAMETER CORRELATIONS WITH SURFACE TENSION (SURFACE FREE ENERGY)

Skaarup was the first to establish a correlation between liquid surface tension and HSP. This correlation with surface tension had been long lost in a report to members of the Danish Paint and Printing Ink Research Laboratory in 1967, as well as in an abstract for a presentation to the Nordic Chemical Congress in 1968.[3,4]

$$\gamma = 0.0688\,V^{1/3}\left[\delta_D^2 + k\left(\delta_P^2 + \delta_H^2\right)\right] \tag{4.1}$$

γ is the surface tension, and k is a constant depending on the liquids involved. This k was reported as 0.8 for several homologous series, 0.265 for normal alcohols, and 10.3 for *n*-alkyl benzenes.

Beerbower independently published essentially the same type of correlation in 1971.[5,6] With the exception of aliphatic alcohols and alkali halides, Beerbower found

$$\gamma = 0.0715\,V^{1/3}\left[\delta_D^2 + 0.632\left(\delta_P^2 + \delta_H^2\right)\right] \tag{4.2}$$

where γ is the surface tension. The constant was actually found to be 0.7147 in the empirical correlation. The units for the cohesion parameters are $(\text{cal/cm}^3)^{1/2}$ and those of the surface tension are dyn/cm in both Equations 4.1 and 4.2. However, values in dyn/cm are numerically equal to those in mN/m. The constant was separately derived as being equal to 0.7152 by a mathematical analysis in which the number of nearest neighbors lost in surface formation was considered, assuming that the molecules tend to occupy the corners of regular octahedra.

The correlations presented by Koenhen and Smolders[7] are also relevant to estimating surface tension from HSP. The author has never explored them in detail, however, so they are not discussed here.

It is interesting to note that δ_P and δ_H have the same coefficient in the surface tension correlations. They also have the same coefficient when solubility is correlated (see Chapter 1, Equation 1.9 or Chapter 2, Equation 2.6). The reason for this is the molecular orientation in the specific interactions derived from permanent dipole–permanent dipole and hydrogen bonding (electron interchange) interactions. The dispersion or London forces arise because of electrons rotating around a positive atomic nucleus. This causes local dipoles and attraction among atoms. This is a completely different type of interaction and requires a different coefficient in the correlations. It is this difference between atomic and molecular interactions which is basic to the entire discussion of similarity between HSP and the Prigogine corresponding states theory in Chapter 2. The finding that the polar and hydrogen bonding (electron interchange) effects require the same coefficient for both bulk and surface correlations suggests that the net effects of the (often mentioned) unsymmetrical nature of hydrogen bonding are no different from the net effects occurring with permanent dipole–permanent dipole interactions. The lack of specific consideration that hydrogen bonding is an unsymmetrical interaction led Erbil[8] to state that HSP has limited theoretical justification, for example. The previous discussion and the contents of Chapters 1 and 2 clearly indicate that the author is not in full agreement with this viewpoint. In fact, it appears that this book presents massive experimental evidence, related both to bulk and surface phenomena, which shows that the geometric mean is valid for estimating interactions between dissimilar liquids. This includes dispersion, permanent dipole–permanent dipole, and hydrogen bonding (electron interchange) interactions.

METHOD TO EVALUATE THE COHESION ENERGY PARAMETERS FOR SURFACES

One can determine the cohesion parameters for surfaces by observing whether or not spontaneous spreading is found for a series of widely different liquids. The liquids used in standard solubility parameter determinations are suggested for this type of surface characterization. It is strongly suggested that none of the liquids be a mixture, as this introduces an additional factor into the evaluations. The liquids in the series often used by the author are indicated in Chapter 3, Table 3.4 or Chapter 5, Table 5.2. Droplets of each of the liquids are applied to the surface and one simply observes what happens. If a droplet remains as a droplet, there is an advancing contact angle and the cohesion energy/surface energy of the liquid is (significantly) higher than that of the surface. The contact angle need not necessarily be measured in this simplified procedure, however. Contact angles have generally been found to increase for greater differences in cohesion parameters between the surface and liquid[9] (see also Figure 4.5). If spontaneous spreading is found, there is presumed to be some "similarity" in the energy properties of the liquid and the surface. The apparent similarity may be misleading. As discussed in greater detail later, the fact of spontaneous spreading for a given liquid does not mean that its HSP are identical with those of the surface being tested. If a given liquid does not spontaneously spread, it can be spread mechanically as a film and observed to see whether it retracts. This can be done according to ASTM D 2578-84 or ISO 8296:1987 (E). This test determines whether or not there is a receding contact angle under the given conditions.

Figure 4.1 shows a complete energy description for an epoxy polymer surface[10,11] based on the testing procedure described previously. The Hansen polar and hydrogen bonding parameters, δ_P and δ_H, are used to report the data. Further explanation of these parameters themselves can be found in

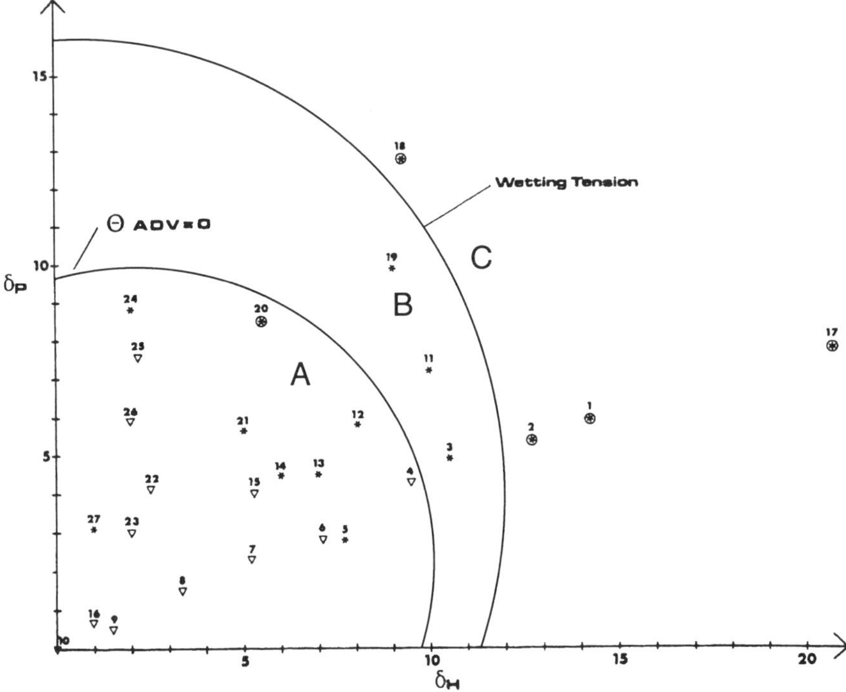

FIGURE 4.1 HSP surface characterization of an epoxy surface showing regions of spontaneous spreading of applied droplets (A), lack of dewetting of applied films (B), and dewetting of applied films (C). Note that this characterization may not be valid for all epoxy surfaces. (From Hansen, C. M. and Wallström, E., *J. Adhes.*, 15(3/4), 281, 1983. With permission.)

Chapter 1. The circular lines can be considered as portraying portions of HSP spheres, but the third Hansen parameter, δ_D, has not been specifically accounted for in the two-dimensional figure.

Figure 4.1 shows two curves which are concave toward the origin. The lower of these divides the test liquids into two groups based on spontaneous spreading or not. Below the line one finds that liquids applied as droplets will spontaneously spread. Liquids which are found in the region above the upper curve will retract when applied as films. A test method to determine this is found in the ASTM and ISO standards given previously, for example, except that one uses a large number of pure liquids instead of the liquid mixtures suggested in the standards. Receding contact angles will generally increase as one progresses to liquids with still higher HSP. Intermediate between the two curves in Figure 4.1 is a region where liquids applied as droplets will remain as droplets, while liquids applied as films will remain as films. This region deserves more attention in future research. The energy properties of these liquids are not as close to those of the surface as are the energy properties of the liquids which spontaneously spread. Spontaneous spreading is more related to adhesion since such liquids want to cover the surface spontaneously. The wetting tension test uses an external force to spread the liquids, after which they may continue to remain as a film. The mobility of the surface layer(s) will play a role in the wetting tension test. Hydrophilic segments can (perhaps) rotate toward a water droplet at some rate, for example, and increase the hydrophilic nature of the surface accordingly. This is discussed more in Chapter 11.

As mentioned earlier, there is still a problem in simplifying these results for easier use and improved understanding. Hexane, for example, does not dissolve an epoxy polymer, but in Figure 4.1 it is almost in the middle of the region describing spontaneous spreading of the liquids. Hexane will not contribute to a "bite" into an epoxy coating for improving intercoat adhesion with a subsequent coating. Hexane is within the region of spontaneous spreading because it has a lower surface energy (surface tension) than the epoxy surface. Nature reduces the free energy level of the surface by requiring hexane to cover the epoxy coating. The result of this is that the center of the normal HSP sphere for describing spontaneous spreading can be assigned sizable negative values.[11] This is both impractical and impossible. A better method of handling this situation is still desired, and until it is found, one must presumably refer to simple plots or other simple comparisons rather than to refined computer techniques, which are more desirable in most cases. In the meantime, interest will still be focused onto the usual test method(s) for determining surface tensions based on the Zisman critical surface tension plots (lack of advancing contact angle) or by using the ASTM procedure for wetting tension (lack of receding contact angle). The following discussion relates these to the HSP-type characterizations discussed earlier.

Additional surface characterization plots for spontaneous spreading and wetting tension using HSP are included in Figure 4.2 for a plasticized polyvinyl chloride (PVC) and in Figure 4.3 for a polyethylene (PE).

A CRITICAL VIEW OF THE CRITICAL SURFACE TENSIONS[12,13]

The Zisman critical surface tension is determined by measuring the extent that affinity is lacking (contact angles) for a surface using pure liquids or liquid mixtures in a series. The surface tension of each of the liquids is known. One can then plot cosine of the contact angle vs. liquid surface tension and extrapolate to the limit where the contact angle is no longer present (see Figure 4.4). Liquids with higher surface tensions than this critical value allow measurement of a contact angle, while liquids with lower surface tensions than the critical value will spontaneously spread. The fact that the liquid with a surface tension just under the critical value spontaneously spreads is often taken as an indication of high affinity. This is difficult to understand and appears to be a misunderstanding. The limiting critical surface tension[12,13] has very little to do with the "best" solvent for the surface. It is more appropriately compared with a very poor solvent which can only marginally dissolve a polymer, for example. This is similar to the condition for a RED number equal to 1.0

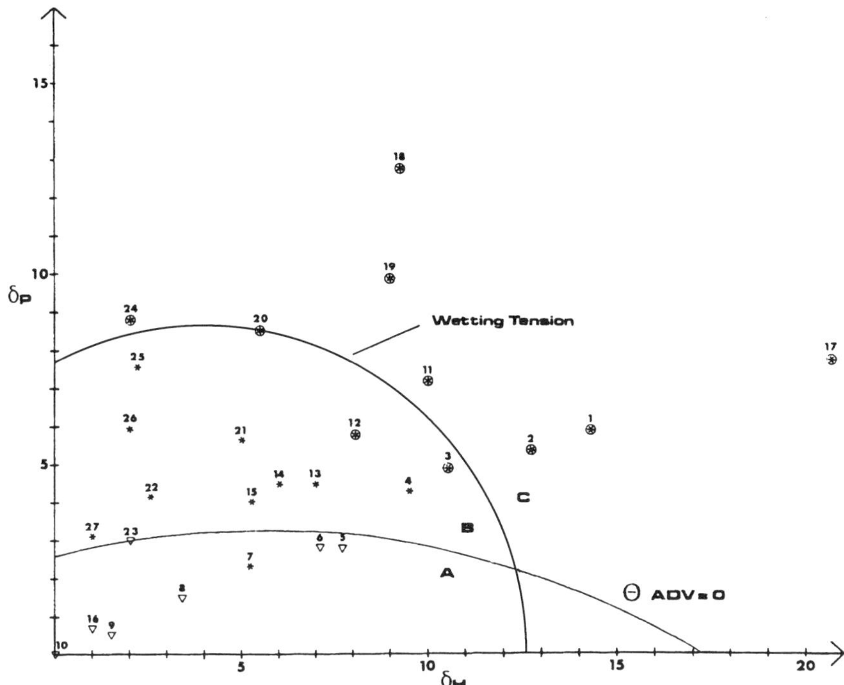

FIGURE 4.2 HSP surface characterization of spontaneous spreading of applied droplets and wetting tension for applied films for plasticized polyvinyl chloride (PVC). Note that these characterizations may not be valid for all PVC surfaces. (From Hansen, C. M. and Wallström, E., *J. Adhes.,* 15(3/4), 280, 1983. With permission.)

discussed in Chapters 1 and 2. Measuring the critical surface tension has been and still will be a useful technique to better understand surfaces, but it should be done with the following in mind.

Who would determine the solubility parameter for a polymer by the following method? One makes up a series of liquids with different, known solubility parameters. The polymer dissolves in some of them, and the degree of swelling of the polymer in question is measured in those liquids which do not dissolve it fully. One subsequently determines the solubility parameter of the polymer by extrapolating the degree of swelling to infinity, this corresponding to total solution. This extrapolation can be done by plotting 1/(degree of swelling) vs. solvent composition (solubility parameter). One now focuses attention upon that liquid which (by extrapolation) just dissolves the polymer. One assumes that there is no better solvent than this one, and consequently assigns the polymer solubility parameters corresponding to those of this boundary solvent. This is exactly what one does when the critical surface tension is measured. This method should clearly never be used to determine solubility parameters for polymers. At the same time, it sheds some light onto the true meaning of the critical surface tension.

If we now consider the region for spontaneous spreading in Figure 4.1, 4.2, or 4.3, it can be seen that the critical surface tension is a point on its boundary. In practice, one finds different critical surface tensions for the same surface depending on which liquids (or liquid mixtures) are used. This is explained by the fact that the cohesion parameter regions of the type shown in Figures 4.1 to 4.3 are not symmetrical around the zero axis. The individual liquid series used to determine the critical surface tension will intersect the cohesion parameter spontaneous spreading boundary at different points. The corresponding total surface tension will vary from intersection to intersection as mentioned earlier. Hansen and Wallström[11] compared the critical surface tension plotting technique with one where a difference in HSP was used instead of liquid surface tension. One arrives at the same general conclusions from both types of plotting techniques. This comparison is made in Figures 4.4 and 4.5.

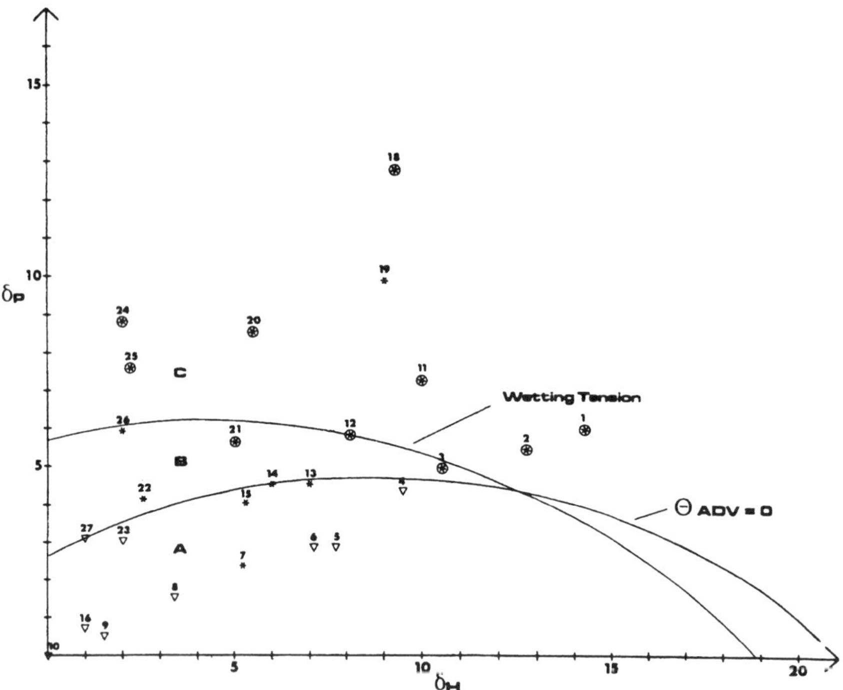

FIGURE 4.3 HSP surface characterization of spontaneous spreading of applied droplets and wetting tension for applied films for a polyethylene (PE) surface. Note that these characterizations may not be valid for all PE surfaces. (From Hansen, C. M. and Wallström, E., *J. Adhes.,* 15(3/4), 279, 1983. With permission.)

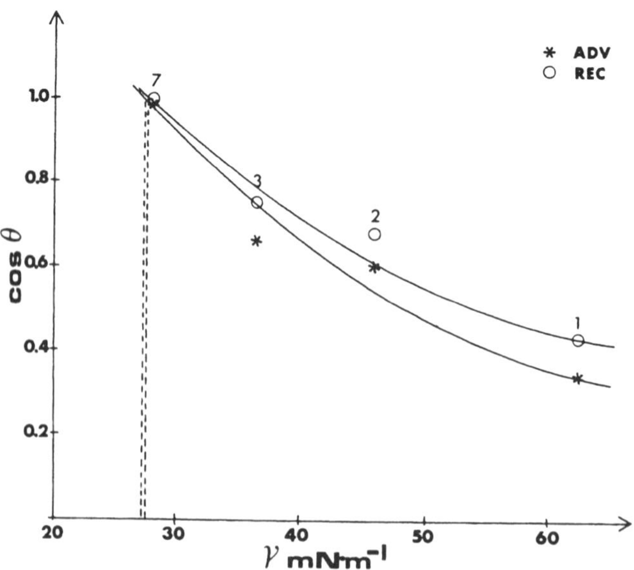

FIGURE 4.4 Zisman critical surface tension plot of cosine of the static advancing and receding contact angles vs. liquid surface tension for low density polyethylene. The same data are used in Figure 4.5. (From Hansen, C. M. and Wallström, E., *J. Adhes.,* 15(3/4), 282, 1983. With permission.)

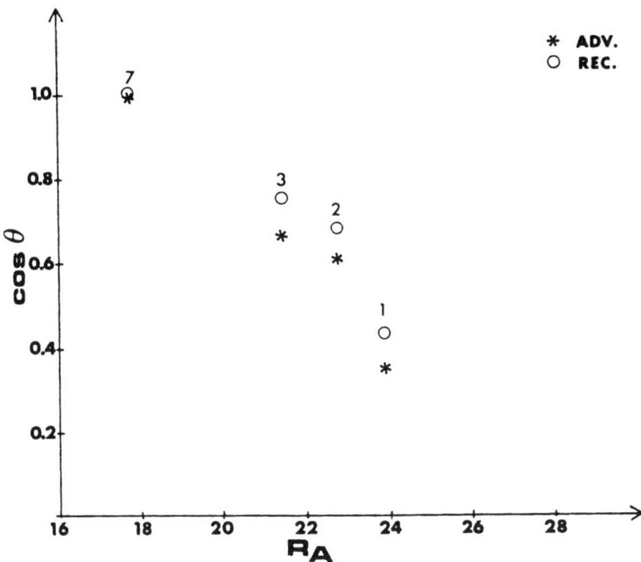

FIGURE 4.5 Critical HSP plot of cosine of the static advancing and receding contact angles vs. the HSP difference as defined by Chapter 1, Equation 1.9 for low density polyethylene. The same data are used in Figure 4.4. (From Hansen, C. M. and Wallström, E., *J. Adhes.*, 15(3/4), 282, 1983. With permission.)

A CRITICAL VIEW OF THE WETTING TENSION

A region larger than that for spontaneous spreading will be found on a δ_P vs. δ_H plot when one plots data for those liquids which remain as films (do not breakup or contract) when they are applied as films. This type of experiment measures the wetting tension. Mixtures of formamide and ethylene glycol monoethyl ether are usually used in practice for these measurements according to ASTM D 2578-84 or ISO 8296:1987 (E). One can also use the same liquids suggested earlier for cohesion parameter determinations and make a plot like that in Figure 4.1. If two different surfaces are to have the same wetting tension behavior, their plots must be the same.

The results of the ASTM test are usually stated in terms of the surface tension of the liquid or liquid mixture which just stays intact as a film for 2 s. This simple single point determination corresponds to determining a single point on the boundary of the HSP plot describing wetting tension for all liquids. A single point determination may not always be sufficient information and certainly neglects the complete picture possible from HSP considerations. Comments identical in principle to those included in the earlier section, A Critical View of the Critical Surface Tensions, on measurement of the critical surface tension are also valid here.

It is hoped the reader now better understands the total energy context of the simple ASTM wetting tension measurements.

ADDITIONAL HANSEN SOLUBILITY PARAMETER SURFACE CHARACTERIZATIONS AND COMPARISONS

Beerbower[14] has reported many other correlations of surface phenomena with HSP. Examples include the work of adhesion on mercury; frictional properties of untreated and treated polyethylene for 2 and 5 min, respectively, with $H_2S_2O_7$; the Joffé effect — effect of liquid immersion on fracture strength of soda-lime glass; and the Rehbinder effect — crushing strength of Al_2O_3 granules under various liquids. Beerbower has also brought cohesion parameters into the discussion of wear and boundary lubrication.[14] It appears that these factors should still have some consideration, even though recent progress and understanding in the area are much more advanced.[15]

Additional surface characterizations using HSP are reported in Chapter 5. These include characterizations of the surfaces of pigments, fillers, and fibers. Both organic and inorganic materials have been characterized. The test method used is to determine sedimentation rates for the materials of interest in the same large number of solvents traditionally used in HSP studies. Adsorption of given liquids onto the particle or fiber surface slows the sedimentation rate, and indeed some (fine) particles with rather high densities suspend for years in organic liquids with rather modest densities. A significant advantage in this testing method is that hexane, for example, is not able to retard sedimentation where it may spontaneously spread, as discussed above. Hexane is not an isolated example of this behavior. The characterizations using standard HSP procedures indicate it is truly high affinity for the surface which is important in these characterizations and not just spontaneous spreading. The reason for this may be the extent (or depth) of the adsorption layer, as well as whether the adsorption occurs at specific sites, or both. Results may be affected when molecules in a surface can orient differently from their original state upon contact with a liquid, for example, with water (see the discussion in Chapter 11).

SELF-STRATIFYING COATINGS

A newer development in the coatings industry is to apply a single coat of paint which separates by itself into a primer and topcoat. A special issue of *Progress in Organic Coatings* was recently devoted to this type of coating.[16] Misev has also discussed formulation of this type of product using HSP concepts.[17] The separation of the binders into primer and topcoat must occur while the coating is still liquid enough to allow the necessary transport processes to occur. The solvent must just dissolve the binders such that they become incompatible when it begins to evaporate. The binder with the lowest energy (surface tension/cohesion parameters) will naturally migrate toward the low energy air interface, and, therefore, this determines which of the binders makes up the topcoat. There are a number of other factors which are important for the process, including polymer molecular weight, rate of solvent evaporation, etc., but these will not be discussed here. This discussion is included because it once more demonstrates how cohesion parameters are coupled with surface energy and also to interfacial energy. The interface between the topcoat and primer is formed from an otherwise homogeneous system. The previous considerations lead to the expectation that the magnitude of the interfacial surface tension between two incompatible polymers is closely related to the difference in their cohesion parameters. Without going into greater detail, it is widely known among those who work with partially compatible polymers that this is indeed the case.[18,19]

Figure 4.6 shows the principles involved for selecting the solvent which can make these work. The polymer with HSP nearest the origin will be the topcoat, since it has the lower (surface or cohesion) energy of the two. A solvent is required which dissolves both polymers, so it will be located in the common region to the spheres portrayed. Mutual solubility of two polymers is promoted when the solvent favors the polymer which is most difficult to dissolve.[20] This is usually the one with the higher molecular weight. It is clear that selection of the optimum solvent for this process of designed generation of an interface is aided by systematic use of HSP.

MAXIMIZING PHYSICAL ADHESION

If one wishes to maximize physical adhesion, the physical similarity (same HSP) of the two interfaces being joined must be as close as possible. The previous discussion suggests that physical similarity can be obtained when two criteria are met. The first criterion is that exactly the same liquids spontaneously spread on each of the surfaces to be joined. The second criterion is that

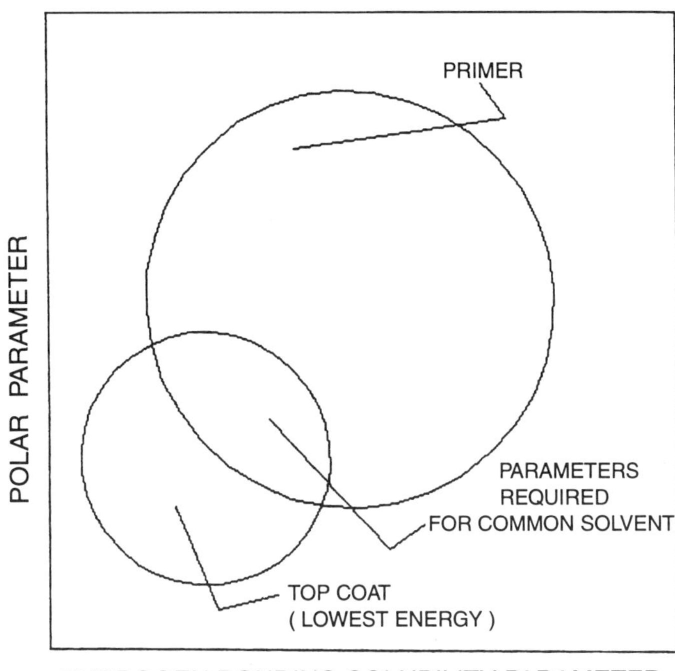

FIGURE 4.6 Sketch of HSP principles used to formulate a self-stratifying coating from an initially homogeneous solution (see discussion in text). (From Birdi, K. S., Ed., *Handbook of Surface and Colloid Chemistry*, CRC Press, Boca Raton, FL, 1997, 324.)

exactly the same liquids maintain films when spread (ASTM method for wetting tension) on each of the surfaces to be joined. Any differences in this spontaneous spreading or wetting tension behavior can be interpreted as being a difference in physical similarity. The differences in the behavior of liquid droplets or films which are observed may suggest which steps can be taken to minimize differences, if this is required. Should one add aliphatic segments to reduce the polar and hydrogen bonding contributions? Should alcohol and/or acid groups be incorporated to increase the hydrogen bonding in the system? This type of approach can be used to establish guidelines for action relative to each of the HSP parameters. Aromatic character and halogens other than fluorine characteristically increase δ_D; nitro and phosphate groups characteristically increase δ_P; and alcohol, acid, and primary amine groups characteristically increase δ_H. Reference can be made to the table of group contributions in Chapter 1 (Table 1.1) for more precise comparisons. The discussion of forming good anchors on pigment and other surfaces found in Chapter 6 is also relevant to the present discussion, since such anchors can also be used to enhance adhesion.

CONCLUSION

Greater insight into the makeup of a product is possible when one not only knows the cohesion parameters, i.e., HSP, for polymers and solvents it contains, but also the HSP for the various surfaces which these encounter. The surfaces of substrates, pigments, fillers, plastics, fibers, and other materials can also be characterized by HSP (see Chapters 3 and 5). This allows mutual interactions to be inferred by comparisons of which materials are similar and which materials are different in terms of their HSP. Similar materials in this context have similar HSP regardless of differences in composition.

The critical surface tension and wetting tension are single point determinations. Cohesion parameters allow a more complete characterization of surfaces than do these single point measurements and, at the same time, allow insight as to how the single point measurements fit into the overall energy picture for the product. Guidelines for systematically changing the affinities of surfaces can also be obtained from HSP concepts.

Both the spontaneous spreading region and the wetting tension region on HSP plots for two different surfaces must be identical if they are to have identical overall surface characteristics.

REFERENCES

1. Flory, P. J., *Principles of Polymer Chemistry*, Cornell University Press, New York, 1953.
2. Barton, A. F. M., Applications of Solubility Parameters and Other Cohesion Energy Parameters, *Polym. Sci. Technol. Pure Appl. Chem.*, 57(7), 905–912, 1985.
3. Skaarup, K. and Hansen, C. M., The Three-Dimensional Solubility Parameter and Its Use, (Det Tredimensionale Opløselighedsparametersystem og dets Anvendelse), Rapport No. 54 (TM 2-67), Lak- og Farveindustriens Forskningslaboratorium, København, 1967 (in Danish).
4. Skaarup, K., Surface Tension and 3-D Solubility Parameters (Overfladespænding og 3-D Opløselighedsparametre), Nordiske Kemikermøde, København, 1968 (in Danish).
5. Beerbower, A., Surface Free Energy: a New Relationship to Bulk Energies, *J. Colloid Interface Sci.*, 35, 126–132, 1971.
6. Hansen, C. M. and Beerbower, A., Solubility Parameters, in *Kirk-Othmer Encyclopedia of Chemical Technology*, Suppl. Vol., 2nd ed., Standen, A., Ed., Interscience, New York, 1971, 889–910.
7. Koenhen, D. N. and Smolders, C. A., The Determination of Solubility Parameters of Solvents and Polymers by Means of Correlation with Other Physical Quantities, *J. Appl. Polym. Sci.*, 19, 1163–1179, 1975.
8. Erbil, H. Y., Surface Tension of Polymers, in *Handbook of Surface and Colloid Chemistry*, Birdi, K. S., Ed., CRC Press, Boca Raton, FL, 1997, 265–312.
9. Hansen, C. M., Characterization of Liquids by Spreading Liquids, *J. Paint Technol.*, 42(550), 660–664, 1970.
10. Hansen, C. M. and Pierce, P. E., Surface Effects in Coatings Processes, *XII Federation d'Associations de Techniciens des Industries des Peintures, Vernis, Emaux et Encres d'Imprimerie de l'Europe Continentale, Congress Book*, Verlag Chemie, Weinheim/Bergstrasse, 1974, 91–99; *Ind. Eng. Chem. Prod. Res. Dev.*, 13(4), 218–225, 1974.
11. Hansen, C. M. and Wallström, E., On the Use of Cohesion Parameters to Characterize Surfaces, *J. Adhes.*, 15(3/4), 275–286, 1983.
12. Zisman, W. A., Relation of the Equilibrium Contact Angle to Liquid and Solid Constitution, in *Contact Angle, Wettability and Adhesion, Advances in Chemistry Series*, No. 43, Gould, R. F., Ed., American Chemical Society, Washington, D.C., 1964, chap. 1.
13. Zisman, W. A., Surface Energetics of Wetting, Spreading, and Adhesion, *J. Paint Technol.*, 44(564), 41, 1972.
14. Beerbower, A., Boundary Lubrication — Scientific and Technical Applications Forecast, AD747336, Office of the Chief of Research and Development, Department of the Army, Washington, D.C., 1972.
15. Krim, J., Friction at the Atomic Scale, *Sci. Am.*, 275(4), 48–56, October 1996.
16. Special issue devoted to Self-Stratifying Coatings, *Prog. Org. Coat.*, 28(3), July 1996.
17. Misev, T. A., Thermodynamic Analysis of Phase Separation in Selfstratifying Coatings — Solubility Parameters Approach, *J. Coat. Technol.*, 63(795), 23–28, 1991.
18. Luciani, A., Champagne, M. F., and Utracki, L. A., Interfacial Tension in Polymer Blends. Part 1: Theory, *Polym. Networks Blends*, 6(1), 41–50, 1996.
19. Luciani, A., Champagne, M. F., and Utracki, L. A., Interfacial Tension in Polymer Blends. Part 2: Measurements, *Polym. Networks Blends*, 6(2), 51–62, 1996.
20. Hansen, C. M., On Application of the Three Dimensional Solubility Parameter to the Prediction of Mutual Solubility and Compatibility, *Färg och Lack*, 13(6), 132–138, 1967.

5 Methods of Characterization for Pigments, Fillers, and Fibers

CONTENTS

ABSTRACT

Cohesion parameters for pigments, fillers, and fibers can often be evaluated by observation of the suspension and/or sedimentation behavior of particulate matter in different liquids. These characterizations are based on relatively stronger adsorption by some of the liquids compared with others. Those liquids with stronger interaction can suspend finer fractions of solids indefinitely or retard sedimentation compared with the other liquids. Data should be interpreted by accounting for differences in the densities and viscosities of the test liquids, such that a relative sedimentation rate can be used for comparisons. The absolute sedimentation rates are generally not of primary interest. Data from such evaluations can be computer processed to assign Hansen cohesion parameters (HSP) to the material in question. Cohesion parameter data are given for some newer pigments and barytes to demonstrate the principles.

INTRODUCTION

The possibilities offered by cohesion parameter characterization of pigments, fillers, and fibers have not been generally recognized, at least not judging from the number of publications appearing on the topic. Pigments and a few fillers were characterized in some of the author's first publications dealing with the solubility parameter.[1,2] These were given δ_D, δ_P, and δ_H parameters (HSP) and a characteristic radius of interaction exactly analogous to polymer characterizations discussed in Chapters 2 and 3. These data together with some more recent pigment characterizations are included in Tables 5.1A and 5.1B. Shareef et al.[3] have also characterized pigment surfaces, including metal oxides. Gardon and Teas,[4] among other things, clearly showed the differences between organic phosphate treated and untreated zinc oxides using a cohesion parameter characterization. Inorganic fibers have also been characterized.[5] All of these characterizations again confirm the universality possible with these parameters. They reflect molecule–molecule interactions whether at surfaces or in bulk.

In the future, more systematic selection of dispersion aids should be possible, since these can also be described with the same energy parameters. Hansen and Beerbower have touched on this topic.[6] Each segment of such molecules requires its own HSP. The discussion in Chapter 9 for the interactions within cell walls in wood demonstrates how this could be done. It has been shown by

TABLE 5.1
HSP Correlations for Older Inorganic Pigments[1,2] and Metal Oxides[3] (units are $MPa^{1/2}$)

Material	δ_D	δ_P	δ_H	Ro
Kronos® RN57 TiO_2[a]	24.1	14.9	19.4	17.2
Aluminum Pulver Lack 80[a]	19.0	6.1	7.2	4.9
Red Iron Oxide[a]	20.7	12.3	14.3	11.5
Synthetic Red Iron Oxide[b]	16.1	8.6	15.0	11.3
Synthetic Yellow Iron Oxide[b]	17.3	6.0	14.5	12.5
	16.1	8.6	15.0	11.3
Zinc Oxide	16.9	7.8	10.6	13.2
	16.2	10.8	12.7	9.8

[a] From Reference 1.
[b] From Reference 3.

TABLE 5.1A
List of Pigments Studied

Pigment	Description
1	*TiO_2, Kronos RN 57,* Titan Co. A/S., Frederikstad, Norway.
2	*Phthalocyanine Blue,* B6, E. I. du Pont de Nemours and Co. (1949).
3	*Isolbonared Nr. 7522, C. I. Pigment Red 48* (C.I. 15865) (MnSalt), Køge Chemical Works, Køge, Denmark.
4	*Peerless Carbon Black*
5	*Isol Fast Yellow IO GX 2505, C.I. Pigment Yellow 3,* Køge Chemical Works, Køge, Denmark.
6	*Reflex Blau TBK Ext.* (No C.I. Index-pigment mixture), Farbwerke Hoechst, Frankfurt (M), Germany.
7	*Isol Ruby BKS 7520, C.I. Pigment Red 57* (C.I. 15850) (Ca Salt), Køge Chemical Works, Køge, Denmark.
8	*Hansagelb 10 G, C.I. Pigment Yellow 3* (C.I. 11710), Farbwerke Hoechst, Frankfurt (M), Germany.
9	*Fanalrosa G Supra Pulver, Pigment Red 81* (C.I. 45160), BASF, Ludwigshafen, Germany.
10	*Heliogenblau B Pulver, C.I. Pigment Blue 15* (C.I. 74160), BASF, Ludwigshafen, Germany.
11	*Heliogengrün GN, C.I. Pigment Green 7,* (C.I. 74260), BASF, Ludwigshafen, Germany.
12	*Permanentgelb H 10 G, C.I. Pigment Yellow 81,* (No C.I. index), Farbwerke Hoechst, Frankfurt (M), Germany.
13	*Permanent Bordeaux FRR, C.I. Pigment Red 12* (C.I. 12385), Farbwerke Hoechst, Frankfurt (M), Germany.
14	*Permanent Violet RL Supra, C.I. Pigment Violet 23,* (C.I. 12505), Farbwerke Hoechst, Frankfurt (M), Germany.
15	*Isol Benzidine Yellow G 2537, C.I. Pigment Yellow 12* (C.I. 21090), Køge Chemical Works, Køge, Denmark.
16	*Brillfast Sky Blue 3862, C.I. Pigment Blue 3* (C.I. 42140), J. W. and T. A. Smith Ltd., London.
17	*Permanent Orange G, C.I. Pigment Orange 13* (C.I. 21110), Farbwerke Hoechst, Frankfurt (M), Germany.
18	*Permanent Red, FGR Extra Pulver, C.I. Pigment Red 112,* (C.I. 12370). Farbwerke Hoechst, Frankfurt (M), Germany.
19	*Isol Fast Red 2G 2516, C.I. Pigment Orange 5,* (C.I. 12075), Køge Chemical Works, Køge, Denmark.
20	*Monolite Fast Blue 3 RS, Powder, C.I. Vat Blue 4* (C.I. 69801), Imperial Chemical Industries.
21	*Heliogenblau LG, Pulver, C.I. Pigment Blue 16* (C.I. 74100), BASF., Ludwigshafen, Germany.
22	*Red Iron Oxide.*
23	*Carbon Black, Printex V* (5519-1), Degussa, Frankfurt (M), Germany.
24	*Aluminum Pulver Lack 80,* Eckart-Werke, 851 Fürth/Bayern, Germany.
25	*Isol Benzidene Yellow GA-PR, 9500, C.I. Pigment Yellow 12,* Køge Chemical Works, Køge, Denmark.

From Reference 1.

TABLE 5.1B
Characteristic Parameters for Various Pigments [units are $(cal/cm^3)^{1/2}$]

Pigment	δ_0	δ_{d_0}	δ_{p_0}	δ_{h_0}	δ_{a_0}	R'_{A_0}	Comments
1	16.8	11.8	7.3	9.5	12.0	8.4	Suspension
2	10.5	9.3	3.1	3.7	4.8	2.3	Few suspending solvents
3	10.0	8.7	3.5	3.5	5.0	2.5	Few suspending solvents
4	13.6	10.3	6.0	6.6	8.9	6.0	Suspension
5	11.9	10.2	4.8	3.8	6.1	4.4	Color only
6	13.2	10.8	3.8	6.6	7.6	7.0	Mixed color and suspension
7	10.5	9.6	3.0	3.2	4.4	3.9	Suspension
8	10.5	9.1	4.0	3.3	5.2	3.3	Color only
9	13.0	9.8	7.0	5.0	8.6	5.2	Color only
10	12.0	10.8	3.5	4.0	5.3	5.2	Suspension
11	12.0	10.0	4.8	4.5	6.6	4.8	Primarily suspension
12	8.8	8.4	1.5	2.3	2.7	2.2	Suspension
13	13.2	10.7	4.8	6.1	7.8	5.2	Color only
14	11.5	9.6	5.2	3.6	6.3	4.4	Mixed color and suspension
15	10.2	9.3	3.0	2.9	4.2	3.9	Mixed color and suspension
16	13.3	9.5	7.2	6.0	9.4	5.1	Suspension
17	11.5	9.7	3.9	4.7	6.1	4.5	Color only
18	11.2	10.0	3.5	3.5	5.0	5.0	Color only
19	14.2	10.9	5.6	7.1	9.0	7.0	Primarily color
20	15.2	10.8	6.5	8.5	10.7	7.0	Suspension
21	13.5	10.7	5.0	6.5	8.2	6.0	Suspension
22	13.7	10.1	6.0	7.0	9.2	5.6	Suspension
23	13.1	10.3	6.0	5.5	8.1	5.5	Suspension
24	10.4	9.3	3.0	3.5	4.6	2.4	Suspension
25	9.1	9.0	2.7	2.3	3.6	2.5	Suspension

From Reference 1.

calculation that hemicelluloses act like surface-active agents with some segments seeking lower energy lignin regions and some segments, those with alcohol groups, orienting toward the higher energy cellulose.

METHODS TO CHARACTERIZE PIGMENT, FILLER, AND FIBER SURFACES

The cohesion parameter (HSP) approach to characterizing surfaces gained impetus by experiments where the suspension of fine particles in pigment powders was used to characterize 25 organic and inorganic pigment surfaces.[1,2] Small amounts of the pigments are shaken in test tubes with a given volume of liquid (10 ml) of each of the test solvents, and one then observes sedimentation or lack of the same. When the solid has a lower density than the test liquid, it will float. Rates of floating have also been noted, but the term sedimentation will be retained here for both sedimentation and floatation. The amounts of solid sample added to the liquids can vary depending on the sample in question, and some initial experimentation is usually advisable. If the pigment or filler particle size is large, say, over 5 μm, the surface effects become less significant compared with a sample where the particle size in only 0.01 μm. Problems arise when the pigments are soluble enough to color the liquid such that sedimentation cannot be evaluated. The larger particle size pigments and fillers may sediment very rapidly. Sedimentation rates have still been used successfully in some of these cases. The sedimentation rate is most easily expressed as the time at which the amount of particles at a given point in the test tubes has fallen to some small amount, perhaps zero. Observations can be made visually, or perhaps instrumentally, in a direction perpendicular to the incidence of a laser light. A visual observation is required in any event, since some samples seem to coat out rapidly on glass surfaces. Some pigments have portions which suspend for years in spite of large density differences and relatively low solvent viscosity. Satisfactory results from this type of measurement require some experience regarding what to look for. This can vary from sample to sample.

A characterization is less certain when there are only 4 or 5 "good" liquids out of the perhaps 40 to 45 tested, although this depends somewhat on which liquids are involved. "Good" in this context means suspension of particulates is prolonged significantly compared with the other test solvents after compensating for differences in density and viscosity. A corrected relative sedimentation time, RST, can be found by modifying the sedimentation time, t_s.

$$\text{RST} = t_s\left(\rho_p - \rho_s\right)/\eta \tag{5.1}$$

The ρs are densities of particle and test liquid, and η is the liquid viscosity.

A prolonged RST implies greater adsorption of the given solvent onto the surface in question. Characterizations based on these techniques tend to place emphasis on the nature of the surfaces for the smaller particle size fractions.

An example of a data sheet used for such studies is included in Table 5.2.

DISCUSSION — PIGMENTS, FILLERS, AND FIBERS

It can be reasoned that one most easily gets the most benefit from a pigment, filler, or fiber when the pigment surface and the binder in question have the same cohesion parameters. There are apparently no publications indicating a systematic modification of pigment surfaces to achieve a given set of cohesion parameters. The characterizations for the organic pigments studied most recently are given in Table 5.3. These data indicate that their respective surfaces are essentially identical. An exception is the first one in the table where the analysis is based on only three "good" solvents which were able to extend sedimentation significantly relative to the other solvents tested.

These results suggest that pigment manufacturers have essentially arrived at the same result, a surface energy compatible with a wide variety of currently used binders. The solvents most frequently appearing as "good" for adsorption onto these surfaces include several chlorinated solvents, toluene, and tetrahydrofuran. Since these solvents dissolve the most commonly used binders, one can conclude that the common binders will adsorb readily onto these pigment surfaces. This will give a good result, provided the solvent is not so good for the binder that it can remove the binder from the pigment surface.

Schröder[7] (BASF) confirms that the optimum polymer adsorption will be found when the binder and pigment surface have the same HSP. He indicates that the solvent should be very poor for the pigment and located in the boundary region for the binder. He prefers the pigment to have HSP values placing it intermediate between the solvent and binder. This is suggested for conditions where the solvent has higher HSP than the pigment, as well as for conditions where the solvent has lower HSP than the pigment. This situation, with the solvent and binder on opposite sides of the pigment, means the composite vehicle has parameters very closely matching those of the pigment.

There is also a relation between how clearly a pigment can be characterized by sedimentation measurements and its zeta potential. Low zeta potential means sedimentation is rapid in all solvents, and this type of characterization becomes difficult or perhaps impossible. The zeta potential reflects the intensity (percentage coverage? how many layers?) of the surface energy characteristics. It does not clearly indicate specific affinity relations of a given binder for the pigment surface as a result

TABLE 5.2
Sedimentation Study

DATE **REF. NR.:**

SAMPLE:

DENSITY OF SAMPLE (Dp):

Solvent	D (Ds) 20°C	Viscosity 20°C	No.	Dp/Ds	Sedimentation Time (min)		Relative Sedimentation Time (RST)	
					From	To	From	To
Acetone	0.79	0.35	4					
Acetophenone	1.03	1.90	6					
Benzene	0.88	0.65	13					
1-Butanol	0.81	4.00	28					
Butyl acetate	0.87	0.74	30					
Butyrolactone	1.29	1.92	37					
Carbon tetrachloride	1.59	0.99	40					
Chlorobenzene	1.10	0.80	41					
Chloroform	1.48	0.37	44					
Cyclohexane	0.78	1.00	47					
Cyclohexanol	0.95	68.00	48					
Diacetone alcohol	0.94	3.20	56					
o-Dichlorobenzene	1.31	1.27	61					
Diethylene glycol	1.12	35.70	75					
Diethyl ether	0.72	0.23	82					
Dimethyl formamide	0.95	0.82	90					
DMSO	1.10	1.98	94					
1,4-Dioxane	1.04	1.31	96					
Dipropylene glycol	1.03	107.0	98					
Ethanol	0.82	1.22	104					
Ethanolamine	0.91	24.10	105					
Ethyl acetate	0.89	0.44	106					
Ethylene dichloride	1.25	0.84	120					
Ethylene glycol	1.12	20.90	121					
Ethyelene glycol monobutyl ether	0.90	2.90	123					
Ethyelene glycol monoethyl ether	0.93	2.05	124					
Ethyelene glycol monomethyl ether	0.96	1.72	126					
Formamide	1.13	3.30	131					
Hexane	0.66	0.33	140					
Isophorone	0.92	2.60	148					
Methanol	0.79	0.59	153					
Methylene dichloride	1.33	0.43	162					
Methyl isobutyl ketone	0.96	0.59	167					
Methyl-2-pyrrolidone	1.03	1.80	172					
Nitrobenzene	1.21	2.03	177					
Nitroethane	1.05	0.55	178					
Nitromethane	1.13	0.62	179					
2-Nitropropane	0.99	0.75	181					
Propylene carbonate	1.20	2.80	204					
Propylene glycol	1.04	56.00	205					
Tetrahydrofurane	0.89	0.55	222					
Toluene	0.87	0.59	225					
Trichloroethylene	1.47	0.58	229					

TABLE 5.3
HSP Correlations for Selected Materials (units are $MPa^{1/2}$)

Material	d_D	d_P	d_H	Ro	FIT	G/T
Organic pigments						
Paliotol® Gelb L1820 BASF	18.9	3.5	10.5	5.4	0.99	3/35
Heliogen® Blau 6930L BASF	18.0	4.0	4.0	4.0	1.00	5/34
Socco Rosso L3855 BASF	17.3	5.7	2.7	4.1	0.99	4/34
Perm Rubin F6B Hoechst	16.7	3.7	3.1	4.8	0.88	6/33
Perm Gelb GRL02 Hoechst	16.7	2.5	3.7	4.5	0.95	5/37
Perm Lackrot LC Hoechst	19.0	5.0	5.0	4.0	1.00	7/28
Inorganic pigments, fillers, etc.						
Cabot Hochdisperse[a]	16.7	9.3	11.5	11.7	—	23/23
Cabot Hochdisperse	19.3	9.5	10.3	12.7	0.79	23/31
Zeta Potential Blanc Fixe[b]	26.5	19.1	14.5	20.4	0.95	5/19

[a] Data analysis which only considers the good solvents to define the least sphere possible. See discussion of the SPHERE1 program in Chapter 1.

[b] Data from Reference 8.

Note: A perfect data fit of 1.0 means that there most probably are other sets of the same parameters which will have a data fit of 1.0 and also define a sphere which surrounds all the good solvents. A data fit of 0.99+ is preferred to define the optimum sphere for this reason. G/T is the number of good (G) liquids and the total (T) number of liquids in a correlation.

of a given surface treatment, for example. This is given by HSP. To obtain a complete picture of the energetics of the surface, one needs an intensity factor, i.e., the zeta potential, as well as a qualitative factor, i.e., the cohesion energy parameters. The latter is generally lacking. One can suspect that some pigments have such high intensity (zeta potential) (at some cost) that even though the cohesion parameters match poorly with a given binder, a system can still function satisfactorily. An HSP correlation for the zeta potential of blanc fixe is given in Table 5.3 using data from Winkler.[8] This is discussed further below.

Acid–base theories have been popular.[9-11] The author has not found it necessary to resort to this type of approach in any activities, although many have clearly found them beneficial. More research is needed to fully understand the successes of the acid–base as well as the HSP approaches. It would seem that the HSP approach allows predictive ability not possible with an acid–base approach. However, the current problem is the lack of data.

Organic pigments normally have a good organic substrate on which to base an organic surface modification. The characterizations may reflect both a surface treatment and the surface of the base particles, depending on how the test liquids interact with these. It should not be too difficult to modify an organic surface to an alternative organic surface with satisfactory properties, if desired. It is conceptually and in practice more difficult to modify an inorganic surface to make it compatible with organic systems. This requires a significant change in surface energy from high to much lower and, presumably, also requires a greater degree of coverage to mask the base inorganic surface. The producer of inorganic pigments and fillers must either give their products suitable surfaces, probably after much effort, or else one needs one or more supereffective additives to be able to achieve a good and stable dispersion. It helps to incorporate given (high cohesion energy) groups in a grinding resin, such as acid, alcohol, amine, etc. The relatively high local cohesion parameters in the binder associated with these groups means that they will have high affinity for the high cohesion energy surface of the inorganic material. At the same time, these local regions of adsorbed polymer segments are not particularly soluble or are insoluble in the cheaper hydrocarbon solvents, for example, which have much lower cohesion parameters. This provides a good, stable anchor on the pigment surface. The solvent will not dissolve that polymer or polymer segment away from the surface. Binders with high acid numbers are frequently used with success in printing inks for the same reason. This is discussed in more detail below.

It is felt that those who understand the use of cohesion parameters are able to more systematically modify surfaces of inorganic materials to optimize or improve their compatibility with organic polymers and binders. This has been done for inorganic Rockwool® fibers which are to be incorporated into polypropylene.[5] It must be presumed that this type of systematic procedure can guide surface treatment of other inorganic materials in a more directed way toward a desired goal.

HANSEN SOLUBILITY PARAMETER CORRELATION OF ZETA POTENTIAL FOR BLANC FIXE

Winkler[8] has recently reported zeta potentials measured for 1% v/v blanc fixe with 0.34% moisture content. Nineteen liquids were included in this careful study. These liquids could easily be divided into two groups. There were 5 systems with zeta potentials greater than about 10 mV, and 14 systems with zeta potentials less than about 5 mV. Table 5.3 includes the results of the correlation of these data with cohesion parameters. The only major "error" was for hexamethylphosphoramide, with a RED of 0.951 and a zeta potential of 1.9 mV. This correlation supports the contention that cohesion parameters are significant for characterization of pigment, filler, and fiber surfaces. This is a good correlation and supports the views presented earlier. According to Winkler, there was no correlation with the acceptor or donor numbers (acid–base).

CONTROLLED ADSORPTION (SELF-ASSEMBLY)

Significant tasks for formulators are to control the surface and interfacial energies of products, especially if they are water reducible. This is required to allow substrate wetting, to maintain stable dispersions, and to provide/ensure adequate and durable adhesion to given substrates. Guidelines for courses of action are frequently available when cohesion energy parameters are referred to. Some guides are discussed in the following.

It is a well-known fact that a small percentage of acid groups (or alcohol groups) on a polymer chain will promote adhesion and adsorption to many surfaces. The cohesion energy parameter of an isolated acid group is high. One can consider the cohesion energy properties of formic acid (δ_D; δ_P; δ_H = 14.3; 11.9; 16.6) as an isolated part of a polymer chain. The polar cohesion energy parameter of an acid group is not so high. It would seem logical to systematically use acid groups for adsorption to high energy surfaces and to make certain that the cohesion energy parameters for the solvent and bulk of the product are much lower, such that isolated acid groups would not be dissolved. This would provide an anchor which the product itself will not be able to remove. This type of adsorption may be called hydrophilic bonding. If, on the other hand, the solvent were too good for the anchor, it could be presumed that even an acid group may be too readily dissolved off the surface, or at least take part in a dynamic equilibrium of adsorption and desorption. Absorbed/adsorbed water can sometimes interfere with such anchors at high energy surfaces.

The reverse of this thinking is systematically used by those designing associative thickeners and by nature itself, such as with hydrophobic bonding in proteins. Certain segments of given molecules have such low cohesive energy parameters that they are no longer soluble in the media, which is usually aqueous, and they either seek out their own kind (associate) or perhaps adsorb on or penetrate into a low energy surface where cohesion energy parameters more suitably match. The positive effects of associative thickeners can be counteracted by the presence of solvents preferentially locating where the hydrophobic bonding is to occur. The hydrophobic bonds lose strength or even may be dissolved away.

A challenge to the creative mind is to derive new uses for high energy groups which are not particularly water soluble/water sensitive. The division of the cohesion energy into at least three parts allows these considerations to be made in a reasonably quantitative manner. One can choose nitro groups or perhaps groups containing phosphorus as examples of species characterized by high polar cohesion energy parameters and low or moderate hydrogen bonding parameters. The total cohesion energy parameters for ethanol and nitromethane are very close, 26.1 and 25.1 $MPa^{1/2}$, respectively. Ethanol is soluble in water, nitromethane is not. Ethanol has a relatively high hydrogen bonding parameter (19.4 $MPa^{1/2}$) compared with nitromethane (5.1 $MPa^{1/2}$). This makes all the difference. Would not the nitro group be a suitable anchor analogous to the previous discussion concerning acid groups? Also, it would not be hydrophilic with the inherent problems of water sensitivity associated with high hydrogen bonding parameters. Several of the pigments reported in Table 5.3 did indeed have moderate affinity for the nitroparafins, for example, but they were included in the lesser interacting group by the arbitrary division into "good" and "bad" groups.

CONCLUSION

Many pigments and fillers have now been characterized by Hansen cohesion parameters (HSP). Many examples are given. A method based on relative sedimentation time and/or suspension is described for doing this. This method has generally allowed useful characterizations, although some experience is helpful. For example, the data are often scattered and not nearly of the quality usually found when observing polymer solution behavior. This scatter may cause some to disregard the method; hopefully, they can develop a better one. The obvious advantages of having solvents, plasticizers, polymers, pigments, fillers, fibers, etc. characterized with the same energy parameters should provide incentive for improving on the present state, both in terms of numbers of characterizations as well as improved methodology.

REFERENCES

1. Hansen, C. M., The Three Dimensional Solubility Parameter and Solvent Diffusion Coefficient, Doctoral Dissertation, Danish Technical Press, Copenhagen, 1967.
2. Hansen, C. M., The Three Dimensional Solubility Parameter — Key to Paint Component Affinities II. *J. Paint Technol.,* 39(511), 505–510, 1967.
3. Shareef, K. M. A., Yaseen, M., Mahmood Ali, M., and Reddy, P. J., *J. Coat. Technol.,* 58(733), 35–44, February 1986.
4. Gardon, J. L. and Teas, J. P., Solubility Parameters, in *Treatise on Coatings, Vol. 2, Characterization of Coatings: Physical Techniques, Part II*, Myers, R. R. and Long, J. S., Eds., Marcel Dekker, New York, 1976, chap. 8.
5. Hennissen, L., Systematic Modification of Filler/Fibre Surfaces to Achieve Maximum Compatibility with Matrix Polymers, Lecture for the Danish Society for Polymer Technology, Copenhagen, February 10, 1996.
6. Hansen, C. M. and Beerbower, A., Solubility Parameters, in *Kirk-Othmer Encyclopedia of Chemical Technology,* Suppl. Vol., 2nd ed., Standen, A., Ed., Interscience, New York, 1971, 889–910.
7. Schröder, J., Colloid Chemistry Aids to Formulating Inks and Paints, *Eur. Coat. J.,* 5/98, 334–340, 1998.
8. Winkler, J., Zeta Potential of Pigments and Fillers, *Eur. Coat. J.,* 1-2/97, 38–42, 1997.
9. Vinther, A., Application of the Concepts Solubility Parameter and Pigment Charge, *Chim. Peint. (England),* 34(10), 363–372, 1971.
10. Soerensen, P., Application of the Acid/Base Concept Describing the Interaction Between Pigments, Binders, and Solvents, *J. Paint Technol.,* 47(602), 31–39, 1975.
11. Soerensen, P., Cohesion Parameters Used to Formulate Coatings (Kohaesionsparametre anvendt til formulering af farver og lak), *Färg och Lack Scand.,* 34(4), 81–93, 1988 (in Danish).

6 Applications — Coatings and Other Filled Polymer Systems

CONTENTS

ABSTRACT

Hansen solubility parameters (HSP) are widely used in the coatings industry to help find optimum solvents and solvent combinations. They also aid in substitution to less hazardous formulations in various other types of products such as cleaners, printing inks, adhesives, etc. The discussion in this chapter includes the physical chemical reasons why solvents function as they do in many practical cases. The behavior of solvents in connection with surfaces of various kinds and the use of HSP to understand and control surface phenomena is especially emphasized. Products where HSP concepts can be used in a manner similar to coatings include other (filled) polymer systems of various types such as adhesives, printing inks, chewing gum, etc. There are many examples of controlled self-assembly.

INTRODUCTION

There are many applications documented in the literature where HSP have aided in the selection of solvents, understanding and controlling processes, and, in general, offering guidance where affinities among materials are of prime importance[1-5] (see also the following chapters and examples below). This chapter emphasizes coatings applications and discusses the practical application of HSP to solvent selection. Computer techniques are helpful, but not necessary. The same principles useful for understanding the behavior of coatings are useful in understanding behavior in a larger number of related products, including adhesives, printing inks, and chewing gum, to mention a few. These contain widely different materials, both liquid and solid, which can be characterized by HSP. This allows their relative affinities to be established. Previous chapters have discussed how to assign HSP to solvents, plasticizers, polymers, and resins, as well as to the surfaces of substrates, pigments, fillers, and fibers. Various additives such as resins, surfactants, flavors, aromas, scents, and the like can also be characterized by HSP to infer how they behave in seemingly complex systems.

SOLVENTS

In order to find the optimum solvent for a polymer, one must have or estimate its HSP. Matching the HSP of an already existing solvent or combination of solvents can be done, but this procedure does not necessarily optimize the new situation. The optimum depends on what is desired of the system. A solvent with the highest possible affinity for the polymer is both expensive and probably not necessary and will rarely be optimum. In more recent years, optimization increasingly includes considerations of worker safety and the external environment. Volatile organic compounds (VOC) are to be reduced to the greatest extent possible.

Whereas hand calculations and plotting of data are still quite useful and at times more rapid than computer processing, it is becoming almost mandatory that computers be used. To this end, most solvent suppliers and many large users of solvents have computer programs to predict solution behavior as well as evaporation phenomena. In spite of these pressures to let the computer do the thinking, an experienced formulator can often arrive at a near-optimum result without recourse to paper or to computers. A major factor in this almost immediate overview is the decrease in the number of solvents useful in coatings. By putting this together with other necessary considerations such as flash point, proper evaporation rate, cost, odor, availability, etc., the experienced formulator who knows the HSP for the relatively few solvents possible in a given situation will be able to select a near-optimum combination by a process of exclusion and simple mental arithmetic. This does not mean the use of HSP is on the way out. The real benefit of this concept is in interpreting more complicated behavior, such as affinities of polymers with polymers and polymers with surfaces as described in the following. Much more work needs to be done in these areas, but the following gives an indication of what might be expected.

As indicated previously, computer techniques can be very useful, but are not always necessary, and simple two-dimensional plots using δ_P and δ_H can often be used by those with limited experience with these techniques to solve practical problems. The nonpolar cohesion parameter, δ_D, cannot be neglected in every case, but, for example, when comparing noncyclic solvents in practical situations, it has been found that their dispersion parameters will be rather close regardless of structure. Cyclic solvents, and those containing atoms significantly larger than carbon, such as chlorine, bromine, metals, etc., will have higher dispersion parameters. The total solubility parameter for aliphatic hydrocarbon solvents is identical with their dispersion parameter and increases only slightly with increased chain length. This same trend is expected for oligomers of a polymer as molecular weight increases. Regardless of the means of processing data, the following examples are intended to illustrate principles on which to base a systematic course of action.

Most coatings applications involve solvents reasonably well within the solubility limit which is indicated by the boundary of a solubility plot such as that shown in Figure 6.1.[1] A maximum of cheaper hydrocarbon solvent is also desired and can frequently be used to arrive at such a situation for common polymers used in coatings. Some safety margin in terms of extra solvency is advised because of temperature changes, potential variations in production, etc. These can lead to a situation where solvent quality changes in an adverse manner. Balance of solvent quality on evaporation of mixed solvents is also necessary. Here again, computer approaches are possible, and calculations of solvent quality can be made at all stages of evaporation. It is usually good practice to include a small or moderate amount of slowly evaporating solvent of good quality and low water sensitivity to take care of this situation. These have frequently been slowly evaporating ketones and esters.

An oxygenated solvent which is frequently added to hydrocarbon solvents and has been cost effective in increasing the very important hydrogen bonding parameter has been *n*-butanol (or sometimes 2-butanol). The mixture of equal parts xylene and *n*-butanol has been widely used in conjunction with many polymers such as epoxies, but a third solvent, such as a ketone, ester, or glycol ether, is often included in small amounts to increase the polar parameter/solvency of the mixture. Neither xylene nor *n*-butanol satisfactorily dissolves an epoxy of higher molecular weight by itself. These are located in boundary regions of the solubility region for epoxies, but on opposite

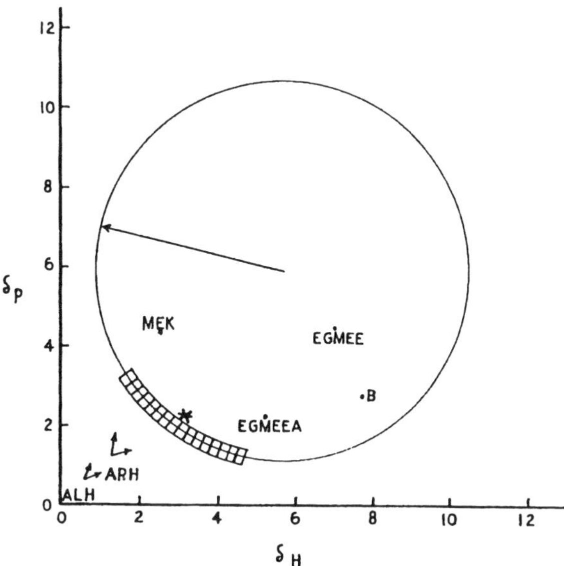

FIGURE 6.1 Sketch showing location of typical solvents relative to the HSP of a binder. Aliphatic hydrocarbons (ALH) and aromatic hydrocarbons (ARH) do not always dissolve well enough so other solvents must be added to bring the mixed solvent composition into the region of solubility for the binder. Ketones (MEK, methyl ethyl ketone), alcohols (B, *n*-butanol), or other solvents such as glycol ethers and their acetates (here ethylene glycol monoethyl ether and ethylene glycol monoethyl ether acetate) can be used to do this. The expected solvent improvement at least cost is discussed in the text as the quantity $(\delta_P^2 + \delta_H^2)^{1/2}$/cost. Units in the figure are in $(\text{cal/cm}^3)^{1/2}$. The choice of solvent today would involve glycol ethers based on propylene glycol as discussed in Chapter 11. (From Hansen, C. M., *Färg och Lack,* 17(4), 75, 1971. With permission.)

sides of the characteristic Hansen spheres (see Figure 6.2).[1] Glycol ethers also can be added to hydrocarbon solvents with advantage, and the polar and hydrogen bonding parameters are higher than if *n*-butanol had been added to the same concentration. There are many possibilities, and a solubility parameter approach is particularly valuable in quickly limiting the number of candidates. The addition of glycol ethers or other water-soluble solvents can have adverse effects, such as increased water sensitivity and poorer corrosion resistance of the final film, since some solvent retention must be anticipated, and the least volatile solvent is enriched and left behind.

Relative costs for improving solvency from a hydrocarbon base solvent can be estimated by the relation $(\delta_P^2 + \delta_H^2)^{1/2}$/cost. This relation has generally pointed to the use of *n*-butanol, for example, as a cost-efficient solvent to increase the hydrogen bonding parameter in particular. Solvents can be ranked in this manner to arrive at the least cost solutions to given solvent selection problems.

Coalescing solvents in water-reducible coatings are often (but not always) those with somewhat higher hydrogen bonding parameters than the polymer, which also means they are water soluble or have considerable water solubility. The distribution between the water phase and the dispersed polymer phase depends on the relative affinities for water and the polymer. Solvents which are not particularly water soluble will preferentially be found in the polymer phase. Such coalescing solvents may be preferred for applications to porous substrates, making certain they are where they are needed when they are needed. Otherwise, a water-soluble coalescing solvent would tend to follow the aqueous phase, penetrating the substrate faster than the polymer particles, which also get filtered out and are not available to do their job in the film when the water evaporates. When water evaporates, the solvent must dissolve to some extent in the polymer to promote coalescence. Of course, this affinity for the polymer is a function of its HSP relative to those of the polymer.

Amines are frequently added to water-reducible coatings to neutralize acid groups built into polymers, thus providing a water-solubilizing amine salt. Amine in excess of that required for total

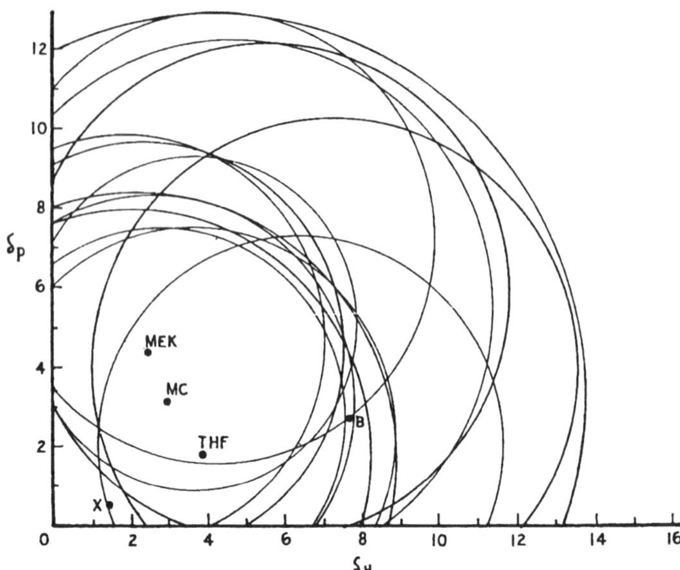

FIGURE 6.2 Sketch showing formulation principles using two relatively poor solvents in combination to arrive at a good solvent. Xylene (X) can be mixed with *n*-butanol (B) to arrive at a mixture which can be improved by additions of tetrahydrofuran (THF), methylene chloride (MC), or methyl ethyl ketone (MEK) among others. These three very volatile solvents have often been used in analytical work, paint removers, and the like because they dissolve all of the typical coatings binders shown in the figure. Labeling requirements have dictated other choices in more recent years. (From Hansen, C. M., *Färg och Lack,* 17(4), 74, 1971. With permission.)

neutralization of the acid groups acts like a solvent. Such amine salts have been characterized separately to demonstrate that they have higher solubility parameters than either (acetic) acid or organic bases.[6] These salts are hydrophilic and have very little affinity for the polymers used in coatings, which means they are to be found in a stabilizing role in the interface in the aqueous phase while still being attached to the polymer. Electrostatic repulsion contributes to stability as well, and the dispersed solubilized polymer can be visualized in terms of a porcupine with raised quills.

Surface-active agents, whether nonionic or ionic, are also to be found where the affinities of the respective parts of their molecules dictate their placement; like seeks like. The hydrophilic end with a high hydrogen bonding parameter will seek the aqueous phase, and the hydrophobic end will seek out an environment where energy differences are lowest (self-assembly). It might be noted here that some solvents have surfactant-like properties as well. Ethylene glycol monobutyl ether, in particular, has been shown to be a good coupling agent, as well as contributing to lowered surface tension.[7] The hydrophobic end of such molecules may reside within the polymer if HSP relations dictate this. Otherwise, if the HSP differences are too great, the hydrophobic portion may be forced to remain in the interfacial region, not being accepted by the aqueous phase either.

Increases in temperature especially lead to lower hydrogen bonding parameters (see Chapter 1). For this reason, solvents with high hydrogen bonding parameters, such as glycols, glycol ethers, and alcohols, become better solvents for most polymers at higher temperatures. This can markedly affect hot-room stability in water-reducible coatings, for example, since more of the solvent will partition to the polymer phase, which swells, becomes more fluid, and has altered affinities for stabilizing surface-active agents. These may dissolve too readily in the swollen, dispersed polymer. When carefully controlled, these temperature effects are an advantage in water-reducible, oven-cured coatings, leading to higher film integrity, since poor solvents at room temperature become good solvents in the oven after the water has evaporated.

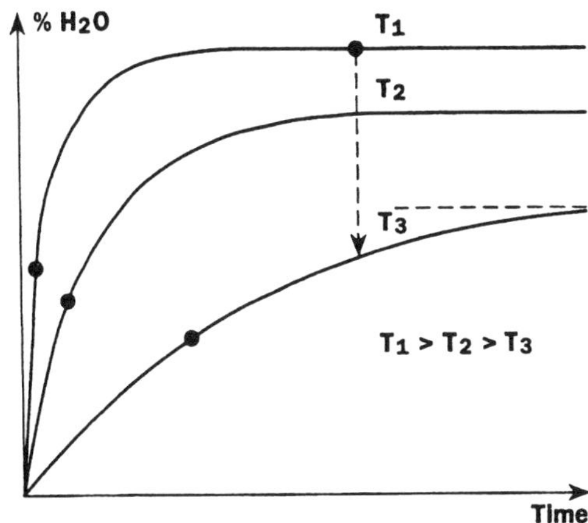

FIGURE 6.3 Sketch of water uptake in a polymer as a function of temperature. Higher temperature leads to more rapid uptake and to higher equilibrium levels. Quenching to a lower temperature (arrow) leads to excess water in the film and possibly to water blisters and delamination (see text for further discussion). (Reprinted from *Prog. Org. Coat.,* 26, Hansen, C. M., New Developments in Corrosion and Blister Formation in Coatings, 115, 1995. With permission from Elsevier Science.)

A very special destructive effect of water is caused by the reduction of its hydrogen bonding parameter with increases in temperature. The solubility of water in most polymers is higher at a higher temperature than it is at a lowered temperature because the HSP for the polymer and water match better at the higher temperature. It has been documented in many cases that a rapid quench from hot water to cold water can cause blisters in coatings.[8] Previously dissolved water within the film now becomes in excess of that soluble in the film. This can be seen in Figure 6.3 where water uptake curves are shown for three temperatures. The amount and rate of uptake is higher for the higher temperatures. Rapid cooling to below the solubility limit at a lower temperature means the system is supersaturated. Excess water freed by this mechanism has been called SWEAT (soluble water exuded at lowered temperatures). If the SWEAT water cannot rapidly diffuse out of the coating, it will appear as a separate phase, perhaps first as clusters, but ultimately at hydrophilic sites or at a substrate. The coating fails by blistering or delamination. This special effect has been noted by the author in coatings (alkyd, polyester, and epoxy), in rigid plastics such as poly(phenylene sulfide) and poly(ether sulfone), and even in EPDM rubber. Examples of measurements of this type are shown in Chapter 7, Figures 7.3 and 7.4 for an EPDM rubber gasket and for a poly(ether sulfone) tensile bar. This effect is not restricted to water; it has also been seen for an epoxy coating which was repeatedly removed from room temperature methanol to measure weight gain. The cooling due to the methanol evaporation was sufficient to produce methanol blisters near the air surface of the coating because of excess amounts of methanol over that soluble at the lower temperature resulting from the methanol evaporation.

The use of supercritical gases as solvents has become possible in recent years. Space limitations prevent going into the details of these developments at this point. It should be noted, however, that when a gas is compressed its cohesive energy density increases. This means that nonpolar gases with their low dispersion solubility parameters can begin to dissolve given organic materials which otherwise have solubility parameters which are too high. Increasing the nonpolar solubility parameter of the gas by increasing the pressure causes a closer match with the corresponding parameter for potential solutes. Similar behavior is found for polar gases such as carbon dioxide. The prevailing pressure and temperature conditions determine its cohesive energy density, and changes in pressure or temperature change solubility relations for this reason. Whereas nonpolar gases are most suitably used for relatively nonpolar solutes, carbon dioxide and, in principle, other polar gases are most suitably used in connection with more polar solutes. The solubility parameters for carbon dioxide have been reported[9] based on the room temperature solubility of the gas in different liquids as being $\delta_D;\delta_P;\delta_H$ equal to 15.3;6.9;4.1. These parameters resemble those of a higher molecular weight ketone. See Chapter 8 for additional examples and data related to carbon dioxide and gas permeability.

Solvent technology has also been used in a wide variety of other products and processes as listed by Barton.[2] One can mention the formulaton of solvent cleaners based on vegetable oils as an additional example.[10]

TECHNIQUES FOR DATA TREATMENT

As mentioned earlier, a simple approach to many practical problems is to make a two-dimensional plot of polar vs. hydrogen bonding parameters with a circle (or estimated circle) for the polymer in question. The circle should encompass the good solvents. One can then plot points for potential solvents and quickly arrive at a starting composition for an experiment. Subsequently, this can be adjusted if necessary. A linear mixing rule based on the volume (or weight) fractions of the solvent components is usually satisfactory. Plasticizers should be included in the calculations. They will be very slow to dissolve rigid polymers, in particular, and are, of course, nonvolatile for all practical purposes.

A special plotting technique for solvent selection developed by Teas[11] is used frequently by those who restore old paintings. The art involved in this stage of the conservation process is to remove the old varnish without attacking the underlying original masterpiece. HSP principles have been used since the late 1960s for selecting solvents and solvent blends for this purpose.[12] The triangular plotting technique uses parameters for the solvents, which, in fact, are modified HSP parameters. The individual Hansen parameters are normalized by the sum of the three parameters. This gives three fractional parameters defined by Equations 6.1, 6.2, and 6.3.

$$f_d = 100\,\delta_D/(\delta_D + \delta_P + \delta_H) \tag{6.1}$$

$$f_p = 100\,\delta_P/(\delta_D + \delta_P + \delta_H) \tag{6.2}$$

$$f_h = 100\,\delta_H/(\delta_D + \delta_P + \delta_H) \tag{6.3}$$

The sum of these three fractional parameters is 100 in the form the equations are written. This allows use of the special triangular technique. Some accuracy is lost, and there is no theoretical justification for this plotting technique, but one does get all three parameters onto a two-dimensional plot with enough accuracy that its use has survived for this type of application (at least). The Teas plot in Figure 6.4 includes an estimate of the solubility/strong attack of older, dried oil paint. A varnish which could be considered for use is Paraloid® B72, a copolymer of ethyl methacrylate and methyl methacrylate from Rohm and Haas. There is a region in the lower, right-hand part of this plot where the varnish is soluble and the dried oil is not. The varnish remover should be in this region. Mixtures of hydrocarbon solvent and ethanol are located in this region and could be considered. HSP correlations for materials of interest in restoration of older paintings are included in Table 6.1.

A helpful simplifying relation to use in solvent selection calculations using solubility parameters is that the resultant values for mixtures can be estimated from volume fraction averages for each solubility parameter component. Solvent quality can be adjusted by the RED number concept, which is discussed in Chapter 1 (Equation 1.10), or graphically as described above.

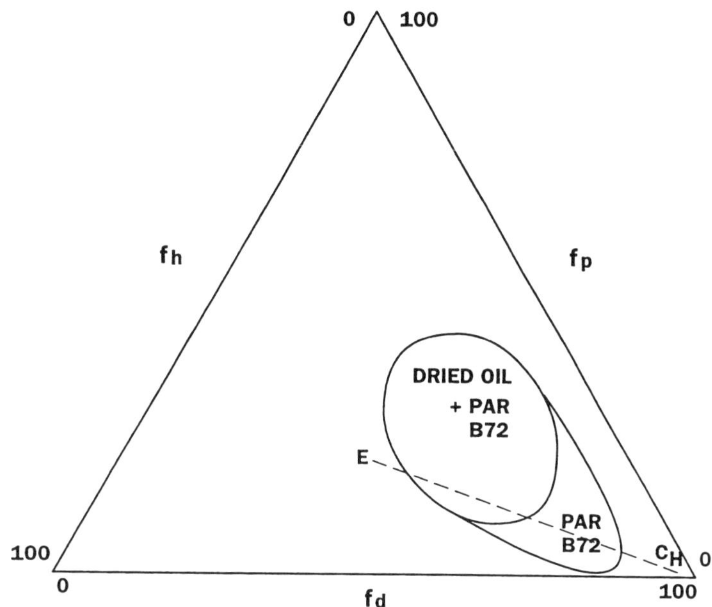

FIGURE 6.4 Teas plot for a typical painting conservation situation where a varnish is to be removed or applied without attacking the underlying original oil painting. Solvents indicated are cyclohexane (C), heptane (H), and ethanol (E) (see text for further discussion).

TABLE 6.1
HSP Correlations for Materials of Interest in the Conservation of Older Paintings

MATERIAL	d_D	d_P	d_H	R_o	FIT	G/T
Paraloid® 22 solubility	17.6	7.4	5.6	9.4	1.000	17/26
Dammar gum dewaxed	18.4	4.2	7.8	8.3	0.915	30/56
Dried oil (estimate)	16.0	6.0	7.0	5.0	1.000	9/22

A computer search with the SPHERE computer program (Chapter 1) for "nearest neighbors" for a given single solvent has been used many times to locate alternates for a wide variety of product types including coatings of various descriptions, cleaners, etc. A similar application is to predict which other solvents will probably be aggressive to a chemically resistant coating where very limited data have indicated a single solvent or two are somewhat aggressive. A nearest neighbor search involves calculation of the quantity Ra (Chapter 1, Equation 1.9) for a whole database, for example, and then arranging the printout in RED number order (Chapter 1, Equation 1.10). The potentially most aggressive liquids are at the top of the list. Solvents with RED less than 1.0 are "good" and therefore easily recognized. Sorting out these possibilities considering toxicity, evaporation rate, cost, etc. leads to the most promising candidates for the substitution.

SOLVENTS AND SURFACE PHENOMENA IN COATINGS (SELF-ASSEMBLY)

Chapters 4 and 5 have been devoted to the characterization of surfaces for substrates, pigments, fillers, and the like. This means the interplay between solvent, polymer, and surfaces can be inferred

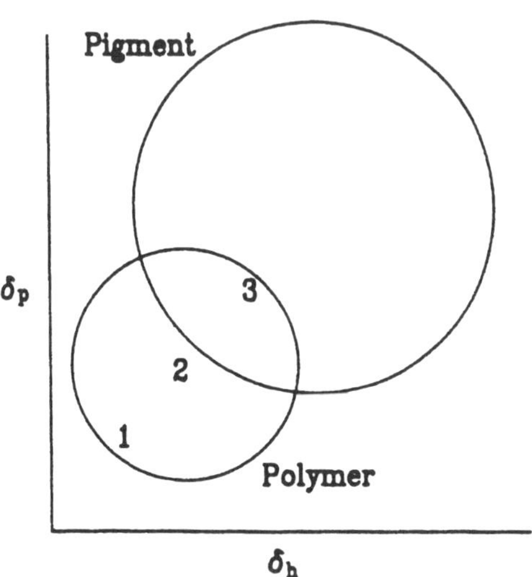

FIGURE 6.5 Sketch showing influence of solvent quality on expected pigment dispersion stability (see text and Figure 1.1 for discussion). (From Hansen, C. M., *Paint and Coating Testing Manual,* 14th ed. of the *Gardner-Sward Handbook,* 1995, 400. Copyright ASTM. Reprinted with permission.)

by their relative affinities. These depend on their HSP relative to each other, and the RED number concept can be quite useful.

As stated previously, the desired solvent quality in many coatings is just slightly better than that of a marginal solvent. This means RED numbers just under 1.0 relative to the polymer will be sought. One reason for the desired marginal solvent quality is that this will ensure that the polymer adsorbed onto pigment surfaces during pigment dispersion has little reason to dissolve away from that surface. The dispersion stabilizing polymer should remain on the pigment surface where it is desired. If this polymer is dissolved away, the result is most likely pigment flocculation which leads to color change, undesired settling, and perhaps even rheological difficulties. The solvent in this case should have a RED number for the pigment surface greater than 1.0, or at least reasonably high, to aid in the planned affinity approach to pigment dispersion stability. Of course the polymer, or some portion of it, and the pigment surface should have high affinity for each other. A sketch of the optimum relations in coatings is given in Figure 6.5 where the marginal solvent is number 1. Solvent 2 would probably be too expensive and, in addition, will probably dissolve the polymer too well.

Schröder[13] (BASF) confirms that the optimum polymer adsorption will be found when the binder and pigment surface have the same HSP. He indicates that the solvent should be very poor for the pigment and located in the boundary region for the binder. He prefers the pigment to have HSP values placing it intermediate between the solvent and binder. This is suggested for conditions where the solvent has higher HSP than the pigment, as well as for conditions where the solvent has lower HSP than the pigment. This situation, with the solvent and binder on opposite sides of the pigment, means the composite vehicle has parameters very closely matching those of the pigment. A very similar type of result was found by Skaarup,[14] who especially emphasized that optimum color strength was found for solvents marginal in quality for the binder and poor for the pigment in question.

In special applications, an extended polymer chain configuration is desirable, but a solid anchor to the pigment surface is also desired. This means a better-than-marginal solvent for the polymer is desired. A good anchor has high affinity for the pigment surface and marginal affinity for the solvent. Solvent 3 (Figure 6.5) would adsorb onto the pigment surface preferentially, and pigment

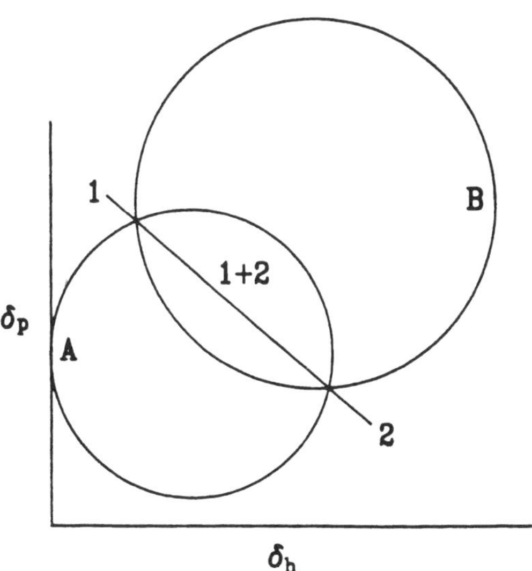

FIGURE 6.6 Sketch showing how two otherwise immiscible polymers can be brought into a homogeneous solution by the use of mixed nonsolvents. (From Hansen, C. M., *Paint and Coating Testing Manual,* 14th ed. of the *Gardner-Sward Handbook,* 1995, 400. Copyright ASTM. Reprinted with permission.)

dispersion stability would be poor. An extension of this thinking may be required for pigment pastes and other very highly filled products. In these cases, there is not much dispersing vehicle relative to the pigment, and the solvent must be considered as being part of the dispersing vehicle. In such cases the solvent may have high affinity for the pigment surface as well as for dispersing polymer. An ideal situation here is where all the ingredients have the same HSP.

POLYMER COMPATIBILITY

In some cases, closer-than-usual matches between solvent and polymer solubility parameters are required. This is true when two polymers are mixed and one of them precipitates. This is most likely the polymer with the larger molecular weight, and it must be dissolved even better. Lower RED numbers with respect to this polymer are desired, while still maintaining affinity for the other polymer. Miscible blends of two polymers have been systematically found using a solvent mixture composed exclusively of nonsolvents.[15] This is demonstrated schematically in Figure 6.6, where it can be seen that different percentage blends of solvents 1 and 2 will have different relative affinities for the polymers. No other alternative theory of polymer solution thermodynamics can duplicate this predictive ability. Polymer miscibility is enhanced by larger overlapping solubility regions for the polymers as sketched in Figure 6.7. Polymers A and B should be compatible, while polymer C would not. Such a systematic analysis allows modification of a given polymer to provide more overlap and enhanced compatibility. The advantages of a copolymer containing the monomers of A or B and C should also be evident. Such a copolymer will essentially couple the system together.

Van Dyk et al.[16] have correlated the inherent viscosity of polymer solutions with HSP. The inherent (intrinsic) viscosity used in this study, [η], is given by Equation 6.4.

$$[\eta] = (\eta_s/\eta_o)/c \tag{6.4}$$

η_s is the solution viscosity, η_o is the solvent viscosity, and c is the solution concentration. The concentration used was about 0.5 g/dl. This is an expression reflecting polymer chain extension in

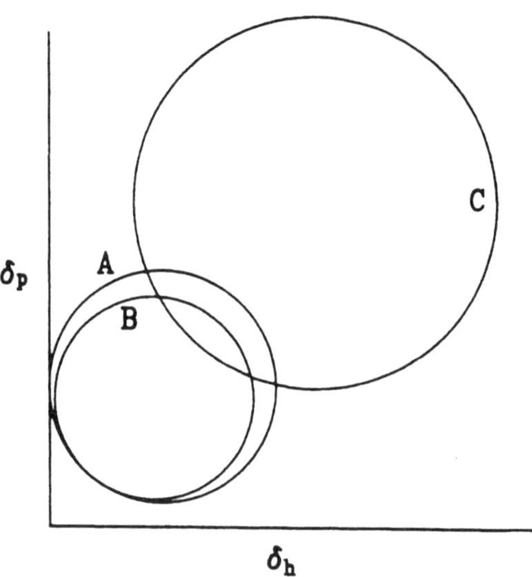

FIGURE 6.7 Sketch describing expected polymer miscibility relations (see text for discussion). (From Hansen, C. M., *Paint and Coating Testing Manual,* 14th ed. of the *Gardner-Sward Handbook,* 1995, 401. Copyright ASTM. Reprinted with permission.)

solution, with higher values reflecting greater chain entanglements because of greater polymer extension. This is interesting in that the solubility parameter is a thermodynamic consideration, while the viscosity is a kinetic phenomena. Higher [η] were found for solvents with HSP nearest those of the polymer.

As stated above additional uses of HSP (and the total solubility parameter) in solvent technology can be found in Barton,[2] but these are too numerous to include here. However, a couple of examples relating to guided polymer compatibility are worthy of special mention. These are the formulation of asymmetric membranes for separations,[17,18] where polymer solutions — having given HSP relations — and at least one solvent soluble in water are used. The solution is immersed in water, the solvent quality becomes bad, and a controlled porous membrane results. Another example of controlled phase relations during a dynamic process is found in the formulation of self-stratifying coatings. This is discussed in Chapter 4 in terms of the creation of interfaces and therefore interfacial surface tension. The HSP principles involved in this type of coating can be seen in Figure 6.8. The solvent must dissolve both the topcoat and primer and allow the lower surface tension topcoat to migrate to the surface during film formation. Formulation principles have been discussed in detail elsewhere.[19,20] Before concluding this section, some of the recent work on miscible polymer blends should also be noted.[21,22] This work used group contribution estimates of the δ_P and δ_H parameters only in an effort to correlate interfacial tension between polymers, assuming that the δ_D parameters would not be too different. While this is a good starting point to prove the procedure has possibilities, further differentiation between the polymers and improved group contribution methods may offer even more improvement.

HANSEN SOLUBILITY PARAMETER PRINCIPLES APPLIED TO UNDERSTANDING OTHER FILLED POLYMER SYSTEMS

Recent characterizations of inorganic fillers and fibers[23] have confirmed that HSP concepts can be applied to engineered fiber-filled systems such as those based on polypropylene.

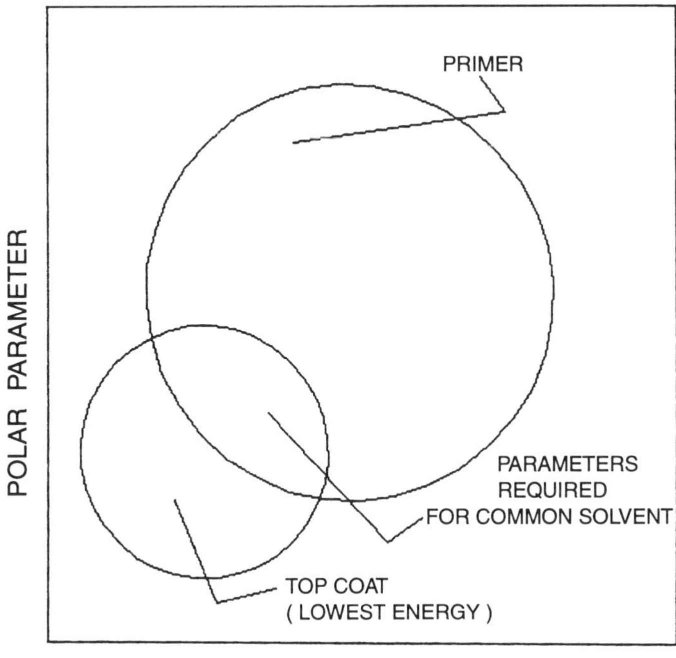

FIGURE 6.8 Sketch illustrating the principles of solvent selection for self-stratifying coatings. (From Birdi, K. S., Ed., *Handbook of Surface and Colloid Chemistry,* CRC Press, Boca Raton, FL, 1997, 324.)

The behavior of chewing gum can also be analyzed in terms of solubility parameter principles.[24] In addition to rheological behavior, appearance, and other performance considerations, a desired product characteristic is that the release of the taste components should be controlled. Greater differences in solubility parameters between flavoring agents and wax-free gum bases lead to enhanced flavor release. Similarity of HSP can lead to stopping the desired release too soon.

Perhaps the most important practical work dealing with solubility parameters and the stability of pigment dispersions is that attributable to Stephen.[25] He concludes that all the (solid) ingredients in a paint formulation should have the same energy characteristics. If they do not, there will be a driving force for this to occur. This can lead to problems. One can just as well make the formulation stable from the start, and then everything will remain stable just where it is because there are no driving forces for anything to move around. While this sounds expensive, obvious, and perhaps too simple, the truth of the matter is well documented in very practical terms.

CONCLUSION

Many practical uses of the solubility parameter concept have been described in detail, including optimizing solvent selection, improving polymer compatibility, and enhancing pigment dispersion. When all of the materials involved in a given product and application can be characterized with the same affinity (solubility/cohesion) parameters, the possibility exists to predict interactions among them, even in complicated situations, such as the formulation of various types of filled systems including coatings, printing inks, adhesives, and other filled polymer systems including chewing gum.

REFERENCES

1. Hansen, C. M., Solubility in the Coatings Industry, *Färg och Lack*, 17(4), 69–77, 1971.
2. Barton, A. F. M., *Handbook of Solubility Parameters and Other Cohesion Parameters*, CRC Press, Boca Raton, FL, 1983.
3. Gardon, J. L. and Teas, J. P., Solubility Parameters, in *Treatise on Coatings, Vol. 2, Characterization of* Coatings*: Physical Techniques, Part II*, Myers, R. R. and Long, J. S., Eds., Marcel Dekker, New York, 1976, chap. 8.
4. Beerbower, A., Boundary Lubrication — Scientific and Technical Applications Forecast, AD747336, Office of the Chief of Research and Development, Department of the Army, Washington, D.C., 1972.
5. Hansen, C. M. and Beerbower, A., Solubility Parameters, in *Kirk-Othmer Encyclopedia of Chemical Technology*, Suppl. Vol., 2nd ed., Standen, A., Ed., Interscience, New York, 1971, 889–910.
6. Hansen, C. M., Some Aspects of Acid/Base Interactions (in German: Einige Aspekte der Säure/Base-Wechselwirkung), *Farbe und Lack*, 83(7), 595–598, 1977.
7. Hansen, C. M., Solvents in Water-Borne Coatings, *Ind. Eng. Chem. Prod. Res. Dev.,* 16(3), 266–268, 1977.
8. Hansen, C. M., New Developments in Corrosion and Blister Formation in Coatings, *Prog. Org. Coat.,* 26, 113–120, 1995.
9. Hansen, C. M., 25 Years with Solubility Parameters, (in Danish: 25 År med Opløselighedsparametrene), *Dan. Kemi,* 73(8), 18–22, 1992.
10. Rasmussen, D. and Wahlström, E., HSP-Solubility Parameters: A Tool for Development of New Products — Modelling of the Solubility of Binders in Pure and Used Solvents, *Surf. Coat. Intl.,* 77(8), 323–333, 1994.
11. Teas, J. P., Graphic Analysis of Resin Solubilities, *J. Paint Technol.,* 40(516), 19–25, 1968.
12. Torraca, G., *Solubility and Solvents for Conservation Problems*, 2nd ed., International Centre for the Study of the Preservation and the Restoration of Cultural Property (ICCROM), Rome, 1978. (13, Via Di San Michelle, 00153 Rome)
13. Schröder, J., Colloid Chemistry Aids to Formulating Inks and Paints, *Eur. Coat. J.,* 5/98, 334–340, 1998.
14. Skaarup, K., The Three Dimensional Solubility Parameter and Its Use. II. Pigmented Systems, Skandinavisk Tidskrift for *Fårg och Lack*, 14(2), 28–42, 1968; 14(3), 45–56, 1968.
15. Hansen, C. M., On Application of the Three Dimensional Solubility Parameter to the Prediction of Mutual Solubility and Compatibility, *Färg och Lack*, 13(6), 132–138, 1967.
16. Van Dyk, J. W., Frisch, H. L., and Wu, D. T., Solubility, Solvency, Solubility Parameters, *Ind. Eng. Chem. Prod. Res. Dev.,* 24(3), 473–478, 1985.
17. Klein, E. and Smith, J. K., Assymetric Membrane Formation, *Ind. Eng. Chem. Prod. Res. Dev.,* 11(2), 207–210, 1972.
18. Chawla, A. S. and Chang, T. M. S., Use of Solubility Parameters for the Preparation of Hemodialysis Membranes, *J. Appl. Polym. Sci.,* 19, 1723–1730, 1975.
19. Misev, T. A., Thermodynamic Analysis of Phase Separation in Selfstratifying Coatings — Solubility Parameters Approach, *J. Coat. Technol.,* 63(795), 23–28, 1991.
20. Special issue devoted to Self-Stratifying Coatings, *Prog. Org. Coat.,* 28(3), July 1996.
21. Luciani, A., Champagne, M. F., and Utracki, L. A., Interfacial Tension in Polymer Blends. Part 1: Theory, *Polym. Networks Blends,* 6(1), 41–50, 1996.
22. Luciani, A., Champagne, M. F., and Utracki, L. A., Interfacial Tension in Polymer Blends. Part 2: Measurements, *Polym. Networks Blends,* 6(2), 51–62, 1996.
23. Hennissen, L., Systematic Modification of Filler/Fibre Surfaces to Achieve Maximum Compatibility with Matrix Polymers, Lecture for the Danish Society for Polymer Technology, Copenhagen, February 10, 1996.
24. Song, J. H. and Reed, M. A., Petroleum Wax-Free Chewing Gums Having Improved Flavor Release, United States Patent 5,286,501, February 15, 1994. Assigned to Wm. Wrigley Jr. Company, Chicago, IL.
25. Stephen, H. G., Parameters Controlling Colour Acceptance in Latex Paints, *J. Oil Colour Chem. Assoc.,* 69(3), 53–61, 1986.

7 Applications — Chemical Resistance

CONTENTS

ABSTRACT

Hansen solubility parameters (HSP) can correlate differences in physical behavior observed in chemical resistance testing of polymers and polymer containing systems when a sufficiently large number of different organic solvents are included in a study. These correlations can then be used to predict the chemical attack expected in systems which have not yet been tested. Examples of HSP correlations included here are for solubility, degree of surface attack, tensile strength reduction, and simple evaluations of chemical resistance of the suitable-for-use or not type. In each case, the molecular size of the liquids used can affect the result and should be considered in some way. A common problem is that tests with larger molecular weight liquids have not reached equilibrium within the timeframe of the exposure. HSP correlations are presented for chemical resistance studies of epoxy and zinc silicate tank coatings, PET, POM, PA6/66, PUR, PPS, PEI, Neoprene®, etc.

INTRODUCTION

HSP are widely used in the coatings industry to select solvents for dissolving polymers and binders. This has been discussed in Chapter 6 and also in References 1 through 12, among other places. The fact of solution is in itself clearly one simple form of chemical attack of the polymers they dissolve. This means that chemical resistance for some polymers can be partly inferred from HSP correlations of their solubility and/or swelling. HSP correlations of this type have been discussed in Chapter 3. An example is that if a chemical does not dissolve an epoxy component or the curing agent, then it is quite unlikely that it will attack a fully crosslinked epoxy coating or glue. The HSP correlations of surface phenomena, which have been discussed in Chapters 4 and 5 and elsewhere,[13-18] can also provide some insight into chemical resistance. Liquids not wetting a surface are not as likely to attack it as those which do wet it, although there are no guarantees. Some surface studies may involve evaluation of a more direct form of chemical attack, such as the attack/whitening of PET coated with "amorphous" PET to improve weldability. This example is discussed in more detail later. Whatever is being correlated, the general considerations of the HSP characterizations discussed in earlier chapters are the same for the HSP correlations of chemical resistance reported here. However, there are certain additional pitfalls to be aware of when correlating chemical resistance. These include (lack of) attainment of equilibrium, the effects of molecular size of the test chemical, difference in local segments of polymers (even in homopolymers), and acid/base reactions.

Once a reliable HSP characterization of chemical resistance is available, it can be used to calculate the behavior of other systems which have not been tested. Obtaining a good HSP correlation of chemical resistance which allows reliable predictions depends very much on careful treatment of the available data or generation of data with such a correlation in mind. Unfortunately, very few studies of chemical resistance have been designed with the purpose of generating HSP correlations. Also, it must be clear that the chemical attack discussed in this chapter does not include true chemical reactions leading to covalent bonding or destruction, such as with acids and bases, whether they are organic or inorganic. Chemical reactions forming new compounds are often found with amines and organic acids. These reactions often lead to discoloration in systematic solubility parameter testing with one or more of the amines used as test solvents. Discolored systems should simply be neglected in an HSP correlation of physical (reversible) solubility. The products of reactions of well-defined organic bases with well-defined organic acids have actually been studied systematically from a solubility parameter point of view.[19] Some of the results of this study are discussed in Chapters 9 and 11.

CHEMICAL RESISTANCE — ACCEPTABLE-OR-NOT DATA

Additional sources of data for solubility parameter characterizations include chemical resistance tables reported by raw material suppliers[20-25] or collected in books[26-28] and other sources such as those supplied on electronic media by the Plastics Design Library.[29] While these data are certainly valuable in themselves, it has been found that given data sets are not always as reliable/consistent/coherent as could be desired for solubility parameter correlations. Attainment of equilibrium may not have been achieved, and this effect is rarely confirmed or even considered. Solvents with low diffusion coefficients will appear to be less aggressive than they might become at longer exposure times. As discussed earlier, true chemical attack with acids and bases must sometimes be sorted out. Likewise, the data are often limited in number and scope, and the chemical reagents have not been chosen for the purpose of solubility parameter correlations. Nevertheless, with the use of due caution, it is often possible to find excellent solubility parameter correlations using chemical resistance data of the acceptable-or-not type, particularly when the list of agents is long. Additional precautions with regard to data of the acceptable-or-not type include whether a molecular size effect is present as discussed in the following. Also, it can be assumed that if a chemical attacks a polymer at, say, 20°C, then it will also attack it at, say, 70°C. It can then be included as data in a 70°C correlation, even though it may not have been tested at that temperature (and in principle the HSP are only valid at room temperature).

EFFECTS OF SOLVENT MOLECULAR SIZE

It has been emphasized in Chapters 1, 2, and 3 that the size of solvent molecules is important for polymer solubility. HSP correlations have confirmed that this effect is even more important when chemical resistance is being considered. Smaller molecules are expected to be better, that is, more aggressive from a thermodynamic point of view than larger ones, all else being equal. This is known from the theories of polymer solubility discussed in Chapters 1 and 2 and also from the discussion of barrier polymers found in Chapter 8. So it is not surprising that solvent molecular size can be an important fourth parameter in correlations of chemical resistance. An appropriate way to check this is sorting output data from a computer (or other) HSP optimization according to the molecular volume of the test solvents. What appear to be errors in the correlation may become systematically arranged. It can easily be seen if the top of the list includes the type of "error" whcre the smaller molecular species are "better" than expected by comparison will all the other solvents. This may take the form of unexpectedly dissolving, being more aggressive than expected, penetrating more rapidly, or reducing mechanical properties more severely. Larger molecular species which are "poorer" than expected by comparison with the data for the other solvents are often seen at the bottom of the list. One can focus upon the molecular size range of greatest interest in such cases and repeat the correlation neglecting those species which are outside of this range of molecular volumes. The correlation is then strictly valid only for the size range specified. Some indication of the behavior of the solvents with V larger than the upper limit is possible if their RED numbers are greater than 1.0. These would not be expected to attack under any circumstances. Likewise, the solvents with V less than the lower limit can be expected to attack if their RED numbers are less than 1.0. Larger numbers of solvents are needed in the study if this is to be done with any benefit.

As stated above, the size and shape of solvent molecules are very important for kinetic phenomena such as diffusion, permeation, and attainment of equilibrium. Chapter 8 discusses correlations of HSP with diffusion phenomena in more detail. However, it will be repeated here that smaller and more linear molecules diffuse more rapidly than larger and more bulky ones. The diffusion coefficient may be so low that equilibrium is not attained for hundreds of years at room temperature in common solvent exposures of rigid polymers like polyphenylene sulfide (PPS) with thicknesses of several millimeters.[30] Such effects lead to comparisons where some systems may have reached an equilibrium uptake, while others have not. Likewise, the second stage in the two-stage drying process in polymer film formation by solvent evaporation can last for many years.[4,31] Polymer samples used for solubility parameter or chemical resistance testing may contain retained solvent or monomer for many years, and this may also affect the evaluations.

Attempts to include the molecular volume into a new composite solubility parameter and size parameter have not been particularly successful.[32,33] This may be because the size effects are not necessarily caused through the thermodynamic considerations on which the solubility parameters are based, but rather through a kinetic effect of diffusion rate.

CHEMICAL RESISTANCE — EXAMPLES

Chemical resistance means different things in different contexts. Various examples of HSP correlations of chemical resistance are included in the following. The HSP data for the correlations discussed are included in Table 7.1.

Experimental data are always preferred over predicted behavior based on a correlation. However, a good HSP correlation can be used to find many chemicals which will clearly attack or which will clearly not attack. There are also situations where the attack is mild, and whether or not satisfactory results are found with a product depends on its use. Data in chemical resistance tables are often of the type +, +/–, –, or satisfactory/questionable/unsatisfactory, recommended/not recommended (R/NR), or something similar. The liquids which attack are clearly good solvents for the material in question and will be located within the HSP spheres with RED numbers being successively lower for more severe attack, all other things being equal. The correlations can include the solvents with mild attack (+/–, questionable) either in the attacking (NR) group or in the non-attacking (R) group. They can also be neglected, not knowing which group to put them into. The treatment used in the individual correlations presented here is indicated in the following. Unless otherwise specified, the results are for room temperature.

TABLE 7.1
Hansen Solubility Parameter Correlations for Selected Materials

Material	d_D	d_P	d_H	R	FIT	G/T
Epoxy tank coat (two component)	18.4	9.4	10.1	7.0	—[a]	—[a]
Epoxy solubility (Epon 1001)	18.1	11.4	9.0	9.1	—[a]	—[a]
Zinc silicate coating	23.5	17.5	16.8	15.6	—[a]	—[a]
PET-amorphous coating	17.0	11.0	4.0	9.0	1.000	7/11
PET-CR (+/– R[b])	18.2	6.4	6.6	5.0	0.865	7/34
PUR-CR (+/– R[b])	18.1	9.3	4.5	9.7	0.981	16/26
POMC/POMH (+/– R[b])	17.1	3.1	10.7	5.2	0.955	2/28
POMC (+/– NR[b])	17.9	5.9	8.3	6.6	0.609	11/28
PA6/PA66 (+/– NR[b])	18.9	7.9	11.7	8.7	0.950	9/31
Halar 300 ECTFE 23°C	16.8	8.4	7.8	2.7	0.993	2/102
Halar 300 ECTFE 50°C	18.1	7.5	8.5	5.2	0.700	18/92
Halar 300 ECTFE 100°C	18.1	7.9	7.9	6.7	0.710	49/91
Halar 300 ECTFE 120/149°C	18.3	8.7	7.9	7.5	0.800	48/74
Neoprene-CR (+/– R[b])	18.1	4.3	6.7	8.9	0.937	30/48
PPS tensile strength <60% 93°C	18.7	5.3	3.7	6.7	0.991	9/16
PEI ULTEM 1000 600 psi	17.3	5.3	4.7	3.3	1.000	3/20
PEI ULTEM 1000 1200 psi	17.0	6.0	4.0	4.0	1.000	4/20
PEI ULTEM 1000 2500 psi	17.4	4.6	9.0	7.2	0.967	9/20
PEI ULTEM 1000 solubility	19.6	7.6	9.0	6.0	0.952	8/45
PES mechanical properties	17.1	9.9	6.3	6.3	0.931	6/25
PES solubility	19.6	10.8	9.2	6.2	0.999	5/41

Note: The symbols G for good solvents and T for total solvents are maintained, with the understanding that G solvents are within the HSP correlation spheres and are not recommended for use.

[a] Data not available.

[b] See text: NR — not resistant, R — resistant.

Tank Coatings

Chemical resistance is important for tank coatings used in the transport of bulk chemicals. The data in Table 7.1 include two older HSP correlations for chemical resistance for two types of tank coatings supplied by Hempel´s Marine Paints. These are for a two component epoxy type and for a zinc silicate type. The data and correlations are about 15 years old. They are included here for purposes of demonstration. Improvements in chemical resistance are known to have been implemented in a newer epoxy tank coating, but no HSP correlation has been made. A HSP correlation of the solubility of a lower molecular weight epoxy, Epon® 1001 (Shell Chemical Corp.), is included for comparison. The numbers are not too different from those of the HSP correlation for chemical resistance for the two component epoxy tank coating. These three correlations have been reported earlier.[8]

The fact of a successful HSP correlation for a completely inorganic type of coating like zinc silicate is surprising. This is still another demonstration of the universality of the applications possible with the HSP concept. While the data fit numbers were not recorded at the time, the two chemical resistance correlations reported here were clearly considered reliable.

PET Film Coating

Another example of chemical resistance, or lack of the same, is the attack of the amorphous, modified PET coating on PET films to improve their weldability. This correlation is based on only

11 well-chosen data points, but clearly shows that attack for many chemicals can be expected. Among those chemicals not attacking are hydrocarbons, glycols, and glycol ethers and higher alcohols which have a reasonably high hydrogen bonding character.

Acceptable or Not — Plastics

Several examples of HSP correlations of data reported in the form acceptable-or-not are included in Table 7.1. The data for these are all found in Reference 20. Other data sources for these are also available. HSP correlations of this type are included for PET, PUR, POMH, POMC, and PA6/PA66 using data from Reference 20. The first three correlations consider reported evaluations of minor attack which will require further evaluation (+/–) as if these systems were suitable for use (Resistant). The last two correlations consider this condition as not suitable for use (Not Resistant). This is to demonstrate that differences are found depending on how the data are considered and that outliers are often found when correlating this type of data. It is for this reason that more extensive tables of HSP correlations of chemical resistance are not reported here. Too much space is required to try to explain why given results are outliers. However, some reasons have been given earlier and others are in the following.

The HSP values for PET based on chemical resistance are somewhat different from those of the amorphous PET coating which is readily attacked by far more solvents. The compositions are also different. The POM correlations have typical problems in that the correlation considering all minor attack as negligible is based on only two severely attacking solvents among the 28 solvents tested. When the solvents demonstrating minor attack are considered as being in the attacking group, the data fit shows that there are many outliers. A systematic analysis of why this is found will not be attempted for reasons of space, even if all the outliers could be explained in one way or another.

The HSP found for a polymer in this type of correlation may not be representative of that polymer in all aspects of its behavior. There is a question as to coverage over the whole range of solvent HSP possible by the test solvents. Additional solvents may be required to make an improved correlation based on the improved coverage possible. There is also a question as to which segments of the polymer may be subject to attack by which solvents. Block copolymers may demonstrate two separate (overlapping) correlations which cannot be reasonably force fitted into a single HSP correlation. Viton® is an example of this. The most severe attack or swelling may occur in one region or another of the polymer or maybe even on a third component, such as a cross-linking segment. Viton® is discussed in Chapters 3 and 8.

In spite of these pitfalls, it is strongly suggested that those generating this type of resistance data should try an HSP correlation to evaluate the consistency of the data before reporting it. Outliers can be reconsidered; whether or not equilibrium has been attained can be inferred; and the possible/probably effects of solvent molecular size may become apparent.

The effects of temperature on the chemical resistance of poly(ethylene co-chlorotrifluoroethylene) (ECTFE) Halar® 300 can be seen in Figure 7.1. The data on which these correlations are based are of the recommended-or-not type and were found in Reference 25. This figure has affectionately been dubbed "bullseye," since there appears to be symmetry about a central point, although this is not strictly true as the HSP data confirm. The radius of the chemical resistance spheres increases with increasing temperature, as expected, since more solvents then become more severe in their attack. The HSP data for these correlations are also included in Table 7.1. The data fits are not particularly good at the higher temperatures.

To complete this section, an HSP correlation of chemical resistance data for Neoprene® rubber (Du Pont)[21] is included. Solvents in the intermediate category, i.e., that of a questionable-for-use recommendation, are considered as being in the nonattacking group for this correlation.

Previously it was indicated why HSP correlations of this type lead most often to guidance rather than to a firm recommendation. There are many pitfalls to be aware of both in generating such correlations as well as in using them, but their usefulness becomes clearer with some experience.

SOLVENT RESISTANCE OF HALAR 300 AT 4 DIFFERENT TEMPERATURES

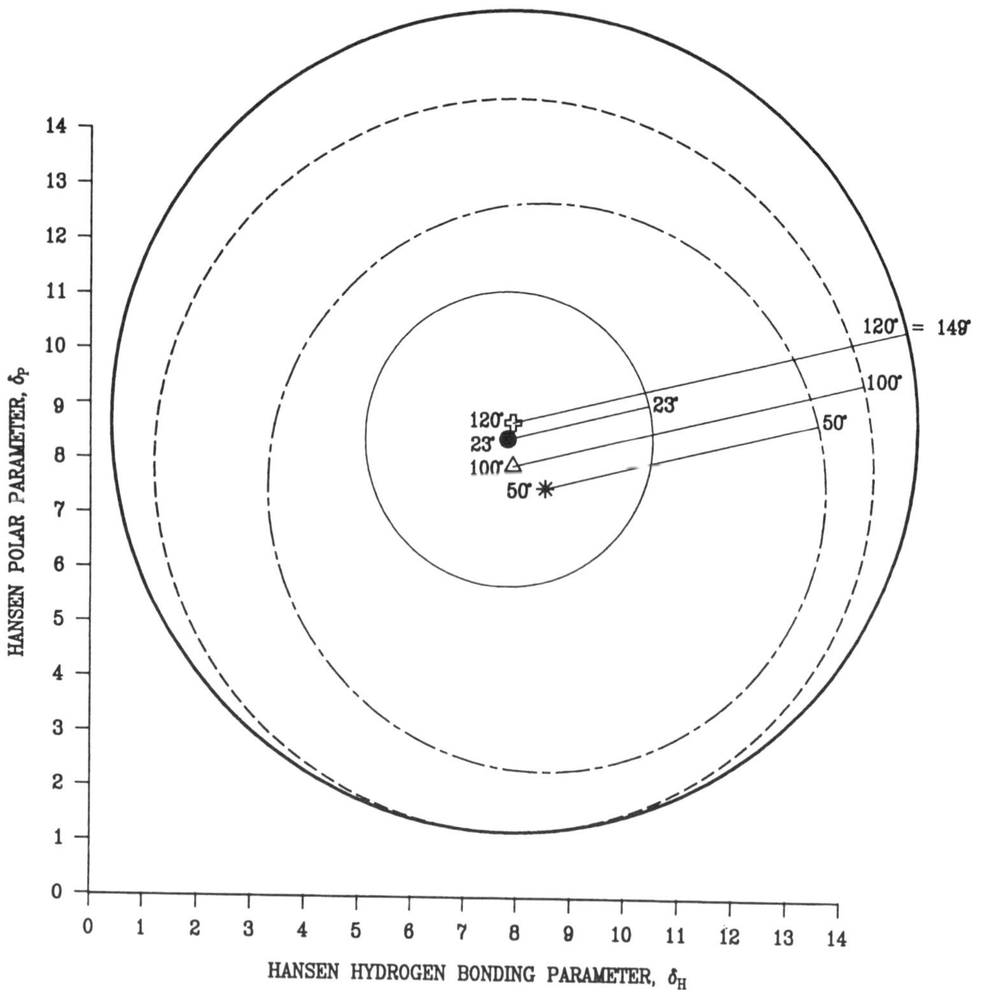

TEMP.	D	P	H	R	FIT	NO(NR)
23°	16.8	8.4	7.8	2.7	0.993	102 (2)
50°	18.1	7.5	8.5	5.2	0.70	92 (18)
100°	18.1	7.9	7.9	6.7	0.71	91 (49)
120/149°	18.3	8.7	7.9	7.5	0.80	74 (48)

FIGURE 7.1 Chemical resistance of Halar® 300 ECTFE at various temperatures. Liquids within the spheres (circles) are not recommended at the given temperatures. HSP data given in Table 7.1.

A suitable goal for a future project is to determine the effective HSP for various media frequently encountered in resistance lists, such as mustard, given detergents, given oils, and the like. This could perhaps be done by composition in some cases. In other cases, one could see whether behavior paralleled that of a known chemical. A third method is to determine these parameters by recognizing a similarity to all materials attacked and a difference from those not attacked. This approach has been used to assign HSP to some liquids when calculations were uncertain. A computer program was developed similar to the SPHERE program as described in Chapter 1, but working in the opposite manner. The solubility of a number of polymers was evaluated in the solvent. The solvent parameters were then systematically varied by the program to reduce the collective error, that is, to locate the best possible set of HSP for the solvent. In general, the data fits for this procedure were comparable to those found for polymer solubility using the SPHERE program. In other words, not all the predictions based on the HSP thus assigned to the solvent agreed with the experimental data, but the errors were small.

Tensile Strength

The long-term exposure of polymers or polymer composites to solvents normally leads to changes in mechanical properties. One of the more direct techniques to measure such effects is to determine the tensile strength. The tensile strength reduction for glass fiber reinforced polyphenylene sulfide (PPS) after exposure to a number of solvents at 93°C for 12 months has been reported.[22] A HSP correlation of these data using the "good" solvents as those which reduce tensile strength under these conditions to less than 60% of the initial value is found in Table 7.1 and Figure 7.2. More extensive correlations for PPS are found in Reference 30.

Additional HSP tensile strength correlations have been generated for polyetherimide, ULTEM® 1000, using data reported by General Electric.[23] It is clear that the chemical resistance is dependent on the stress level. Higher stress levels lead to more severe attack by a larger number of chemicals. The solvents considered as being those which attack led to cracking during the study which lasted 336 h. Some led to earlier cracking than others, which could be treated in a separate correlation, but this has not been done. A more rapid attack is expected from the better solvents with the smallest size and shape. The correlations all have high data fits.

The next entry in Table 7.1 is for true solubility of ULTEM 1000. These data were generated in a standard solubility parameter study. This correlation is not directly comparable with the previous ones for ULTEM 1000 since the number and range of solubility parameters included in the test solvents are different. The parameters for the polymer found in this correlation are those expected to reflect its (thermodynamic) affinities most closely. The previous study[23] did not include a sufficient number of solvents having widely different HSP to give a true total picture of the ULTEM 1000.

A final HSP correlation of the suitable-for-use or not type is presented in Table 7.1. This is for polyether sulfone (PES) based on mechanical properties after exposure to various liquids. The HSP correlation for the recommendation from the supplier[24] for glass fiber reinforced PES (Ultrason® E, BASF) can be compared with the HSP correlation for simple solubility of the polymer in standard test liquids, also given in Table 7.1. There is a difference, but it is not large.

SPECIAL EFFECTS WITH WATER

As stated elsewhere in this book (Chapters 1 and 9, in particular), the seemingly unpredictable behavior of water has often led to its being an outlier in HSP correlations. For this reason, it is suggested that data for water used as a test solvent not be included in HSP correlations. Water can be a very aggressive chemical. Water uptake in most polymers increases with increased temperature.

Solvent Resistance of Polymer Composites - Glass Fiber Reinforced Polyphenylene Sulfide

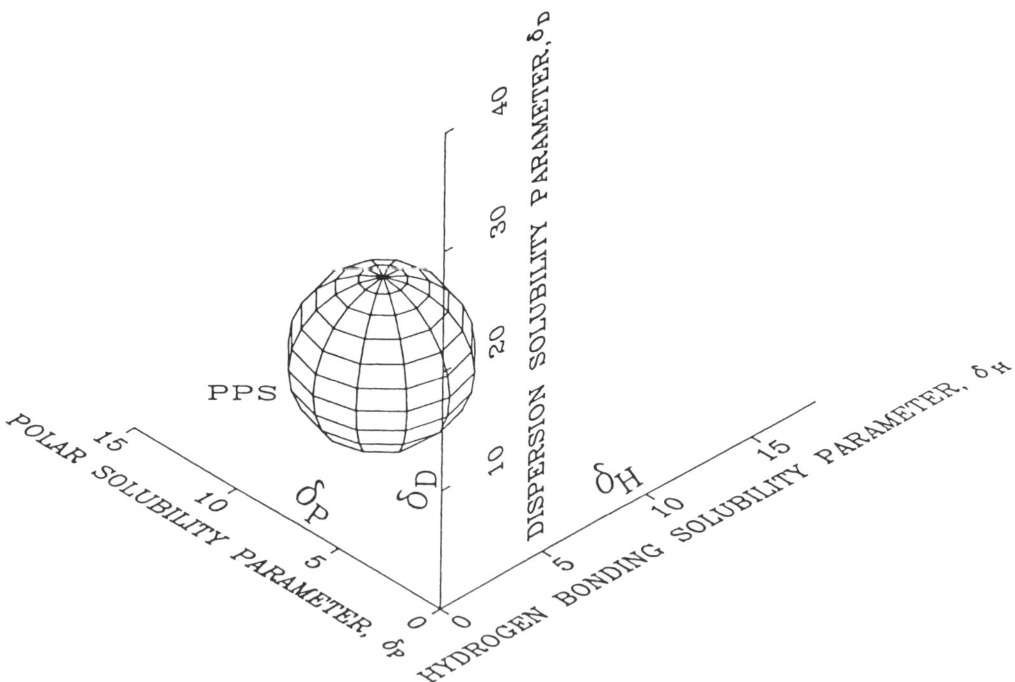

FIGURE 7.2 HSP correlation of the tensile strength reduction of Ryton® PPS. Within the sphere are liquids which reduce the tensile strength to less than 60% of the original value after exposure for 1 year at 93°C. (From FORCE Institute, *Solvent Resistance of Polymer Composites, Glass Fibre Reinforced Polyether Sulphone (PES),* 1st ed., Center for Polymer Composites, 1994, 31. With permission.)

This is because the solubility parameters of the water and polymer are closer at higher temperatures. The very high δ_H parameter for water decreases more rapidly with increasing temperature than the δ_H parameter for most polymers. This has been discussed in Chapters 1 and 6, but it is repeated here with examples for those interested in chemical resistance.

Water is an exceptionally good plasticizer because of its small molecular size. The presence of water not only softens (reduces the glass transition temperature) a polymer as such, but it also means diffusion rates of other species will be increased. Therefore, the presence of water in a film can influence the uptake of other materials, with hydrophilic materials, in particular, being more prone to enter the film.

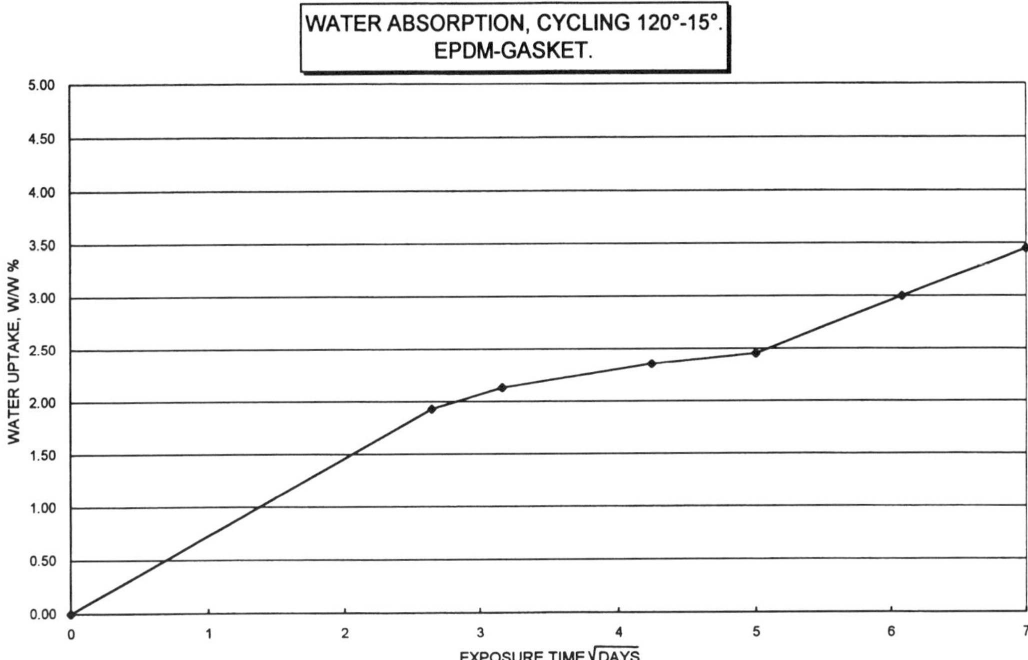

FIGURE 7.3 A rapid quench to a lower temperature can free water already dissolved in a polymer in the form of SWEAT. SWEAT can lead to blistering, cracking, and delamination. The data in the figure are weight gain for EPDM with cycling in water between 120 and 15°C. Water in excess of the equilibrium value at longer cycling times is SWEAT.

The increase of water uptake with increased temperature can cause special problems with blistering if the temperature of a water-saturated polymer falls rapidly to a lower temperature. The previously soluble water can no longer be truly dissolved. Some of the water already in the polymer is now in excess and suddenly appears as small clusters or droplets of freed liquid water within the polymer itself (see Chapter 6, Figure 6.3). These droplets can quickly collect into blisters, especially if there are hydrophilic sites in the polymer or at an interface to which water will rapidly diffuse. This special type of failure has been discussed in more detail elsewhere[31] (see also Chapters 1 and 6). The phase separated water has been called SWEAT (soluble water exuded at lowered temperatures). The author has observed this phenomenon as a mechanism of failure for epoxies, polyesters, alkyds, polyethersulfone (PES), polyphenylene sulfide (PPS), and even EPDM rubber. This mechanism can be confirmed experimentally by cycling samples continually exposed to water between two relevant temperatures using a quench from the higher one to the lower one. One follows weight gain by rapidly weighing samples which are surface dry. Typical results for the SWEAT phenomena for EPDM are seen in Figure 7.3 and for PPS in Figure 7.4. Control samples which are not cycled reach equilibrium and stay there, whereas the cycled samples suddenly begin to gain weight well beyond the equilibrium value. The extra weight is phase separated water within the samples. This has been discussed in detail in Reference 34 for PPS and PES.

A related problem can be encountered in chemical resistant coatings for tanks which have been in contact with methanol. If a coated tank has been used to store methanol, and perhaps hot methanol in particular, the coating is more than likely saturated with methanol. It may take several days of exposure to fresh air (to reduce the amount of methanol to acceptable levels) before subsequent direct contact with water or seawater can be tolerated. If there is too much methanol retained in

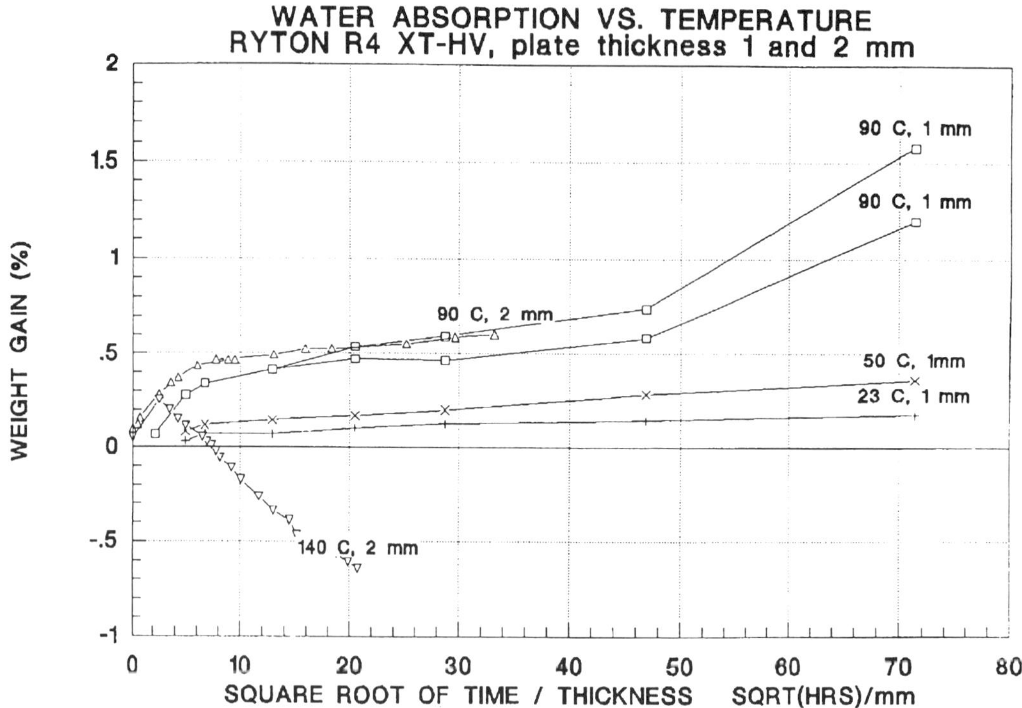

FIGURE 7.4 A rapid quench to a lower temperature can free water already dissolved in a polymer in the form of SWEAT. SWEAT can lead to blistering, cracking, and delamination. The data in the figure are weight gain for PPS with cycling in water between 90 and 23°C. Water in excess of the equilibrium value is SWEAT. (From Hansen, C. M., *The Resistance of PPS, PES and PA Polymer Composites to Temperature Cycling During Water Exposure*, Center for Polymer Composites (Denmark), Danish Technological Institute, Taastrup, 1994, 11. With permission.)

the coating, the water diffusing into the coating will associate with the methanol. The increasing water content in the mixture of methanol and water will ultimately cause the solubility parameters of the mixture to be sufficiently high so that it becomes incompatible with the coating. Blisters form and total delamination can occur. These blisters are often near the substrate, since this is where the retained methanol will be found at highest concentrations.

CONCLUSION

HSP can correlate differences in physical behavior observed in chemical resistance testing of polymers and polymer containing systems when a sufficiently large number of different organic solvents are included in a study. HSP correlations including systematic consideration of the solvent molar volume (or other suitable size parameter[s]) should be an inherent part of all future studies of chemical resistance. These correlations aid in the determination of whether equilibrium has been attained, as well as provide insight into the behavior expected from nontested solvents whose HSP are stored in a solvent database or can be calculated.

Examples include HSP correlations of true solubility and swelling, degree of surface attack, tensile strength reduction, and correlations for simple evaluations of chemical resistance of the suitable-or-not type. It is reemphasized that, in each case, the molecular size of the liquids evaluated will affect the result, and this should be considered in some way.

Data for true acidic or basic chemical attack must not be included in HSP correlations, since HSP correlations reflect physical attack and not chemical attack. It is strongly suggested that data for water not be included in these correlations as well. Its behavior is too unpredictable compared with other test liquids, and if it is included as an outlier, this fact will force a correlation with less predictive ability than had it been neglected.

REFERENCES

1. Hansen, C. M., The Three Dimensional Solubility Parameter — Key to Paint Component Affinities I. Solvents, Plasticizers, Polymers, and Resins, *J. Paint Technol.,* 39(505), 104–117, 1967.
2. Hansen, C. M., The Three Dimensional Solubility Parameter — Key to Paint Component Affinities II. Dyes, Emulsifiers, Mutual Solubility and Compatibility, and Pigments, *J. Paint Technol.,* 39(511), 505–510, 1967.
3. Hansen, C. M. and Skaarup, K., The Three Dimensional Solubility Parameter — Key to Paint Component Affinities III. Independent Calculation of the Parameter Components, *J. Paint Technol.,* 39(511), 511–514, 1967.
4. Hansen, C. M., The Three Dimensional Solubility Parameter and Solvent Diffusion Coefficient, Their Importance in Surface Coating Formulation, Doctoral Dissertation, Danish Technical Press, Copenhagen, 1967.
5. Hansen, C. M., The Universality of the Solubility Parameter, *Ind. Eng. Chem. Prod. Res. Dev.,* 8(1), 2–11, 1969.
6. Hansen, C. M. and Beerbower, A., Solubility Parameters, in *Kirk-Othmer Encyclopedia of Chemical Technology,* Suppl. Vol., 2nd ed., Standen, A., Ed., Interscience, New York, 1971, 889–910.
7. Hansen, C. M., 25 Years with Solubility Parameters (in Danish: 25 År med Opløselighedsparametrene), *Dan. Kemi,* 73(8), 18–22, 1992.
8. Hansen, C. M., Solubility Parameters, in *Paint Testing Manual,* Manual 17, Koleske, J. V., Ed., Americal Society for Testing and Materials, Philadelphia, 1995, 383–404.
9. Barton, A. F. M., *Handbook of Solubility Parameters and Other Cohesion Parameters,* CRC Press, Boca Raton, FL, 1983.
10. Barton, A. F. M., *Handbook of Polymer-Liquid Interaction Parameters and Solubility Parameters,* CRC Press, Boca Raton, FL, 1990.
11. Anonymous, Brochure: Co-Act — A Dynamic Program for Solvent Selection, Exxon Chemical International Inc., 1989.
12. Dante, M. F., Bittar, A. D., and Caillault, J. J., Program Calculates Solvent Properties and Solubility Parameters, *Mod. Paint Coat.,* 79(9), 46–51, 1989.
13. Hansen, C. M., Characterization of Surfaces by Spreading Liquids, *J. Paint Technol.,* 42(550), 660–664, 1970.
14. Hansen, C. M., Surface Dewetting and Coatings Performance, *J. Paint Technol.,* 44(570), 57–60, 1972.
15. Hansen, C. M. and Pierce, P. E., Surface Effects in Coatings Processes, *Ind. Eng. Chem. Prod. Res. Dev.,* 13(4), 218–225, 1974.
16. Hansen, C. M. and Wallström, E., On the Use of Cohesion Parameters to Characterize Surfaces, *J. Adhes.,* 15(3/4), 275–286, 1983.
17. Shareef, K. M. A., Yaseen, M., and Reddy, O. J., Suspension Interaction of Pigments in Solvents. Characterization of Pigments Surfaces in Terms of Three-dimensional Solubility Parameters of Solvents, *J. Coat. Technol.,* 58(733), 35–44, Feb. 1986.
18. Beerbower, A., Surface Free Energy: a New Relationship to Bulk Energies, *J. Colloid Interface Sci.,* 35, 126–132, 1971.
19. Hansen, C. M., Some Aspects of Acid/Base Interactions (in German: Einige Aspekte der Säure/Base-Wechselwirkung), *Farbe und Lack,* 7, 595–598, 1977.
20. Anonymous, Plastguide, SCS Dukadan AS, Randers, Denmark, 1990.
21. Anonymous, Fluid Resistance of Viton®, Du Pont Company, Polymer Products Department, Elastomers Division, Wilmington, DE, 1989.
22. Anonymous, RYTON® PPS Polyphenylene Sulfide Resins — Corrosion Resistance Guide, Philips Petroleum Co. (U.S.)

23. Anonymous, Ultem® Resin Design Guide, GE Plastics, Pittsfield, MA, 1989.
24. Anonymous, Verhalten von Ultrason® gegen Chemikalien — BASF Technische Information TI-KTE/TH-01 d 82132, October 1991.
25. Anonymous, Expanded List — Chemical Resistance of Halar® Fluoropolymer, Ausimont, USA, Inc.
26. Carlowitz, B., Thermoplastic Plastics (in German), *Thermoplastische Kunststoffe*, Zechner & Hüthig, Speyer am Rhein, 1980.
27. Anonymous, Chemical Resistance Data Sheets Volume 1 Plastics; Volume 2 Rubbers, New Edition — 1993, Rapra Technology Limited, Shawbury, Shrewsbury, Shropshire.
28. Anonymous, Chemical Resistance of Plastics and Elastomers used in Pipeline Construction, George Fischer +GF+, 1992.
29. Plastics Design Library, various collections of data, William Andrew, Inc., Norwich, NY 13815.
30. Hansen, C. M., *Solvent Resistance of Polymer Composites — Glass Fibre Reinforced Polyphenylene Sulfide*, Centre for Polymer Composites (Denmark), Danish Technological Institute, Taastrup, 1993, 1–62, ISBN 87-7756-286-0.
31. Hansen, C. M., New Developments in Corrosion and Blister Formation in Coatings, *Prog. Org. Coat.*, 26, 113–120, 1995.
32. Van Dyk, J. W., Paper presented at the Fourth Chemical Congress of America, New York, August 25–30, 1991.
33. Anonymous (Note: This was, in fact, Van Dyk, J. W., but this does not appear on the bulletin), Using Dimethyl Sulfoxide MSO) in Industrial Formulations, Bulletin #102, Gaylord Chemical Corp., Slidell, LA, 1992.
34. Hansen, C. M., *The Resistance of PPS, PES and PA Polymer Composites to Temperature Cycling During Water Exposure*, Centre for Polymer Composites (Denmark), Danish Technological Institute, Taastrup, 1994. ISBN 87-7756-357-3.

8 Applications — Barrier Polymers

CONTENTS

ABSTRACT

The permeation coefficient, **P**, of a liquid or a gas through a polymer is given by the product of the diffusion coefficient, **D**, and the solubility coefficient, **S**: **P** = **DS**. **S** correlates with the Hansen solubility parameter(s) (HSP). At low permeant concentrations **D** is a constant. However, as the permeant concentration increases, its plasticizing effect on the polymer becomes significant, and the diffusion coefficient increases markedly. This effect can be very significant. The successful correlations of permeation phenomena with HSP are thought to be largely a result of this exceptional dependence of **D** on the dissolved permeant. Since the amount of permeant being dissolved increases with closer matches of the HSP for permeant and barrier polymer, the end result is that both **S** and **D**, and therefore **P**, are functions of the HSP match. HSP correlations are given for breakthrough times in chemical protective clothing, permeation rates through barrier polymers, and barrier polymer swelling. Both liquids and gases are treated.

INTRODUCTION

The permeation of a liquid or a gas through a polymer can be described by the relation

$$\mathbf{P} = \mathbf{DS} \tag{8.1}$$

P, the permeation coefficient, is the product of the diffusion coefficient, **D**, and the solubility coefficient, **S**. The diffusion coefficient indicates how fast the permeant molecules can move through the polymer. The solubility coefficient indicates how much of the permeant can be dissolved in the polymer. The amount dissolved in the polymer determines the concentration gradient over a film, and the concentration gradient is the driving force for mass transport. When solubility is higher, the concentration gradient is correspondingly higher, and, assuming the same diffusion coefficient, mass transport will be proportionately higher. **S** will be lower when the HSP of the barrier film and a solvent are very different.

A significant factor affecting **D** is the molecular size and shape of the permeant molecules. Larger molecular size and more complex and bulky molecular shape are major factors which lead to lower diffusion coefficients. The diffusion coefficient for oxygen in polyvinyl chloride (PVC) is well over a million times greater than that of *n*-hexane (at low concentrations) in the same polymer.[1] This difference in diffusion coefficients is a result of differences in molecular size. Likewise, it has been found that the rate of diffusion at the same concentration is about the same for different solvents with approximately the same size and shape, even though they may have different solubility parameters (but not so different that both are able to dissolve in the polymer at the level of comparison).[2-4] The polymers in these studies were a copolymer of 87% vinyl chloride and 13% vinyl acetate, polyvinyl acetate, and polymethyl methacrylate.

CONCENTRATION-DEPENDENT DIFFUSION

Low molecular weight liquids are plasticizers for polymers if they can be dissolved in them. Water, for example, can significantly soften many polymers even though it is dissolved to only a few percent. The low molecular weight materials can greatly reduce the glass transition temperature of their mixtures with a polymer since they have considerably more free volume associated with them than the polymers themselves. This extra free volume allows easier polymer segmental motion. The diffusion of the smaller species (and other species) becomes faster as their local concentration and plasticizing effect become greater.

The solvent diffusion coefficient data in Figure 8.1 were first presented in Reference 3. This figure shows diffusion coefficients for several solvents in polyvinyl acetate (PVAc) at 25°C. The diffusion coefficient for water shown in the figure was found by absorption and desorption experiments in thin films where a correction for the surface resistance was also required.[5] It can be seen in this figure that for moderate solvent concentrations in this rigid polymer, the local diffusion coefficient increases by a factor of about 10 for an increase in solvent concentration of about 3 to 4 vol%. Since this behavior is general for solvents in polymers, a rule of thumb indicates that the local diffusion coefficient for solvents in rigid polymers can increase by a factor of about one million when about 20 vol% solvent is present compared with the solvent-free state.[3-7] This rule of thumb assumes that the polymer behaves as a rigid polymer over the concentration range being considered. This difference corresponds to the speed of a snail in the woods compared with a modern jet airliner.

Concentration-dependent diffusion coefficients are also found for elastomers. Here, the rule of thumb is that the diffusion coefficient increases by a factor of about 10 for an increase in solvent concentration of about 15 vol%.[7] This shows that liquid contact with chemical protective clothing, for example, leads to concentration-dependent diffusion coefficients because the amount taken up at the contact surface on liquid contact is very often more than 15%.

Concentration-dependent diffusion has been discussed at length by Crank.[8] It is also discussed here because it is a major factor in the success of HSP correlations of permeation phenomena. The Crank-Nicholsen finite difference treatment for concentration-dependent diffusion[8] was extended by Hansen[3] and used to describe film formation by solvent evaporation,[4] to explore what is termed anomalous diffusion,[5] to develop an easy method to evaluate data leading to concentration-dependent diffusion coefficients,[6] and to account for the effects of concentration-dependent diffusion and surface boundary resistance simultaneously.[5,7] Klopfer[9] developed analytical solutions involving concentration-dependent diffusion for many situations found in practical building applications, particularly with respect to transport of water in building materials. Concentration-dependent diffusion can be handled properly without great difficulty for most situations of practical interest.

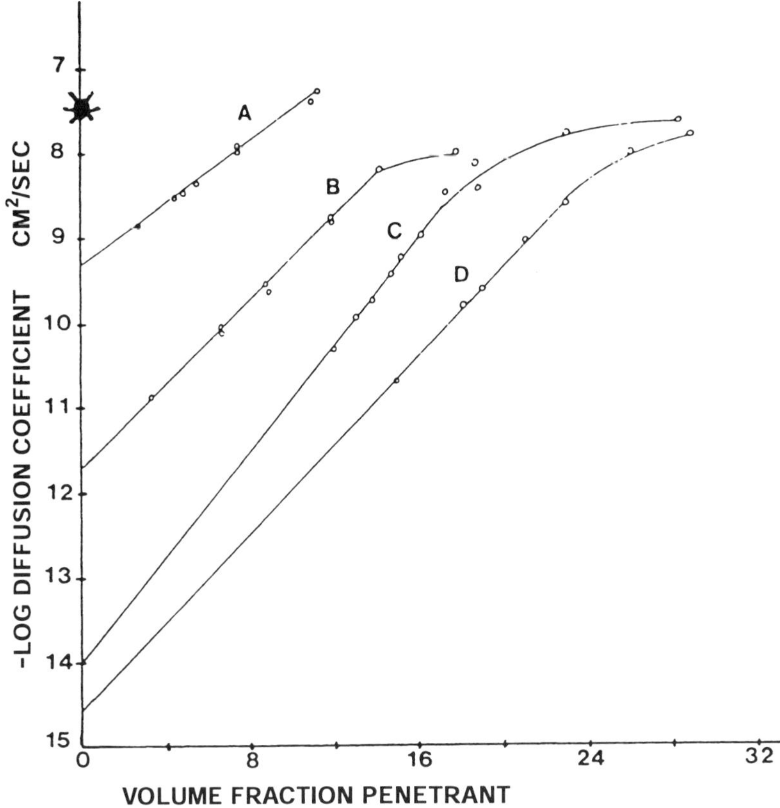

FIGURE 8.1 Diffusion coefficients in polyvinyl acetate at 25°C for methanol, ethylene glycol monomethyl ether, chlorobenzene, and cyclohexanone. Original data are in Reference 3. The data point for water (*) is included for comparison. (From Hansen, C. M., Permeability of polymers, *Pharmaceutical and Medical Packaging 98*, 1998, 7.12 [ISBN 87-89753-24-0]. With permission.)

Neglect of this effect can lead to errors, the significance of which will increase with increasing amounts of the dissolved materials.

In addition to demonstrating concentration dependence, the diffusion coefficient data for PVAc in Figure 8.1 also show the well-established relations that those solvents with larger and more complicated chemical structures are those with lower diffusion coefficients. Water has one "significant" atom, methanol has two, and ethylene glycol monomethyl ether (EGMME) has five. The diffusion coefficient for water in PVAc at low concentration, Do, is 10,000 times larger than that for the latter. An example of how to estimate diffusion coefficients in PVAc for other liquids, such as methylene chloride, is as follows. The diffusion coefficients in PVAc for methylene chloride, with three significant atoms, can be expected to be somewhat lower than those for methanol, but much higher than those for EGMME. Planar chlorobenzene diffuses more rapidly than nonplanar cyclohexanone, even though the number of significant atoms is the same. Another type of comparison which is possible is to state that the diffusion coefficients for toluene are expected to be close to those for chlorobenzene because of a similarity in molecular size and shape. This was confirmed by solvent retention studies where toluene and chlorobenzene were retained in identical amounts in a film of VYHH® (87 wt% vinyl chloride, 13 wt% vinyl acetate, Union Carbide). Toluene, which does not dissolve this polymer, was introduced by placing a completely dry polymer film in a closed container over toluene vapors.

Diffusion can be expected to be slower in more rigid polymers, i.e., those with higher glass transition temperatures, unless the rigidity is such as to allow decided holes of suitable size to enable quite rapid diffusion of much smaller molecules. These considerations lead to the best combination of properties for a barrier polymer as being one with a high glass transition temperature and with HSP far removed from those of the permeant. If, in practice, this leads to water sensitivity, an alternate strategy, such as a laminated system, may be required.

SOLUBILITY PARAMETER CORRELATIONS BASED ON PERMEATION PHENOMENA

SOLUBILITY PARAMETER CORRELATIONS OF BREAKTHROUGH TIMES

Extensive permeation studies and collections of permeation data are available within the chemical protective clothing industry.[10,11] Such data can also be used to establish correlations with HSP. A list of HSP for barrier polymers used in chemical protective clothing has been published[12] based on data by Forsberg and Olsson.[10] Some of these correlations have been improved in most instances by correlating the more extensive data of Forsberg and Keith.[11] The definition of a "good" solvent which was used for these correlations was that the breakthrough time was less than some selected value, either 20 min, 1 h, or 4 h. Table 8.1 includes some of these improved HSP correlations based on a 1-h breakthrough time.

HSP alone cannot always correlate barrier properties unless comparisons are limited to solvent molecules with approximately the same size (and shape). This, of course, means that the diffusion coefficients at the reasonably low concentrations expected in better barrier polymers do not vary too greatly from each other. In many cases, satisfactory correlations could only be found when the differences in HSP between the permeant and the barrier polymer were combined with a size (and shape) parameter(s). The molecular volume, V, was found to be a reasonably successful single parameter for this purpose. Printing the correlation data arranged in increasing order of permeant V clearly showed whether the molecular size was important. With regard to the protective ability of the different garments, it was found that, in general, and as expected, the solvents with larger, more complicated structures required much longer times for breakthrough for a given protective membrane type than comparison with other solvents would indicate. Outliers were usually these larger molecular species and permeants with smaller or more linear structure, where diffusion is much

TABLE 8.1
HSP Correlations of Breakthrough Times for Barrier Polymers Typically Used in Chemical Protective Clothing

Type	δ_D	δ_P	δ_H	Ro	V Limits	FIT	No.
Neoprene®	16.0	8.8	4.0	10.1	None	0.574	66
Neoprene	19.0	8.0	0.0	13.2	71.0/172	1.000	50
Butyl	17.0	1.5	0.0	7.3	71.0/175.8	0.902	86
Viton®	15.6	9.6	7.8	7.1	72.6/148.9	0.896	77
Nitrile	19.8	13.3	2.2	13.6	84.3/177.2	0.907	58
Challenge 5100®	16.6	7.0	3.8	2.3	None	0.925	116
PE	16.5	2.7	6.1	7.9	None	0.969	32

Note: "Good" solvents in these correlations have breakthrough times of less than 1 h.

more rapid than expected in average comparisons. This size effect is in agreement with what has been known about solvent retention in coatings[2-4,12] and what has been discussed previously. An excellent example of this type of improved correlation is included in Table 8.1 for the breakthrough times of less than 1 h for neoprene rubber used in chemical protective clothing. The first correlation for this material listed in Table 8.1 gives a very poor data fit (0.574). There were 46 out of 66 liquids which had breakthrough times shorter than 1 h. It is clear from closer analysis of the details of the correlation that the outliers are methanol, carbon disulfide, and allyl alcohol with shorter breakthrough times than predicted and the phthalate plasticizers which have longer breakthrough times than predicted. A perfect fit is found when the molecular volume range of the permeants included in the correlation is abbreviated to between 71 and 172 cc/mol. This eliminates these "outliers." This correlation is based on 39 liquids with breakthrough times of less than 1 h out a total of 50.

The HSP correlations for 1-h breakthrough times for other barrier polymers discussed in Table 8.1 give polymer HSP in the range of those expected from their composition. This includes butyl rubber, Viton® (The Du Pont Company), nitrile rubber, Challenge® 5100 (Chemical Fabrics Corporation, Merrimack, NH), and polyethylene (PE). The thicknesses of all of the films discussed here are those commonly used in chemical protective clothing. HSP correlations of the swelling of Viton are discussed here as well as in Chapters 2 and 3 and by Zellers.[14,15]

The RED number (Chapters 1 and 2) is a key parameter to judge solvent quality. This is given in HSP correlations using Chapter 1, Equation 1.10 as Ra (Chapter 1, Equation 1.9), the difference in HSP between a solvent and a polymer, divided by the radius of the correlating sphere, Ro. The radius of the sphere is actually determined as the difference in HSP of the "worst" good solvent and the HSP for the polymer (which is the center point of the sphere). RED numbers near zero indicate very good solvents (rapid breakthrough). RED numbers increase as the solvent quality decreases. For RED numbers greater than 1.0, the solvent quality is considered "bad."

Figure 8.2 shows one way to graphically use RED numbers to present data from HSP correlations of permeation phenomena. In this correlation, "good" permeants have breakthrough times of less than 3 h. The data are plotted using V vs. solvent–polymer affinity, i.e., the RED number.[16] The barrier material, Challenge 5100, is a fluoropolymer supported by fiberglass. The abbreviations for the permeants in Figure 8.2 are explained in Table 8.2. This correlation shows that molecules with molar volumes greater than about 100 cc/mol will not have breakthrough times of less than 3 h, regardless of RED number. Molecules with molar volumes greater than about 75 cc/mol require a terminal double bond and lower RED numbers to breakthrough under these conditions. Molecules with still lower molar volumes appear to come through with only a slight dependence on the RED number. The effect with the terminal double bonds clearly indicates the preferential direction of motion for this type of molecule. The molecules in effect worm their way through the barrier polymer.

Solubility Parameter Correlation of Permeation Rates

Permeation rates for different permeants in a polymer can also be correlated to find HSP for the polymer. This is done by dividing a data set into two groups. The "good" solvents will have permeation rates greater than an arbitrarily selected value, and the "bad" solvents will have permeation rates lower than this value. Such a correlation based on permeation coefficients for various liquids in PE is included in Table 8.3. The permeation coefficient data [(g × mm)/(m^2 × d)] are reported by Pauly[17] for low density polyethylene (LDPE). "Good" solvents are arbitrarily considered as those which have permeation coefficients in these units which are greater than 1.5 at 21.1°C. The parameters reported correlate the data well, but are somewhat different from those which might be expected for a polyolefin. Reasons for this are not evident, but may include additives in the polymer, local oxidation, or some other local variation in the composition of the polymer.

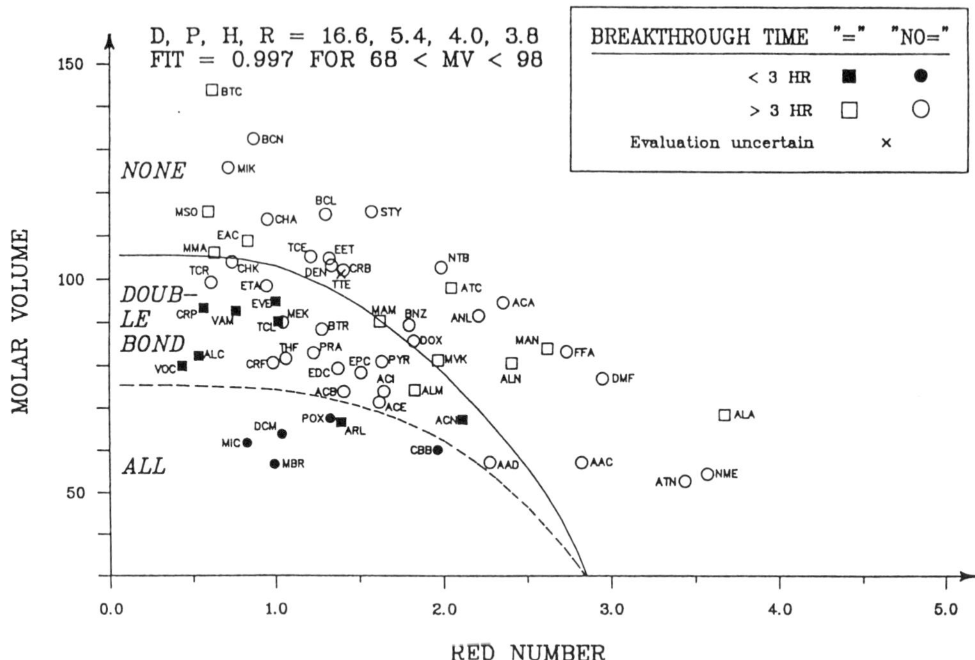

FIGURE 8.2 Graphical method to present HSP correlations. The data are plotted using permeant molar volume vs. RED number. HSP correlation for breakthough times of less than 3 h in Challenge 5100. Symbols used are explained in Table 8.2. (From Hansen, C. M. et al., *The Performance of Protective Clothing: Fourth Volume, ASTM STP 1133,* McBriarty, J. P. and Henry, N. W., Eds., American Society for Testing and Materials, Philadelphia, 1992, 906. Copyright ASTM. Reprinted with permission.)

It should be remembered that permeation occurs in the amorphous regions only. This is why high density PE is a better barrier polymer than low density PE; the higher densities are attributable to a higher percentage of crystallinity.

A problem of some concern is the permeation through buried water pipes by chemicals or oil products which somehow reach them, either by general pollution or by gasoline or oil spills. One clearly expects more extensive permeation by chemicals which have HSP not too different from those of the polymer from which the pipe is made, all other things being equal. These pipes are often made from polyolefins.

A HSP correlation has been possible in a very special case of polymer permeability where the barrier polymer is viable human skin.[18] This is discussed in more detail in Chapter 9. Human skin is a polymeric barrier with several functions, one of which is to help keep undesirable chemicals out of the body. Some chemicals readily permeate this boundary, and this fact has been used to establish a tentative HSP correlation for the permeation rate of viable human skin. This correlation also has a relation to the HSP correlation for the swelling of psoriasis scales,[19] which is also discussed in Chapter 9.

Solubility Parameter Correlation of Polymer Swelling

Solubility is a major factor in the equation **P** = **DS**, so correlations of solvent uptake in polymers are important to understand their barrier properties. The correlation for swelling of polypropylene reported in Table 8.3 is based on solvent uptake data reported by Lieberman and Barbe.[20] The limit of 0.5% weight gain was arbitrarily set to differentiate "good" solvents from "bad" ones. A different limit might give different parameters. The HSP found in a given correlation of swelling depends

TABLE 8.2
List of Symbols Used in Figure 8.2

Symbol	Compound	Symbol	Compound
AAC	Acetic acid	EBR	Ethyl bromide
AAD	Acetaldehyde	EDC	Ethylene dichloride
ACA	Acetic anhydride	EET	Diethyl ether
ACB	Acetyl bromide	EIM	Ethyleneimine
ACE	Acetyl chloride	EPC	Epichlorohydrin
ACI	Acetone	ESH	Ethanethiol
ACN	Acrylonitrile	ETA	Ethyl acetate
ALA	Allyl alcohol	EVE	Ethyl vinyl ether
ALC	Allyl chloride	EVK	Ethyl vinyl ketone
ALM	Allyl amine	F12	Dichlorodifluoromethane (Freon 12)
ALN	Allyl cyanide	FFA	Furfural
ANL	Aniline	HXA	Hexane
ARL	Acrolein	MAL	Methanol
ATC	Allyl isothiocyanate	MAM	Methyl acrylate
ATN	Acetonitrile	MAN	Methacrylonitrile
BCM	Bromochloromethane	MAT	Methyl acetate
BCN	Butyl acetate	MBR	Methyl bromide
BNZ	Benzene	MEK	Methyl ethyl ketone
BTC	Butyl acrylate	MES	Methyl sulfide
BTR	Butyraldehyde	MIC	Methyl isocyanate
BUT	Butane	MIK	Methyl isobutyl ketone
BZN	Benzonitrile	MMA	Methyl methacrylate
CAC	Chloroacetylchloride	MSO	Mesityl oxide
CBB	Carbon disulfide	MVK	Methyl vinyl ketone
CBT	Carbon tetrachloride	NEE	Nitroethane
CCF	Dichloromonofluoromethane (Freon 21)	NME	Nitromethane
CHA	Cyclohexylamine	NTB	Nitrobenzene
CHK	Cyclohexanone	POX	Propylene oxide
CLA	Chloroacetone	PRA	Propylamine
CLB	1-Chlorobutane	PYR	Pyridine
CRB	Chlorobenzene	STY	Styrene
CRF	Chloroform	TCE	1,1,2,2,-Tetrachloroethylene
CRP	Chloroprene	TCL	Trichloroethylene
DCM	Dichloromethane	TCR	1,1,1-Trichloroethane
DEN	Diethylamine	THF	Tetrahydrofuran
DMF	Dimethyl formamide	TOL	Toluene
DOX	1,4-Dioxane	TTE	Tetrachloroethylene
DSO	Dimethyl sulfoxide	VAM	Vinyl acetate
EAC	Ethyl acrylate	VDC	1,1-Dichloroethylene

From Hansen, C. M. et al., *The Performance of Protective Clothing: Fourth Volume, ASTM STP 1133,* McBriarty, J. P. and Henry, N. W., Eds., American Society for Testing and Materials, Philadelphia, 1992, 903.

on which polymer segments the smaller amounts of permeant prefer to associate with. The predictive ability of the correlation will depend on the number of test liquids used in the study and their given HSP values. How different are the HSP of the test liquids? What are their values compared with the predictions desired? The parameters reported in Table 8.3 for polypropylene seem to accurately

TABLE 8.3
HSP Correlations Related to Barrier Polymers

Material	d_D	d_P	d_H	Ro	FIT	G/T
LDPE permeation coefficient 21.1°C	16.3	5.9	4.1	8.2	1.000	26/47
Permeation viable skin[a]	17.6	12.5	11.0	5.0	1.000	4/13
PP swelling	18.0	3.0	3.0	8.0	1.000	13/21
ACLAR® 22 >5 wt% swelling	14.7	3.9	6.7	6.8	1.000	6/26
ACLAR® 22 2%<swelling<5%	18.0	1.0	2.0	4.0	1.000	4/21
Psoriasis scales swelling[b]	24.6	11.9	12.9	19.0	0.927	35/50
Viton® swell[c] >10 wt% 20°C	13.1	13.7	3.9	14.7	0.742	36/57
EVOH sol/swell[d]	20.5	10.5	12.3	7.3	0.925	5/24
Polyvinyl chloride swell[e]	18.7	9.7	7.7	6.4	1.000	13/47
Cellophane — >25% swell	16.1	18.5	14.5	9.3	0.955	4/22
PETP chemical resistance (+/– OK)[f]	18.2	6.4	6.6	5.0	0.865	7/34
PA6/PA66 chemical resistance (+/– OK)[g]	17.0	3.4	10.6	5.1	0.984	2/31
PA6/PA66 chemical resistance (+/– bad)[g]	18.9	7.9	11.7	8.0	0.950	9/31

Note: Data fit and the number of good liquids (G) and total number of liquids (T) in the correlations are also indicated.

[a] This correlation is discussed in more detail in Chapter 9 and is based on limited data.[18]
[b] This correlation is based on data by Lieberman and Barbe[19] and is discussed in more detail in Chapter 9.
[c] This correlation is discussed in Chapter 3. Swelling data is from Reference 21.
[d] Ethylene vinyl alcohol copolymer (EVOH), four liquids dissolved and one (morpholine) swelled very strongly.
[e] Visual observation of very strong swelling and/or solubility.
[f] Polyethylene terephthalate-P (PETP)chemical resistance based on rather uncertain data[22] (see discussion in Chapter 7). Recommendation of uncertain-for-use is considered okay (OK) for use. Attacking solvents are within correlating HSP sphere.
[g] Polyamide 6/66 chemical resistance based on rather uncertain data[22] (see discussion in Chapter 7). Recommendation of uncertain-for-use is used as indicated. Attacking solvents are within the correlating HSP sphere.

From Hansen, C. M., Permeability of polymers, *Pharmaceutical and Medical Packaging 98*, 7.6, 1998 (ISBN 87-89753-24-0). With permission.

reflect what is expected in terms of low polarity and low hydrogen bonding properties for this type of polymer.

As stated previously, a problem of some significance in any study of solvents at low concentrations in polymers is that the smaller amounts of solvent relative to the polymer can lead to preferential association of solvent with those local regions/segments/groups in the polymer which have energies (HSP) most similar to their own. Like seeks like. These local regions may not necessarily reflect the same affinities as the polymer as a whole, such as are indicated by the totally soluble-or-not approach most commonly used in HSP evaluations. These local association effects can influence results on swelling studies at low solvent uptakes in both good and bad solvents, for example. Copolymers, such as Viton, are particularly susceptible to this problem. Swelling data for Viton[21] are correlated by the HSP values included in Table 8.3. A poor data fit can be anticipated when a single HSP sphere is used to describe what should be represented by (at least) two overlapping HSP spheres (see also Figure 8.3). Zellers[14,15] also had difficulty correlating the swelling of Viton. Other types of studies carried out at low solvent concentrations can also be influenced by this segregation/association phenomena. An extension of this type of situation can be cited in

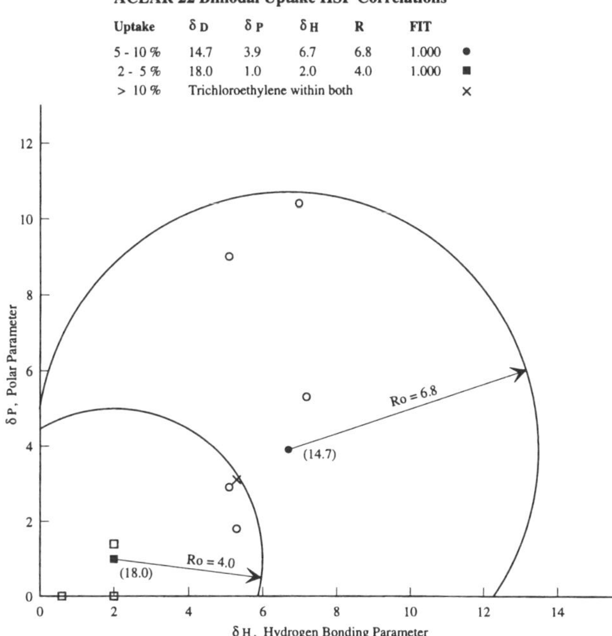

FIGURE 8.3 Bimodal HSP correlation(s) for uptake of liquids in ACLAR® 22. Trichloroethylene uptake is the largest among the test solvents because it is the only solvent found within both regions. It has RED numbers of 0.999 for the >5% correlation and 0.978 for the correlation of uptake between 2 and 5%. Since the data fit is 1.0 for both correlations, other sets of parameters can also give data fits of 1.0. However, the numbers are approximately correct.

the tendencies of water to associate with itself as well as with local hydrophilic regions within polymers. The amount of water taken up at equilibrium is not reflected by an overall HSP correlation of polymer solubility or swelling. As little as 1% of hydrophilic additive can effectively destroy the water barrier properties of a polymer film, but this small amount cannot be measured in swelling or solubility studies leading to HSP correlations. This fact, among other things, has made simple predictions of the behavior of water very difficult.

Correlations of polymer solubility and swelling have led to several of the HSP data sets reported in Table 8.3 (see also the data reported in Chapter 3). The HSP correlations for chemical resistance based on data of the acceptable-for-use or recommended/not recommended type are not as reliable as those usually found for solubility and swelling where a suitably large number of liquids are used in the testing. The reasons for this are discussed in depth in Chapter 7. The data used in the chemical resistance correlations reported in Table 8.3 were taken from Reference 22.

SOLUBILITY PARAMETER CORRELATION OF PERMEATION COEFFICIENTS FOR GASES

Gases can also be assigned δ_D, δ_P, and δ_H parameters. For strictly nonpolar gases, the values of δ_P and δ_H will be zero, but other gases, such as carbon dioxide, hydrogen sulfide, etc., will have significant values for all three parameters. Table 8.4 gives the δ_D, δ_P, and δ_H parameters for a number of gases. It is not surprising that there are HSP correlations of permeation coefficients for gases in different polymers as a function of their solubility parameter differences. One such correlation using the total solubility parameter has been given by König and Schuch,[23] who showed that the better barrier polymers for oxygen, i.e., those with low oxygen permeation coefficients, were those

TABLE 8.4
HSP for Common Gases of Interest in Permeation Phenomena

Gas	δ_D	δ_P	δ_H
Water	15.5	16.0	42.3
Ammonia	13.7	15.7	17.8
Chlorine	17.3	10.0	0.0
Sulfur dioxide	15.8	8.4	10.0
Carbon dioxide	14.7	3.9	6.7
Carbon monoxide	11.5	4.9	0
Ethane	15.6	0	0
Ethylene	15.0	2.7	2.7
Helium	1.0	0	0
Hydrogen	5.1	0	0
Hydrogen sulfide	17.0	6.0	10.2
Methane	14.0	0	0
Nitrogen oxide	11.5	20.0	0
Nitrogen	11.9	0	0
Nitrous oxide	12.0	17.0	0
Oxygen	14.7	0	0
Acetylene	14.4	4.2	11.9

whose solubility parameters were most different from the solubility parameters of oxygen. The better barrier polymers for oxygen include polyacrylonitrile and polyvinyl alcohol, while the poorer barriers include polyolefins and polytetrafluoroethylene (PTFE). The amounts of gases dissolved at low pressures are usually low, and constant diffusion coefficients are expected. This may not be true at higher pressures where solubility parameters for the gases increase more rapidly than those of the polymers and polymers can absorb them to a greater extent.

An example of how HSP principles can be applied to interpreting the behavior of gas barrier films can be found in the performance of poly(chlorotrifluoroethylene). The data on which the example is based are taken from the commercial literature supplied by Allied Signal concerning their barrier films under the tradename of ACLAR®.[24] These films are excellent barriers for water and oxygen, and various laminating possibilities exist, including polyethylene, polyvinyl chloride, and polyethylene terephthalate. The barrier properties of films made from this material are not nearly as good for carbon dioxide as they are for nitrogen or oxygen. A contributing factor in this is that the HSP of the polymer is somewhat different from the HSP of oxygen and nitrogen, but identical with the HSP of carbon dioxide. A HSP correlation for the swelling of ACLAR 22 to greater than 5 wt% is included in Table 8.3. The RED numbers for water, nitrogen, oxygen, and carbon dioxide based on this correlation are 5.5, 1.4, 1.1, and 0, respectively. Nitrogen has slower permeation than oxygen, and both are much slower than carbon dioxide, in general agreement with this ranking. One might have expected the permeation rate of carbon dioxide to be lower than that of nitrogen and oxygen since it is a larger molecule, but the enhanced solubility of carbon dioxide in the polymer overrides this expectation.

Figure 8.3 shows that there are two distinct spherical characterizations possible for ACLAR 22. The first of these, as discussed earlier, is for liquid uptake to greater than 5% by weight. The second of these is for uptake between 2 and 5% by weight. This second correlation is also reported in Table 8.3. There is one liquid in the data which is common to both of the HSP regions pictured in Figure 8.3. This is trichloroethylene (which was assumed to be a good solvent in both of the perfect correlations in spite of being absorbed to over 10% by weight). Even though trichloroethylene has high RED numbers in both correlations, this solvent is absorbed more than any of the other solvents tested because of this property. The primary uptake region has HSP in what might be expected from a fluoropolymer, while the secondary HSP region is what might be expected from a chlorinated species. Such secondary regions can potentially allow higher permeation rates and greater absorption of unpredictable materials based on a single HSP correlation. Searching a database of solvents, plasticizers, aromatic compounds, etc. would show which of these could behave in an unexpected manner.

Sometimes an indirect approach allows prediction of the uptake of a gas in a polymer. This involves determining the uptake of the gas in a liquid having solubility parameters which are similar to those of the polymer. This approach expands the usefulness of gas–liquid equilibrium data. Correlations of gas–liquid solubility with the solubility parameter are included in Figure 8.4 for the equilibium values for water[25] and in Figure 8.5 for the equilibrium values for nitrogen.[26] The

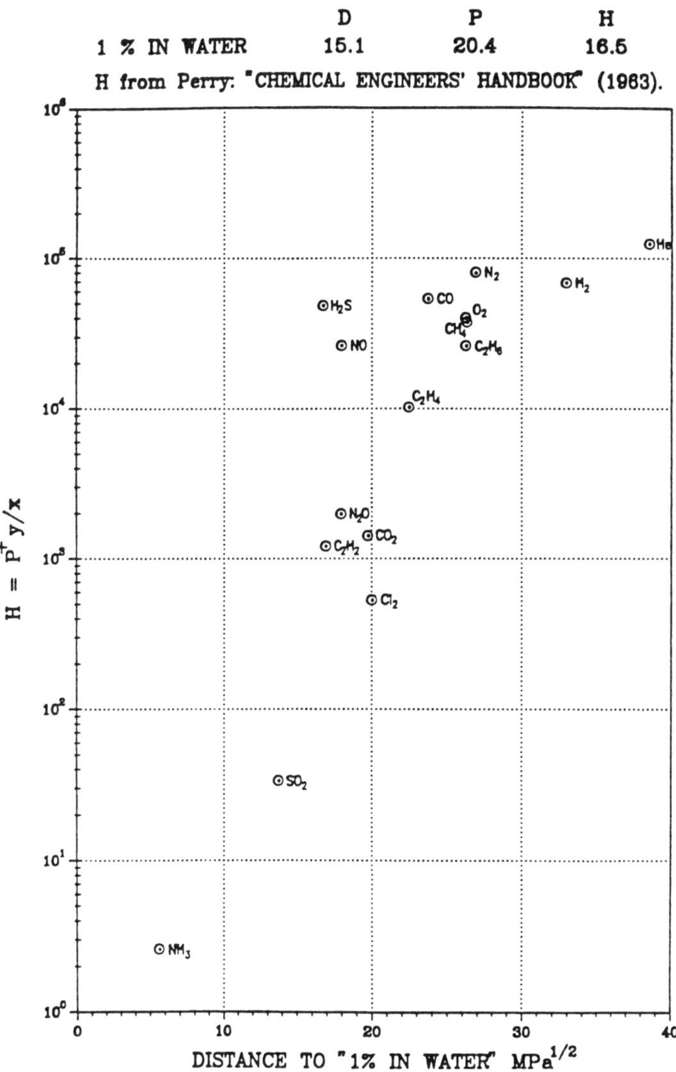

FIGURE 8.4 HSP correlation of gas–water equilibrium data using water HSP values found with a correlation using a limit of >1% liquid soluble in water as a "good" solvent.[19] and gas–water equilibrium data from Perry et al.[25] P* is the total pressure, y is the mole fraction in the gas phase, and x is the mole fraction in the liquid phase. (From Hansen, C. M., *Dan. Kemi,* 73(8), 21, 1992.)

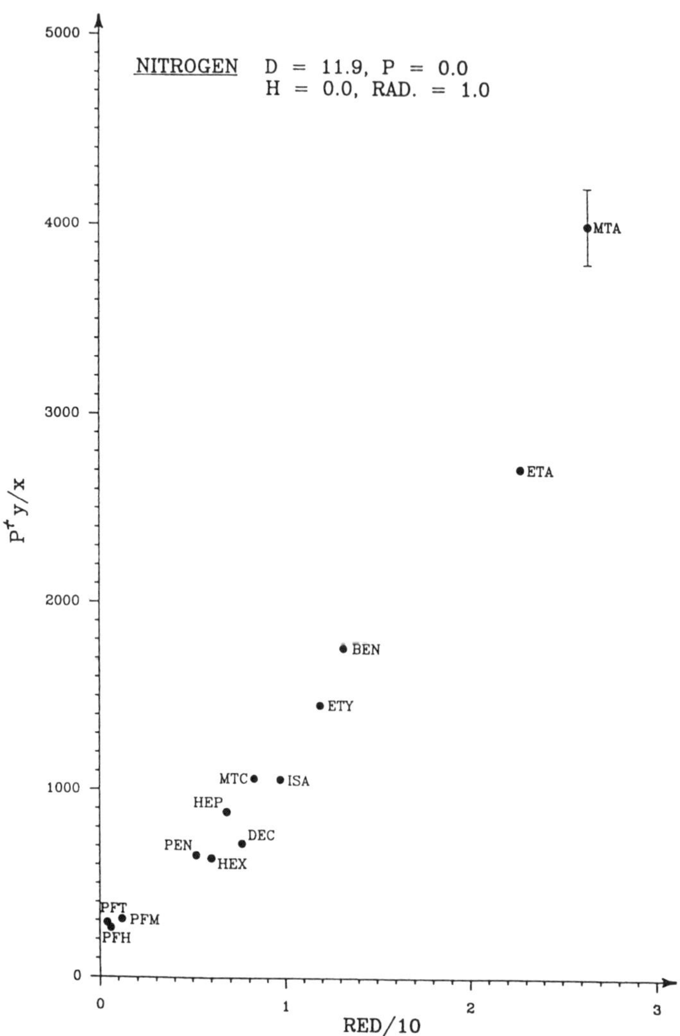

FIGURE 8.5 HSP correlation of nitrogen–liquid equilibrium data at temperatures near 25°C and low pressure. P* is the total pressure, y is the mole fraction in the gas phase, and x is the mole fraction in the liquid phase. (From Hansen, C. M., *Dan. Kemi,* 73(8), 20, 1992. With permission.)

quantity P*y/x is given by the total pressure, P*, the mole fraction in the gas phase, y, and the mole fraction in the liquid phase, x. The abbreviations used in Figure 8.5 are explained in Table 8.5.

The solubility parameters for gases not found in Table 8.4 may be found in standard references.[27-29] HSP for gases can be calculated using the procedures outlined in Chapter 1 with the special figure for gases included in Chapter 11 (Figure 11.2).

It might be noted that the scale in Figure 8.4 for the uptake of water in various liquids is exponential with data covering almost five decades in concentration. The phenomena correlated in this figure confirm the expectation that nonpolar polymers, with solubility parameters far different from those of water, will be good barrier polymers for water because of low water solubility. This is generally true, of course. As mentioned earlier, such polymers include the polyolefins as well as chlorinated and fluorinated polymers. These comments and generalities are not necessarily valid for polymers containing additives. Depending on the nature of the additive and the amounts present, some of these can totally change the barrier performance of the base polymer.

TABLE 8.5
Key to Symbols Used in Figure 8.5

Symbol	Compound
PFH	Perfluoroheptane
PTF	Perfluorotributylamine
PFM	Perfluoromethylcyclohexane
PEN	Pentane
HEX	Hexane
HEP	Heptane
MYC	Methyl cyclohexane
ISA	Isobutyl acetate
ETY	Ethyl acetate
BEN	Benzene
ETA	Ethanol
MTA	Methanol

Laminates

Laminated barrier polymer systems are designed to make the best use of the properties of each of the individual layers, as well as to optimize cost with performance. The most common type has a polyolefin on the exterior surfaces to protect the inner barrier polymer from water. These interior barrier polymers often have relatively high solubility parameters, such as ethylene vinyl alcohol copolymers (EVOH), polyamides (PA), or polyethylene terephthalate (PET). If the inner barrier polymer takes up water, it will be plasticized, and its barrier properties will be reduced. A polyolefin laminated to such a potentially water-sensitive barrier film can significantly delay this uptake and loss of barrier properties and maintain reasonable costs. Depending on the performance desired, various combinations of laminates can be systematically designed using HSP considerations as one of the design parameters.

GENERAL CONSIDERATIONS

HSP correlations have been possible for many phenomena where differences in behavior on contact with different solvents have been studied. The HSP correlations are preferably based on systems in thermodynamic equilibrium, although the correlations presented previously on breakthrough times are an exception to this. These correlations were possible because of the exceptional dependence of the permeation phenomena on the amount of permeant being dissolved.

There is a strong dependence of the diffusion coefficient of permeants in polymers on their size and shape. This can clearly affect HSP correlations of permeation coefficients, since two permeants with identical HSP will have different **D** if their sizes and shapes are significantly different. This differential in diffusion rate based on solvent size and shape can also give apparent errors in HSP correlations of polymers for their chemical resistance, for example, where not enough exposure time has been allowed for attainment of equilibrium. This is clearly a problem in the determination of equilibrium degree of swelling and low amounts of uptake. This problem has also been found, for example, for exposures of thick samples (3 to 4 mm) of rigid polymers used for tensile testing after solvent exposure for given times. Crystalline polymers also have a tendency to be more readily soluble in solvents with lower V, all other parameters being equal, but this is explained by thermodynamic considerations rather than a relatively faster diffusion process. In all of these cases, the majority of the outliers in the correlations are the test liquids with higher V. The time required for attainment of equilibrium with the larger diffusing molecules can be so long as to be prohibitive for their reasonable inclusion in HSP correlations. It is suggested that diffusion rates be carefully considered when liquids with very high V are outliers in HSP correlations. Neglecting such data for the sake of a correlation can be justified, but this is a warning that the diffusion process for these liquids has not yet achieved equilibrium and that the effects of such liquids can be expected to be more severe at still longer times than those used in the study. HSP correlations can be used in this way to find those exposure liquids which have not reached equilibrium at the exposure time chosen for evaluations.

CONCLUSION

Successful HSP correlations for the permeation and solubility behavior of selected barrier polymers have been presented to demonstrate the use of simple principles to arrive at optimum barrier systems, as well as to determine reliable HSP for the polymers studied. Selection of polymer–permeant combinations with widely different solubility parameters will ensure low solubility of the permeant in the polymer. This reduces the concentration gradient and prevents significant self-plasticization of the polymer. The self-plasticization leads to concentration-dependent diffusion coefficients, an effect which becomes more significant with increasing amounts of permeant being dissolved, i.e., a closer HSP match.

HSP correlations have been presented for breakthrough times in chemical protective clothing, permeation rates in barrier polymers, and swelling of various types of polymers. Both gases and liquids are treated.

REFERENCES

1. Rogers, C. E., Permeation of Gases and Vapours in Polymers, in *Polymer Permeability*, Comyn, J., Ed., Elsevier Applied Science, London, 1985, 11–73.
2. Hansen, C. M., Some Aspects of the Retention of Solvents in High Polymer Films, *Färg och Lack*, 10(7), 169–186, 1964.
3. Hansen, C. M., The Three Dimensional Solubility Parameter and Solvent Diffusion Coefficient, Doctoral Dissertation, Danish Technical Press, Copenhagen, 1967.
4. Hansen, C. M., A Mathematical Description of Film Drying by Solvent Evaporation, *J. Oil Colour Chem. Assoc.*, 51(1), 27–43, 1968.
5. Hansen, C. M., Diffusion in Polymers, *Polym. Eng. Sci.*, 20(4), 252–258, 1980.
6. Hansen, C. M., The Measurement of Concentration-Dependent Diffusion Coefficients — The Exponential Case, *Ind. Eng. Chem. Fundam.*, 6(4), 609–614, 1967.
7. Hansen, C. M., Diffusion Coefficient Measurements by Solvent Absorption in Concentrated Polymer Solutions, *J. Appl. Polym. Sci.*, 26, 3311–3315, 1981.
8. Crank, J., *The Mathematics of Diffusion*, Oxford University Press, Oxford, 1956.
9. Klopfer, H., *Water Transport by Diffusion in Solid Materials* (in German: *Wassertransport durch Diffusion in Feststoffen*), Bauverlag GMBH, Wiesbaden, 1974.
10. Forsberg, K. and Olsson, K. G., *Guidlines for Selecting Chemical Protective Gloves,* (in Swedish: *Riktlinjer för val av kemiskyddshandskar*), Förening Teknisk Företagshälsovård (FTF), Stockholm, 1985.
11. Forsberg, K. and Kieth, L. H., *Chemical Protective Clothing Performance Index*, 4th Ed., Instant Reference Sources, Austin, TX, 1991.
12. Hansen, C. M. and Hansen, K. M., Solubility Parameter Prediction of the Barrier Properties of Chemical Protective Clothing, *Performance of Protective Clothing: Second Symposium. ASTM STP 989*, Mansdorf, S. Z., Sager, R., and Nielsen, A. P., Eds., American Society for Testing and Materials, Philadelphia, 1988, 197–208.
13. Hansen, C. M., The Free Volume Interpretation of Plasticizing Effectiveness and Diffusion in High Polymers, *Off. Dig.*, 37(480), 57–77, 1965.

14. Zellers, E. T., Three-Dimensional Solubility Parameters and Chemical Protective Clothing Permeation. I. Modelling the Solubility of Organic Solvents in Viton® Gloves, *J. Appl. Polym. Sci.,* 50, 513–530, 1993.
15. Zellers, E. T., Three-Dimensional Solubility Parameters and Chemical Protective Clothing. II. Modelling Diffusion Coefficients, Breakthrough Times, and Steady-State Permeation Rates of Organic Solvents in Viton® Gloves, *J. Appl. Polym. Sci.,* 50, 531–540, 1993.
16. Hansen, C. M., Billing, C. B., Jr., and Bentz, A. P., Selection and Use of Molecular Parameters to Predict Permeation Through Fluoropolymer-Based Protective Clothing Materials, in *The Performance of Protective Clothing: Fourth Volume, ASTM STP 1133*, McBriarty, J. P. and Henry, N. W., Eds., American Society for Testing and Materials, Philadelphia, 1992, 894–907.
17. Pauly, S., Permeability and Diffusion Data, in *Polymer Handbook*, 3rd ed., Branderup, J. and Immergut, E. H., Eds., Wiley-Interscience, New York, 1989, VI/445–446.
18. Ursin, C., Hansen, C. M., Van Dyk, J. W., Jensen, P. O., Christensen, I. J., and Ebbehoej, J., Permeability of Commercial Solvents Through Living Human Skin, *Am. Ind. Hyg. Assoc. J.,* 56, 651–660, 1995.
19. Hansen, C. M. and Andersen, B. H., The Affinities of Organic Solvents in Biological Systems, *Am. Ind. Hyg. Assoc. J.,* 49(6), 301–308, 1988.
20. Lieberman, R. B. and Barbe, P. C., Polypropylene Polymers, in *Encylopedia of Polymer Science and Engineering*, 2nd ed., Mark, H. F., Bikales, N. M., Overberger, C. G., Menges, G., and Kroschwitz, J. I., Eds., Wiley-Interscience, New York, 1988, 13, 482–3.
21. Anonymous, Fluid Resistance of Viton®, Du Pont Company, Polymer Products Department, Elastomers Division, Wilmington, DE, 1989.
22. Anonymous, Plastguide, SCS Dukadan AS, Randers, Denmark, 1990.
23. König, U. and Schuch, H., Structure and Permeability of Polymers (in German: Konstitution und Permeabilität von Kunststoffen), *Kunststoffe,* 67(1), 27–31, 1977.
24. Anonymous, ACLAR® Barrier Films, AlliedSignal — Advanced Materials, Allied Signal Inc., Morristown, NJ.
25. Perry, J. H., Chilton, C. H., and Kirkpatrick, S. D., Eds., *Chemical Engineers' Handbook*, 4th ed., McGraw-Hill, New York, 1963, 2–7.
26. Hansen, C. M., 25 Years with Solubility Parameters (in Danish: 25 År med Opløselighedsparametrene), *Dan. Kemi,* 73(8), 18–22, 1992.
27. Hildebrand, J. and Scott, R. L., *The Solubility of Nonelectrolytes,* 3rd ed., Reinhold, New York, 1950.
28. Hildebrand, J. and Scott, R. L., *Regular Solutions,* Prentice-Hall Inc., Englewood Cliffs, NJ, 1962.
29. Barton, A. F. M., *Handbook of Polymer-Liquid Interaction Parameters and Solubility Parameters*, CRC Press, Boca Raton, FL, 1990.

9 Hansen Solubility Parameters — Biological Materials

CONTENT

ABSTRACT

The Hansen solubility parameters (HSP) of many biological materials can be found from correlations of how they interact with well-defined liquids. Examples included in this chapter are cholesterol, chlorophyll, wood chemicals and polymers, human skin, nicotine, lard, and urea. The often quoted "Like Dissolves Like" has been expanded to "Like Seeks Like" (self-association) to discuss the implications of these correlations. The ability of HSP to correlate surface phenomena has made this change mandatory.

The adsorption of polymers onto pigment surfaces is known to be influenced by solvent quality. It seems reasonable to assume that similar relations exist in biological systems. The "solvent quality" in a given environment is expected to determine whether a protein is dissolved or not and also to control the way it adsorbs onto other materials or interacts with itself. Controlled changes in solvent quality can lead to controlled changes in conformation.

INTRODUCTION

HSP have been used to characterize many biological materials.[1-7] Most of the materials discussed in these references are also included in the present discussion, but many more can be added by experiment or calculation.

There are many simple experimental methods to determine the HSP for materials. These involve contacting a material of interest with a series of well-chosen liquids. The fact of solubility, differences in degree of equilibrium swelling, rapid permeation or not, significant surface adsorption or not, or other measurable quantity significantly influenced by physical affinity relations can be

Hansen Solubility Parameters

	δD	δP	δH	R
Lignin	21.9	14.1	16.9	13.7

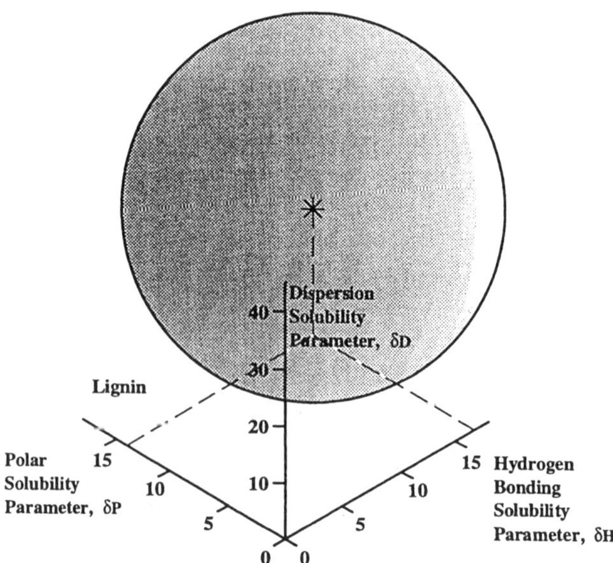

FIGURE 9.1 HSP correlation showing the solubility of lignin. Good solvents are located within the sphere. (From Hansen, C. M. and Björkman, A., The Ultrastructure of Wood from a Solubility Parameter Point of View, *Holzforschung,* 52, 339, 1998. With permission.)

observed and used to find the HSP for a material being studied. These methods have been discussed in more detail in earlier chapters. The HSP for simpler compounds can be calculated according to the methods given in Chapter 1. HSP calculations for nicotine, skatole, wood chemicals, etc. are discussed in this chapter. Figure 9.1 shows a typical HSP sphere correlating solubility data for lignin.[1] The good solvents are located within the sphere which is based on Chapter 1, Equation 1.9. Again, as stated in previous chapters, this equation is in agreement with the Prigogine corresponding states theory of polymer solutions as discussed in Chapter 2.

Table 9.1 contains HSP data for several biologically interesting materials. These are discussed in the following in more detail with an indication of how such data may be used. The data included in this table are the δ_D, δ_P, and δ_H parameters; the radius of interaction for the HSP correlation, Ro; if appropriate, the data fit (where 1.000 is perfect as discussed in Chapter 1); and the number of "good" solvents, G, and the total number of solvents, T, used in the correlation. The units for the solubility parameters and Ro are $MPa^{½}$. Table 9.2 includes data from the correlation of the solubility of lignin reported in Figure 9.1. The RED numbers indicate solvent quality with lower values being indicative of better solvents (see Chapter 1, Equation 1.10). The correlations reported here are a result of data processing with the SPHERE program described in Chapter 1. This is discussed in more detail later. The output is arranged in order of expected solvent quality with the best solvent at the top of the list.

A HSP correlation can, of course, be used to predict the behavior of solvents not included in the experimental work which lead to it. It is convenient to print the solvent database in order from best solvent to worst solvent to aid in finding alternatives. This is a quantitative application of the

TABLE 9.1
Hansen Solubility Parameter Correlations for Biologically Interesting Materials, MPa½

Material	d_D	d_P	d_H	R_o	FIT	G/T
Cholestcrol solubility	20.4	2.8	9.4	12.6	1.000	25/41
Lard 37°C solubility	15.9	1.2	5.4	12.0	1.000	29/50
Lard 23°C solubility	17.7	2.7	4.4	8.0	1.000	21/50
Olive oil solubility	15.9	1.2	5.4	12.0	1.000	29/50
Psoriasis scales swelling	24.6	11.9	12.9	19.0	0.927	35/50
Human skin — permeation	17.6	12.5	11.0	5.0	1.000	4/13
Nicotine — calculation	18.8	7.8	6.4	—	—	—
Skatole — calculation	20.0	7.1	6.2	—	—	—
Chlorophyll — solubility	20.2	15.6	18.2	11.1	0.864	7/35
Sinapyl alcohol calculation	19.2	7.3	16.1	—	—	—
Coniferyl alcohol calculation	19.0	7.0	16.3	—	—	—
p-Coumaryl alcohol calculation	19.1	7.0	17.3	—	—	—
Lignin — solubility	21.9	14.1	16.9	13.7	0.990	16/82
Dextran C (= amorphous cellulose)	24.3	19.9	22.5	17.4	0.999	5/50
Sucrose solubility	23.4	18.4	20.8	16.0	0.981	6/50
N-methyl-morpholine-*N*-oxide calculation	19.0	16.1	10.2	—	—	—
Blood serum — swelling	25.5	10.3	22.1	17.8	0.980	4/51
Zein — solubility	22.4	9.8	19.4	11.9	0.964	4/50
Urea — solubility	22.9	14.9	21.3	16.2	0.984	14/50
Water — >1% soluble in	15.1	20.4	16.5	18.1	0.856	88/167
Water — totally miscible	18.1	17.1	16.9	13.0	0.880	47/166
Water — single molecule	15.5	16.0	42.3	—	—	—

Note: The units for the solubility parameters and R_o are MPa½. G/T represents the number of good liquids (G) and the total number of liquids (T) in the correlation.

generally used statement "Like Dissolves Like" (self-association). In the following discussion, this concept is expanded to "Like Seeks Like" (self-association). This implies that segments of molecules seek regions of similar HSP if this is possible. This may result in solutions or in selective orientation of segments of molecules in more complicated systems. Examples include surface active agents which usually are described as having a hydrophilic end and a hydrophobic end. The hydrophilic end seeks water, and the hydrophobic end seeks a hydrophobic environment. These general concepts of hydrophilic and hydrophobic can also be quantified using HSP. Examples are included in the following.

HYDROPHOBIC BONDING AND HYDROPHILIC BONDING (SELF-ASSOCIATION)

The concept of "Like Seeks Like" offers a general explanation of hydrophobic bonding. An aliphatic hydrocarbon chain on a protein, for example, is not soluble in water and ultimately finds another aliphatic hydrocarbon chain with which to associate. This same type of process leads to micelle formation when the solubility limit of surface active agents is exceeded. Hydrophobic bonding is found when the HSP for the associating segments are too low to allow solubility in the continuous phase.

Hydrophilic (hyperphilic?) bonding is found when the HSP for the associating segments are too high to allow solubility in the continuous phase. If the continuous phase is a hydrocarbon liquid, the associating segments may be characterized by high δ_H, for example, because of the presence

TABLE 9.2
Calculated Solubility Sphere for Zein

D = 22.4 P = 9.8 H = 19.4 RAD. = 11.9 FIT = 0.964 NO = 50

No.	Solvent	D	P	H	SOLUB	RED	V
14	1,3-Benzenediol	18.0	8.4	21.0		0.761	87.5
17	Benzyl alcohol	18.4	6.3	13.7		0.876	103.6
70	Diethanolamine	17.2	10.8	21.2		0.891	95.9
197	Phenol	18.0	5.9	14.9		0.893	87.5
154	*o*-Methoxyphenol	18.0	8.2	13.3		0.910	109.5
135	Furfuryl alcohol	17.4	7.6	15.1		0.933	86.5
139	Hexamethylphosphoramide	18.5	8.6	11.3		0.950	175.7
45	3-Chloro-1-propanol	17.5	5.7	14.7		0.976	84.2
27	1,3-Butanediol	16.6	10.0	21.5	0*	0.991	89.9
205	Propylene glycol	16.8	9.4	23.3	0*	0.997	73.6
75	Diethylene glycol	16.6	12.0	20.7	1	0.998	94.9
118	Ethylenediamine	16.6	8.8	17.0		0.999	67.3
46	*m*-Cresol	18.0	5.1	12.9	1*	1.001	104.7
10	Aniline	19.4	5.1	10.2	0	1.004	91.5
98	Dipropylene glycol	16.5	10.6	17.7	0	1.004	130.9
218	1,1,2,2-Tetrabromoethane	22.6	5.1	8.2		1.021	116.8
105	Ethanolamine	17.0	15.5	21.2	0	1.037	59.8
217	Succinic anhydride	18.6	19.2	16.6		1.043	66.8
213	2-Pyrolidone	19.4	17.4	11.3		1.061	76.4
8	Allyl alcohol	16.2	10.8	16.8		1.068	68.4
121	Ethylene glycol	17.0	11.0	26.0	1*	1.068	55.8
126	Ethylene glycol monomethyl ether	16.2	9.2	16.4	1*	1.073	79.1
48	Cyclohexanol	17.4	4.1	13.5	0	1.087	106.0
81	Diethylenetriamine	16.7	13.3	14.3		1.090	108.0
15	Benzoic acid	18.2	6.9	9.8		1.099	100.0
236	Triethyleneglycol	16.0	12.5	18.6		1.101	114.0
219	1,1,2,2-Tetrachloroethane	18.8	5.1	9.4		1.108	105.2
104	Ethanol	15.8	8.8	19.4	0	1.112	58.5
199	1-Propanol	16.0	6.8	17.4		1.117	75.2
174	Morpholine	18.8	4.9	9.2		1.127	87.1
124	Ethylene glycol monoethyl ether	16.2	9.2	14.3	0	1.128	97.8
90	Dimethylformamide	17.4	13.7	11.3	0	1.130	77.0
210	Propylene glycol monophenyl ether	17.4	5.3	11.5		1.136	143.2
214	Quinoline	19.4	7.0	7.6		1.137	118.0
141	Hexylene glycol	15.7	8.4	17.8		1.140	123.0
93	Dimethyl sulfone	19.0	19.4	12.3		1.155	75.0
94	Dimethyl sulfoxide	18.4	16.4	10.2	0	1.165	71.3
117	Ethylene cyanohydrin	17.2	18.8	17.6		1.166	68.3
28	1-Butanol	16.0	5.7	15.8	0	1.169	91.5
200	2-Propanol	15.8	6.1	16.4		1.179	76.8
119	Ethylene dibromide	19.2	3.5	8.6		1.180	87.0
224	Tetramethylurea	16.7	8.2	11.0		1.198	120.4
136	Glycerol	17.4	12.1	29.3		1.198	73.3
80	Diethylene glycol monomethyl ether	16.2	7.8	12.6		1.200	118.0
79	Diethylene glycol monoethyl ether	16.1	9.2	12.2		1.221	130.9
89	*N,N*-Dimethylacetamide	16.8	11.5	10.2		1.226	92.5
23	Bromoform	21.4	4.1	6.1		1.228	87.5
29	2-Butanol	15.8	5.7	14.5		1.232	92.0

TABLE 9.2 (continued)
Calculated Solubility Sphere for Zein

No.	Solvent	D	P	H	SOLUB	RED	V
187	1-Octanol	17.0	3.3	11.9		1.233	157.7
130	Ethyl lactate	16.0	7.6	12.5		1.236	115.0
173	Methyl salicylate	16.0	8.0	12.3		1.239	129.0

of an alcohol, acid, or amide group. This concept is used frequently in the coatings industry. Improved pigment dispersion is found when such groups are incorporated into binders. They adsorb onto high HSP pigment and filler surfaces because of a lack of solubility in the continuous media. This anchor is solid, and the dispersions are stable. Another example of hydrophilic bonding from the coatings industry is found with thixotropic (drip-free) alkyd paints. The HSP relations are sketched in Figure 9.2. Polyamide-type polymers are incorporated as block segments into an alkyd binder. The alkyd is soluble in mineral spirits, but the polyamide is not. These segments have HSP which are too high. They associate and give the structure characteristics of thixotropic paints. This structure can be broken down by spraying, brushing, or rolling and quickly builds up again to prevent dripping, sagging, and the like. Hydrophilic bonding can be influenced by the presence of water, since it, too, has high affinity for alcohol, acid, and amide groups. The presence of excess water can be a problem for stability of performance for an otherwise successful thixotropic alkyd paint. The hydrophilic bonding between the polyamide segments of different molecules can be

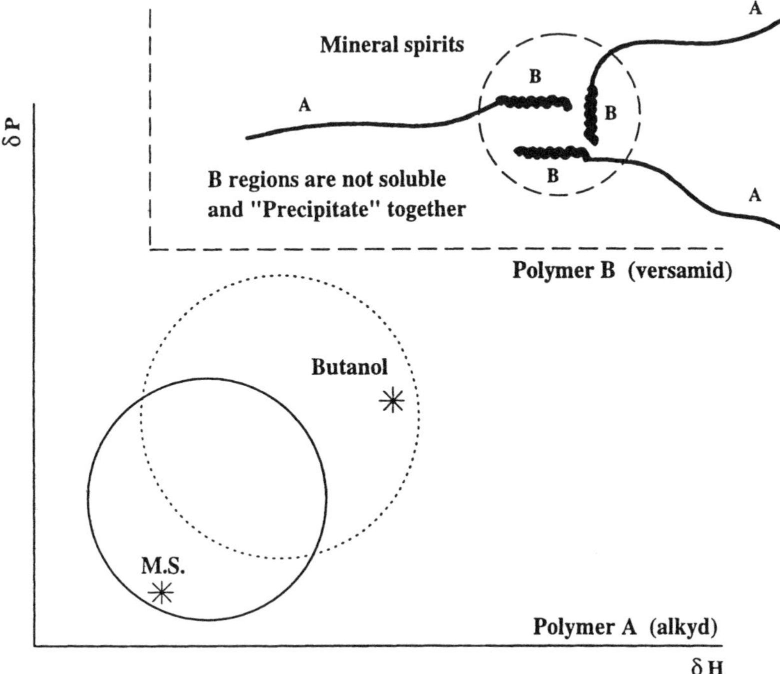

FIGURE 9.2 HSP relations for establishing thixotropy in an alkyd-type paint. The Versamid segments associate because they are not soluble in mineral spirits. Addition of *n*-butanol destroys the thixotropic effect, since the solvent then becomes too good. Similar relations exist for the true solution of proteins by additions of urea to water.

reduced by the addition of a solvent which dissolves these polyamides, such as *n*-butanol. If enough *n*-butanol is mixed into the mineral spirits usually used in this type of product, the thixotropic behavior will disappear, since all segments of the copolymer molecules are now truly dissolved and there is no association between them.

Hydrophilic bonding is also found in biological systems. An example is the helical structure of proteins which is traditionally attributed to hydrogen bonding. This hydrogen bonded structure is based on the lack of solvency in the continuous media, water, because of the HSP of these segments being too high. Additions of urea, as discussed later in more detail, increase the HSP of the continuous media to such an extent that it can now dissolve the hydrogen bonded segments. The protein is denatured, which in fact means these segments are dissolved in a good solvent. Additions of salts can also improve solvency for a given material or segments of materials. Additions of salts can also reduce solvency. These phenomena must also have their explanation in the "Like Seeks/Dissolves Like" phenomena, but more research is required to quantify them. Such mechanisms of controlling solvent quality can be expected to be used in many biological systems to control adsorption and/or transport of various types of materials as in self-association.

CHOLESTEROL

Cholesterol has been characterized with HSP based on its solubility in a large number of solvents. $\delta_D;\delta_P;\delta_H$ and Ro for cholesterol solubility were found as 20.4;2.8;9.4 and 12.6, all values having units of $MPa^{1/2}$. The test method involved placing 0.5 g of cholesterol in test tubes together with 5 ml of each of 41 different solvents. The temperature was 23°C. Total solution or not at this concentration was evaluated visually. The 25 "good" solvents dissolved the entire amount of cholesterol added. These data were analyzed by the SPHERE computer program described in Chapter 1 to find the HSP for cholesterol. This has also been reported in Reference 8. Figure 9.3 shows this HSP correlation for cholesterol. This figure also includes several solvents which are discussed in the following.

The data fit of 1.0 indicates that there are other sets of parameters for spheres which can be expected to give a perfect separation of the good solvents from the bad ones by a "spherical" HSP correlation. Continued testing with additional test solvents located in the boundary region of the sphere is possible to define it more precisely. This was not warranted under the present circumstances, but is recommended if more extensive use of these data is planned

A general confirmation of the HSP correlation for cholesterol was done by studying mixtures of nonsolvents. Many mixtures of two nonsolvents which dissolve polymers when admixed have been reported in the literature.[1] Such synergistic mixtures can be predictably found when they are pairwise on opposite sides of an HSP sphere. The 50:50 vol mixtures of *n*-hexane with 2-nitropropane and *n*-hexane with ethanol predictably dissolved cholesterol at 0.5 g/5 ml.

During the course of this study, it also became obvious that the solubility of cholesterol in hydrocarbons was limited and quite temperature dependent, being considerably higher at slightly elevated temperatures. This behavior in hydrocarbon solvents relates to the interactions of cholesterol in the hydrocarbon (hydrophobic) portions of lipid layers. The limited solubility in hydrocarbon media and very low solubility in water favors a location at an aqueous interface with the alcohol group of the cholesterol molecule oriented toward the high energy aqueous phase, where it is more compatible, and the hydrocarbon portions oriented into the lipid layer. Changes toward lower temperature will tend to force more cholesterol out of a hydrocarbon matrix. The δ_H parameter of alcohol solvents decreases relatively more rapidly with increasing temperature than for solvents where the δ_H parameter is low (or zero), such as with the hydrocarbon solvents. This brings the HSP of the "alcohol" solvent closer to the HSP of the hydrocarbon solvents, and miscibility improves markedly.

	MATERIAL	δ_D	δ_P	δ_H	R
⊗	CHOLESTEROL	20.4	2.8	9.4	12.6
✳	2-NITROPROPANE	16.2	12.1	4.1	
✳	HEXANE	14.9	0.0	0.0	
✳	ETHANOL	15.8	8.8	19.4	

● DISSOLVING MIXTURES OF NON-SOLVENTS

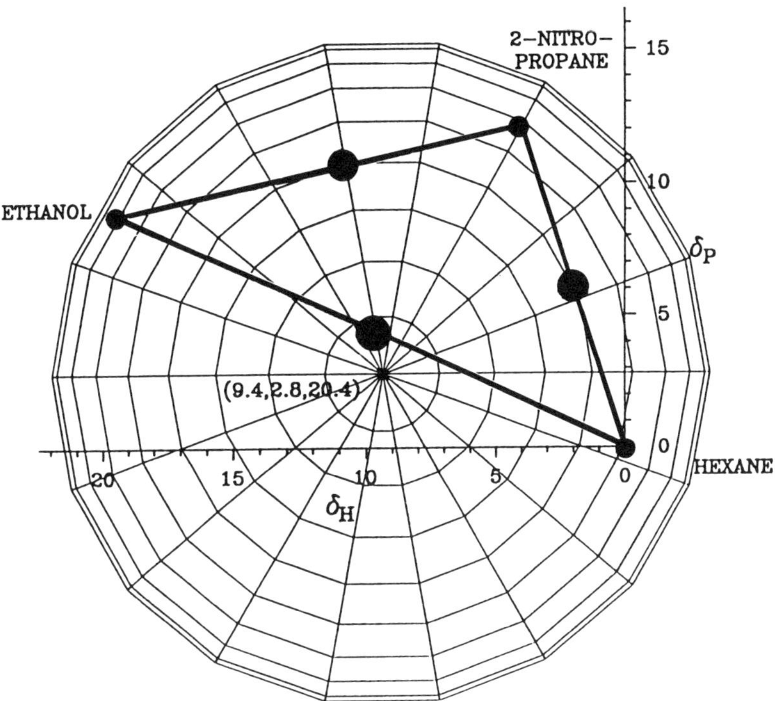

FIGURE 9.3 HSP sphere correlating the solubility of cholesterol. Mixtures of nonsolvents which synergistically interact to become solvents when mixed are indicated. These can predictably be found by selecting pairs located on opposite sides of the HSP solubility parameter sphere.

One can also surmise what might happen when ethanol or other organic solvent is present in the body. Organic solvents with HSP resembling those of the lipid layer may be found due to occupational exposure or for other reasons, such as drinking alcohol-containing beverages. The presence of ethanol or other organic solvent in the lipid layer allows greater cholesterol miscibility in its hydrocarbon portions. The reason for this is the synergistic effect of ethanol and hydrocarbon segments described earlier. The simple experiments described previously indicate that the cholesterol uptake in hydrocarbon portions of a lipid layer will be greatly enhanced when ethanol is present. This, of course, preferentially removes some of the cholesterol from the blood stream.

The solubility of cholesterol in an essentially nonsolvent such as water can be enhanced by additions of a solvent improver such as ethanol. The average HSP for these mixtures are closer to those of cholesterol itself. Therefore, those persons with alcohol in their blood can anticipate a slightly higher solubility of cholesterol in their blood because the continuous phase has solubility parameters closer to those of cholesterol. This effect and that discussed earlier should help to reduce cholesterol levels in the blood and blood vessels of those who ingest small to moderate amounts of alcohol on a regular basis.

LARD[2]

Experimental data and HSP correlations for the solubility of refined lard at 23 and 37°C have been reported.[2] The criterion for a good solvent is that it totally dissolve the sample at the given temperature. The concentrations chosen were 10%. The results of the correlations are given in Table 9.1. The refined lard is a semisolid with a melting point of 42°C.

The composition of refined lard is very similar to that of human depot fat, so the conclusions drawn for the solubility of lard will also be generally valid for depot fat. Olive oil is a convenient material to use at room temperature to study the behavior of depot fat (lard), since the same solvents that dissolve it at room temperature also dissolve lard at 37°C. This is reported in Table 9.1.

The best room temperature solvents for lard include trichloroethylene, styrene, toluene, and methyl methacrylate. Octyl alcohol does not have a strong affinity for lard at room temperature with a RED number (see Chapters 1 and 2) of 0.96. The good solvents reflect the crystalline nature of the lard, since toluene, for example, is an excellent swelling solvent for partly crystalline polyethylene. Esters are among the best solvents for lard at 37°C, reflecting the presence of the ester groups in the lard, which is very nearly a liquid at this temperature.

HUMAN SKIN

A first attempt to characterize human skin with HSP was made by visually evaluating the swelling of psoriasis scales immersed for a prolonged time in different solvents.[2] Uptake could clearly be seen by dimensional changes and a marked enhancement of clarity. It was anticipated that the solubility parameter correlation for the psoriasis scales (keratin) would to some extent reflect permeation in human skin, but that other factors, such as the presence of water and lipids, for example, would also be important. The data fit for this correlation (0.927) indicates that a reasonably reliable correlation for swelling of the psoriasis scales (keratin) has been found. However, the δ_D parameter is thought to be too high.

Permeation data generated in an extensive study allowed placement of the tested solvents into groups according to actual permeation rates through viable human skin.[4] Figure 9.4 graphically shows the HSP correlation which resulted. There are too few data to establish a reliable correlation, but a sphere with center at $\delta_D;\delta_P;\delta_H$ of 17.6;12.5;11.0, which has a radius of 5.0, encompasses the parameters for the four solvents with the highest permeation rates while excluding the others. The units for these parameters are $MPa^{1/2}$ This correlation cannot be considered precise because of insufficient data, and there are, in fact, numerous spheres with somewhat similar but different combinations of the parameters which also can accomplish this. Nevertheless, there is a good guideline for future work, whether it be an expanded correlation or formulation of products designed for a prescribed compatibility with human skin. Calculations for skatole and nicotine predict that moderate rates of skin permeation can also be expected for these.

It might be noted that the four solvents with high permeation rates also have very high affinity for psoriasis scales according to the correlation previously noted. Likewise, the cyclic solvents propylene carbonate, gamma-butyrolactone, and sulfolane have, or are predicted to have, high affinity for psoriasis scales, but they are placed in the low permeation rate group for actual permeation through viable human skin. This reflects the complexity of actual skin permeation and the importance of using viable skin for testing. The cyclic nature of the solvents, however, is also expected to slow the rate of permeation relative to linear solvents of comparable affinity. Factors affecting permeation have been discussed at length in Chapter 8. Of course, the presence of water and/or other skin components can also have an effect on the permeation rate. Finally, the swelling of the psoriasis scales involved equilibrium swelling of the individual systems, while the permeation rate studies did not have this uniformity. Concentration gradients are required for permeation to occur.

SOLUBILITY PARAMETER PLOT FOR SKIN PERMEATION RATE

	δ_D	δ_P	δ_H	M_V	PERMEATION RATE
DMSO	18.4	16.4	10.2	71.3	
DMF	17.4	13.7	11.3	77.0	● HIGH
DMAC	16.8	11.5	10.2	92.5	
NMP	18.0	12.3	7.2	96.5	
MCL	18.2	6.3	6.1	63.9	○ MODERATE
MEK	16.0	9.0	5.1	90.1	
ETH	15.8	8.8	19.4	58.5	
BAC	15.8	3.7	6.3	132.5	
PPC	20.0	18.0	4.1	85.0	× LOW
TOL	18.0	1.4	2.0	106.8	
BTA	19.0	16.6	7.4	76.8	
SUL	18.4	16.6	7.4	95.3	
OAC	15.8	2.9	5.1	196.0	✥ "0"

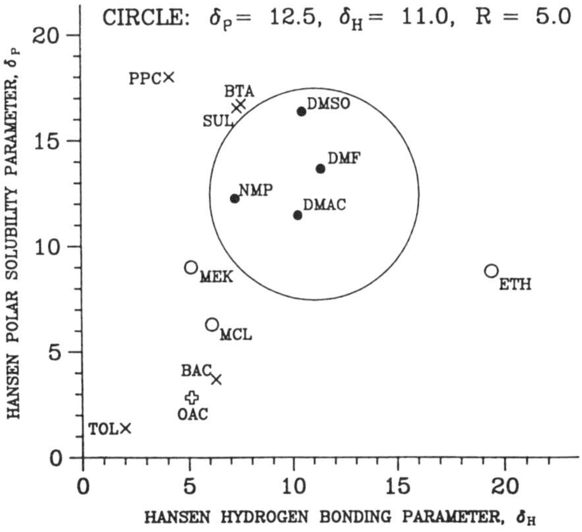

CIRCLE: δ_P = 12.5, δ_H = 11.0, R = 5.0

HANSEN POLAR SOLUBILITY PARAMETER, δ_P

HANSEN HYDROGEN BONDING PARAMETER, δ_H

FIGURE 9.4 Permeation rates of selected solvents through viable human skin show a correlation with the HSP,[4] although the data are not extensive.

PROTEINS — BLOOD SERUM AND ZEIN

HSP correlations for the swelling of blood serum and for the solubility of Zein, a protein derived from corn, are included in Table 9.1. The data used in these correlations are found in Reference 3. Solvents with the lowest RED numbers in the correlation for the solubility of Zein are listed in Table 9.2. The HSP parameters for blood serum and Zein are not too different. The blood serum data are based on visual observation of swelling, while the Zein data are for visual observation of true solution. It is noteworthy that there are only four "good" solvents in the data set reported in Table 9.2 and that the HSP parameters for the proteins are much higher than for any liquid which can be used in such testing. These HSP parameters are found by a form of extrapolation, where all of the good solvents are located in the boundary region of the respective spheres. The values are very much dependent on the mathematical model which includes the coefficient "4" (see Chapters 1 and 2). The saturated solution of urea and water is also a (predictably) good solvent in that it swells blood serum and dissolves Zein, but it was not included as a data point in the correlations as such. Mixtures of solvents, water, and mixtures of solvents with water have been avoided as test solvents to the extent possible because of too many interactions which are apparently not always predictable by these simple considerations. The general prediction that additions of urea to water will improve solvency of proteins is discussed below.

CHLOROPHYLL AND LIGNIN[5]

The results of HSP correlations of solubility for lignin and chlorophyll are given in Table 9.1. More specific information on the lignin correlation is found in Tables 9.3 and 9.4. It can be seen that these indeed have high affinity/physical resemblance to each other, with the HSP values not being too different. A major difference is that chlorophyll is soluble in ethanol, while lignin is not. This indicates a higher hydrophilicity, of course, and gives a higher δ_H parameter to chlorophyll compared with lignin.

It can be presumed that the HSP for these materials are the result of natural selection by nature for optimum compatibility relations with immediate surroundings and function. A discussion of this is beyond the scope of this work, but this point has been studied in more detail for the relations among wood chemicals and wood polymers as outlined in the next section. Here, the HSP for lignin have a demonstrated clear importance with regard to compatibility relations.

WOOD CHEMICALS AND POLYMERS

The results of HSP calculations and correlations for several wood chemicals and polymers are given in Table 9.1. These results are part of a study considering the ultrastructure of wood from a solubility parameter point of view.[6] The study is based on the principle of "Like Seeks Like" and leads to a proposed configuration of the ultrastructure. The HSP for amorphous cellulose are presumed to be similar to those of Dextran (Dextran C, British Drug Houses). The crystallinity in cellulose will require that good solvents have higher affinity/HSP than most of those dissolving Dextran, however. *N*-methyl-morpholine-*N*-oxide is an example. The HSP for Dextran are higher than those of sucrose (which values are similar to the other sugars as well). It is common for polymers to have higher HSP than the monomers from which they are made. It is also common that the solubility of crystalline polymers requires good solvents to have higher HSP than otherwise expected and that smaller molecular volume is an advantage.

The relatively high HSP for cellulose, which also includes a large number of –OH groups, provides a proper energetic environment for the backbones of hemicelluloses, as well as those of their side groups which contain –OH groups. The hemicellulose side groups with acetyl and ether linkages can be expected to orient toward the lower HSP lignin. Neither lignin nor hemicelluloses are compatible with cellulose in the usual sense, but the hemicelluloses can form oriented configurations in connection with cellulose and with lignin. The monomers for lignin, sinapyl alcohol, coniferyl alcohol, and *p*-coumaryl alcohol all have HSP which are on the boundary of the solubility sphere for solubility of Dextran (amorphous cellulose), so their affinities indicate they will seek the lower HSP domain of the lignin. Hemicelluloses act like surfactants, with some side groups favoring the cellulose environment and others favoring the lignin environment. If one considers the HSP for higher ketones, esters, and ethers in Table 9.3, it can be seen that none of these simple liquids will dissolve lignin. This indicates that the acetyl- and ether-containing side groups on the hemicelluloses may not penetrate lignin as such, but prefer to remain on its surface, probably finding a local (interface) site with closest possible HSP. A sketch of these predicted relations is found in Figure 9.5. This is a clear example of self-association in nature.

TABLE 9.3
Calculated Solubility Sphere for Lignin Solubility

D = 21.9 P = 14.1 H = 16.9 RAD. = 13.7 FIT = 0.990 NO = 82

No.	Solvent	D	P	H	SOLUB	RED	V
2	Acetic acid	14.5	8.0	13.5	0	1.195	57.1
3	Acetic anhydride	16.0	11.7	10.2	0	1.006	94.5
4	Acetone	15.5	10.4	7.0	0	1.212	74.0
5	Acetonitrile	15.3	18.0	6.1	0	1.277	52.6
6	Acetophenone	19.6	8.6	3.7	0	1.096	117.4
10	Aniline	19.4	5.1	10.2	0	0.897	91.5
12	Benzaldehyde	19.4	7.4	5.3	0	1.044	101.5
13	Benzene	18.4	0.0	2.0	0	1.582	89.4
24	1-Bromonaphthalene	20.3	3.1	4.1	0	1.254	140.0
27	1,3-Butanediol	16.6	10.0	21.5	1	0.895	89.9
28	1-Butanol	16.0	5.7	15.8	0	1.060	91.5
30	Butyl acetate	15.8	3.7	6.3	0	1.403	132.5
34	Butyl lactate	15.8	6.5	10.2	0	1.158	149.0
36	Butyric acid	14.9	4.1	10.6	0	1.337	110.0
37	gamma-Butyrolactone	19.0	16.6	7.4	1	0.833	76.8
38	Butyronitrile	15.3	12.4	5.1	0	1.298	87.3
39	Carbon disulfide	20.5	0.0	0.6	0	1.586	60.0
40	Carbon tetrachloride	17.8	0.0	0.6	0	1.683	97.1
41	Chlorobenzene	19.0	4.3	2.0	0	1.369	102.1
42	1-Chlorobutane	16.2	5.5	2.0	0	1.506	104.5
44	Chloroform	17.8	3.1	5.7	0	1.293	80.7
46	*m*-Cresol	18.0	5.1	12.9	1	0.917	104.7
47	Cyclohexane	16.8	0.0	0.2	0	1.761	108.7
48	Cyclohexanol	17.4	4.1	13.5	0	1.013	106.0
49	Cyclohexanone	17.8	6.3	5.1	0	1.193	104.0
51	Cyclohexylchloride	17.3	5.5	2.0	0	1.424	118.6
56	Diacetone alcohol	15.8	8.2	10.8	0	1.085	124.2
61	*o*-Dichlorobenzene	19.2	6.3	3.3	0	1.210	112.8
62	2,2-Dichlorodiethyl ether	18.8	9.0	5.7	0	1.006	117.2
71	Diethylamine	14.9	2.3	6.1	0	1.552	103.2
75	Diethylene glycol	16.6	12.0	20.7	1	0.836	94.9
78	Diethylene glycol monobutyl ether	16.0	7.0	10.6	0	1.105	170.6
80	Diethylene glycol monomethyl ether	16.2	7.8	12.6	1*	1.001	118.0
82	Diethyl ether	14.5	2.9	5.1	0	1.605	104.8
86	Diethyl sulfide	16.8	3.1	2.0	0	1.543	107.4
87	Di(isobutyl) ketone	16.0	3.7	4.1	0	1.480	177.1
90	Dimethylformamide	17.4	13.7	11.3	1	0.774	77.0
94	Dimethyl sulfoxide	18.4	16.4	10.2	1	0.727	71.3
96	1,4-Dioxane	19.0	1.8	7.4	0	1.211	85.7
97	Dipropylamine	15.3	1.4	4.1	0	1.631	136.9
98	Dipropylene glycol	16.5	10.6	17.7	1	0.831	130.9
104	Ethanol	15.8	8.8	19.4	1	0.988	58.5
105	Ethanolamine	17.0	15.5	21.2	1	0.788	59.8
106	Ethyl acetate	15.8	5.3	7.2	0	1.306	98.5
109	Ethylbenzene	17.8	0.6	1.4	0	1.615	123.1
111	2-Ethyl-1-butanol	15.8	4.3	13.5	0	1.169	123.2
121	Ethylene glycol	17.0	11.0	26.0	1*	1.002	55.8
123	Ethylene glycol monobutyl ether	16.0	5.1	12.3	0	1.134	131.6

TABLE 9.3 (continued)
Calculated Solubility Sphere for Lignin Solubility

No.	Solvent	D	P	H	SOLUB	RED	V
124	Ethylene glycol monoethyl ether	16.2	9.2	14.3	1	0.925	97.8
125	Ethylene glycol monoethyl ether acetate	15.9	4.7	10.6	0	1.204	136.1
126	Ethylene glycol monomethyl ether	16.2	9.2	16.4	1	0.906	79.1
133	Furan	17.8	1.8	5.3	0	1.372	72.5
136	Glycerol	17.4	12.1	29.3	0	1.128	73.3
140	Hexane	14.9	0.0	0.0	0	1.904	131.6
142	Isoamyl acetate	15.3	3.1	7.0	0	1.447	148.8
145	Isobutyl isobutyrate	15.1	2.9	5.9	0	1.516	163.0
146	Isooctyl alcohol	14.4	7.3	12.9	0	1.237	156.6
148	Isophorone	16.6	8.2	7.4	0	1.125	150.5
151	Mesityl oxide	16.4	6.1	6.1	0	1.268	115.6
153	Methanol	15.1	12.3	22.3	0	1.076	40.7
157	Methylal	15.0	1.8	8.6	0	1.479	169.4
164	Methyl ethyl ketone	16.0	9.0	5.1	0	1.274	90.1
165	Methyl isoamyl ketone	16.0	5.7	4.1	0	1.411	142.8
166	Methyl isobutyl carbinol	15.4	3.3	12.3	0	1.279	127.2
167	Methyl isobutyl ketone	15.3	6.1	4.1	0	1.464	125.8
171	Morpholine	18.8	4.9	9.2	1	0.986	87.1
177	Nitrobenzene	20.0	8.6	4.1	0	1.054	102.7
178	Nitroethane	16.0	15.5	4.5	0	1.254	71.5
179	Nitromethane	15.8	18.8	5.1	0	1.286	54.3
181	2-Nitropropane	16.2	12.1	4.1	0	1.260	86.9
193	1-Pentanol	15.9	4.5	13.9	0	1.143	108.6
199	1-Propanol	16.0	6.8	17.4	0	1.117	75.2
204	Propylene carbonate	20.0	18.0	4.1	0	1.513	85.0
205	Propylene glycol	16.8	9.4	23.3	0*	0.944	73.6
212	Pyridine	19.0	8.8	5.9	1	0.987	80.9
216	Styrene	18.6	1.0	4.1	0	1.421	115.6
222	Tetrahydrofuran	16.8	5.7	8.0	0	1.163	81.7
223	Tetrahydronaphthalene	19.6	2.0	2.9	0	1.392	136.0
225	Toluene	18.0	1.4	2.0	0	1.538	106.8
228	1,1,1-Trichloroethane	16.8	4.3	2.0	0	1.500	99.3
229	Trichloroethylene	18.0	3.1	5.3	0	1.298	90.2
236	Xylene	17.6	1.0	3.1	0	1.524	123.3

In addition to those previously mentioned, one can deduce which chemicals are most prone to penetrate directly through wood. These will dissolve lignin. Included are chlorinated phenols and other wood impregnation materials. It is known that pentachlorophenol, for example, readily diffuses into and through wood specimens. Still another question is how wood transports its own chemicals at various stages of the life of a tree. The same principles are valid. A preferred pathway is where HSP are similar. This can be made possible by molecular rotation and orientation. This can perhaps change with time and local environment.

Other types of predictions are possible from comparisons of the HSP correlations in Table 9.1. For example, it can be determined that all the solvents dissolving lignin are also predicted to swell psoriasis scales. This generality then suggests special care is in order when handling wood-impregnating chemicals. The protective clothing chosen should have HSP quite different from the HSP of the chemical involved, as discussed in Chapter 8.

TABLE 9.4
Calculated Solubility Sphere for Lignin Solubility

D = 21.9 P = 14.1 H = 16.9 RAD. = 13.7 FIT = 0.990 NO = 82

No.	Solvent	D	P	H	SOLUB	RED	V
213	2-Pyrolidone	19.4	17.4	11.3		0.599	76.4
217	Succinic anhydride	18.6	19.2	16.6		0.609	66.8
93	Dimethyl sulfone	19.0	19.4	12.3		0.665	75.0
94	Dimethyl sulfoxide	18.4	16.4	10.2	1	0.727	71.3
139	Hexamethylphosphoramide	18.5	8.6	11.3		0.758	175.7
154	*o*-Methoxyphenol	18.0	8.2	13.3		0.761	109.5
14	1,3-Butanediol	18.0	8.4	21.0		0.766	87.5
117	Ethylene cyanohydrin	17.2	18.8	17.6		0.769	68.3
90	Dimethyl formamide	17.4	13.7	11.3	1	0.774	77.0
81	Diethylenetriamine	16.7	13.3	14.3		0.785	108.0
105	Ethanolamine	17.0	15.5	21.2	1	0.788	59.8
70	Diethanolamine	17.2	10.8	21.2		0.792	95.9
17	Benzyl alcohol	18.4	6.3	13.7		0.800	103.6
135	Furfuryl alcohol	17.4	7.6	15.1		0.821	86.5
98	Dipropylene glycol	16.5	10.6	17.7	1	0.831	130.9
37	gamma-Butyrolactone	19.0	16.6	7.4	1	0.833	76.8
75	Diethylene glycol	16.6	12.0	20.7	1	0.836	94.9
197	Phenol	18.0	5.9	14.9		0.839	87.5
118	Ethylenediamine	16.6	8.8	17.0		0.865	67.3
8	Allyl alcohol	16.2	10.8	16.8		0.866	68.4
236	Triethyleneglycol	16.0	12.5	18.6		0.878	114.0
27	1,3-Butanediol	16.6	10.0	21.5	1	0.895	89.9
10	Aniline	19.4	5.1	10.2	1	0.897	91.5
45	3-Chloro-1-propanol	17.5	5.7	14.7		0.902	84.2
126	Ethylene glycol monomethyl ether	16.2	9.2	16.4	1	0.906	79.1
89	*N,N*-Dimethyl acetamide	16.8	11.5	10.2		0.911	92.5
243	Trimethylphosphate	16.7	15.9	10.2		0.913	115.8
15	Benzoic acid	18.2	6.9	9.8		0.915	100.0
46	*m*-Cresol	18.0	5.1	12.9	1	0.917	104.7
172	Methyl-2-pyrrolidone	18.0	12.3	7.2		0.918	96.5
218	1,1,2,2-Tetrabromoethane	22.6	5.1	8.2		0.919	116.8
124	Ethylene glycol monoethyl ether	16.2	9.2	14.3	1	0.925	97.8
214	Quinoline	19.4	7.0	7.6		0.929	118.0
205	Propylene glycol	16.8	9.4	23.3	0*	0.944	73.6
238	Triethylphosphate	16.7	11.4	9.2		0.965	171.0
219	1,1,2,2-Tetrachloroethane	18.8	5.1	9.4		0.968	105.2
224	Tetramethylurea	16.7	8.2	11.0		0.973	120.4
79	Diethylene glycol monoethyl ether	16.1	9.2	12.2		0.981	130.9
174	Morpholine	18.8	4.9	9.2	1	0.986	87.1
212	Pyridine	19.0	8.8	5.9	1	0.987	80.9
104	Ethanol	15.8	8.8	19.4	1	0.988	58.5
134	Furfural	18.6	14.9	5.1		0.989	83.2
141	Hexylene glycol	15.7	8.4	17.8		0.998	123.0
210	Propylene glycol monophenyl ether	17.4	5.3	11.5		1.000	143.2
80	Diethylene glycol monomethyl ether	16.2	7.8	12.6	1*	1.001	118.0
121	Ethylene glycol	17.0	11.0	26.0	1	1.002	55.8
62	2,2-Dichlorodiethyl ether	18.8	9.0	5.7	0	1.006	117.2
3	Acetic anhydride	16.0	11.7	10.2	0	1.006	94.5

TABLE 9.4 (continued)
Calculated Solubility Sphere for Lignin Solubility

No.	Solvent	D	P	H	SOLUB	RED	V
232	Tricresyl phosphate	19.0	12.3	4.5		1.008	316.0
48	Cyclohexanol	17.4	4.1	13.5	0	1.013	106.0
199	1-Propanol	16.0	6.8	17.4	0	1.013	75.2
204	Propylene carbonate	20.0	18.0	4.1	0	1.015	85.0
234	Triethyanolamine	17.3	22.4	23.3		1.018	133.2
184	Nonyl phenoxy ethanol	16.7	10.2	8.4		1.021	275.0
173	Methyl salicylate	16.0	8.0	12.3		1.026	129.0
92	Dimethyl phthalate	18.6	10.8	4.9		1.028	163.0
130	Ethyl lactate	16.0	7.6	12.5		1.034	115.0
12	Benzaldehyde	19.4	7.4	5.3	0	1.044	101.5
239	Trifluoroacetic acid	15.6	9.9	11.6		1.044	74.2
66	Di-(2-Chloro-isopropyl) ether	19.0	8.2	5.1		1.052	146.0
177	Nitrobenzene	20.0	8.6	4.1	0	1.054	102.7
119	Ethylene dibromide	19.2	3.5	8.6		1.059	87.0
28	1-Butanol	16.0	5.7	15.8	0	1.060	91.5
200	2-Propanol	15.8	6.1	16.4		1.066	76.8
153	Methanol	15.1	12.3	22.3	0	1.076	40.7
23	Bromoform	21.4	4.1	6.1		1.077	87.5
56	Diacetone alcohol	15.8	8.2	10.8	0	1.085	124.2
116	Ethylene carbonate	19.4	21.7	5.1		1.088	66.0
102	Epichlorohydrin	19.0	10.2	3.7		1.090	79.9
29	2-Butanol	15.8	5.7	14.5		1.095	92.0
6	Acetophenone	19.6	8.6	3.7	0	1.096	117.4
78	Diethylene glycol monobutyl ether	16.0	7.0	10.6	0	1.105	170.6
162	Methylene dichloride	18.2	6.3	6.1		1.112	63.9
18	Benzyl butyl phthalate	19.0	11.2	3.1		1.113	306.0
7	Acrylonitrile	16.4	17.4	6.8		1.116	67.1
132	Formic acid	14.3	11.9	16.6		1.121	37.8
148	Isophorone	16.6	8.2	7.4	0	1.125	150.5
187	1-Octanol	17.0	3.3	11.9		1.125	157.7
136	Glycerol	17.4	12.1	29.3	0	1.128	73.3
131	Formamide	17.2	26.2	19.0		1.129	39.8
123	Ethylene glycol monobutyl ether	16.0	5.1	12.3	0	1.134	131.6
120	Ethylene dichloride	19.0	7.4	4.1		1.136	79.4
193	1-Pentanol	15.9	4.5	13.9	0	1.143	108.6
180	1-Nitropropane	16.6	12.3	5.5		1.144	88.4
21	Bromobenzene	20.5	5.5	4.1		1.144	105.3
127	Ethylene glycol monomethyl ether acetate	15.9	5.5	11.6		1.145	121.6
115	Ethyl cinnamate	18.4	8.2	4.1		1.149	166.8
209	Propylene glycol monomethyl ether	15.6	6.3	11.6		1.149	93.8
84	Diethyl phthalate	17.6	9.6	4.5		1.149	198.0
85	Diethyl sulfate	15.7	14.7	7.1		1.154	131.5
34	Butyl lactate	15.8	6.5	10.2	0	1.158	149.0
77	Diethylene glycol hexyl ether	16.0	6.0	10.0		1.160	204.3
207	Propylene glycol monoethyl ether	15.7	6.5	10.5		1.160	115.6
202	Propylamine	16.9	4.9	8.6		1.162	83.0
222	Tetrahydrofuran	16.8	5.7	8.0	0	1.163	81.7
188	2-Octanol	16.1	4.9	11.0		1.163	159.1
111	2-Ethyl-1-butanol	15.8	4.3	13.5	0	1.169	123.2
144	Isobutyl alcohol	15.1	5.7	15.9		1.169	92.8

TABLE 9.4 (continued)
Calculated Solubility Sphere for Lignin Solubility

No.	Solvent	D	P	H	SOLUB	RED	V
171	2-Methyl-1-propanol	15.1	5.7	15.9		1.169	92.8
55	1-Decanol	17.5	2.6	10.0		1.171	191.8
211	Propylene glycol monopropyl ether	15.8	7.0	9.2		1.174	130.3
58	Dibutyl phthalate	17.8	8.6	4.1		1.180	266.0
99	Dipropylene glycol methyl ether	15.5	5.7	11.2		1.192	157.4
49	Cyclohexanone	17.8	6.3	5.1	0	1.193	104.0
2	Acetic acid	14.5	8.0	13.5	0	1.195	57.1
128	Ethyl formate	15.5	8.4	8.4		1.196	80.2
127	Trichlorobiphenyl	19.2	5.3	4.1		1.200	187.0
11	Anisole	17.8	4.1	6.7		1.202	119.1
25	Ethylene glycol monoethyl ether acetate	15.9	4.7	10.6	0	1.204	136.1
61	*o*-Dichlorobenzene	19.2	6.3	3.3	0	1.210	112.8
96	1,4-Dioxane	19.0	1.8	7.4	0	1.211	85.7
4	Acetone	15.5	10.4	7.0	0	1.212	74.0
183	Nonyl phenol	16.5	4.1	9.2		1.212	231.0
1	Acetaldehyde	14.7	8.0	11.3		1.212	57.1
91	1,1-Dimethyl hydrazine	15.3	5.9	11.0		1.213	76.0
198	*bis*-(*m*-Phenoxyphenyl) ether	19.6	3.1	5.1		1.224	373.0
192	2,4-Pentanedione	17.1	9.0	4.1		1.226	103.1
114	Ethyl chloroformate	15.5	10.0	6.7		1.232	95.6
57	Dibenzyl ether	17.3	3.7	7.3		1.232	192.7
129	2-Ethyl hexanol	15.9	3.3	11.8		1.236	156.6
146	Isooctyl alcohol	14.4	7.3	12.9	0	1.237	156.6
220	Tetrachloroethylene	19.0	6.5	2.9		1.237	101.1
72	2-(Diethylamino) ethanol	14.9	5.8	12.0		1.241	133.2
19	Benzyl chloride	18.8	7.1	2.6		1.247	115.0
16	Benzonitrile	17.4	9.0	3.3		1.247	102.6
110	Ethyl bromide	16.5	8.0	5.1		1.250	76.9
178	Nitroethane	16.0	15.5	4.5	0	1.254	71.5
24	1-Bromonaphthalene	20.3	3.1	4.1	0	1.254	140.0
176	Naphthalene	19.2	2.0	5.9		1.257	111.5
181	2-Nitropropane	16.2	12.1	4.1	0	1.260	86.9
155	Methyl acetate	15.5	7.2	7.6		1.260	79.7
242	2,2,4-Trimethyl 1,3-pentanediol monoisobutyrate	15.1	6.1	9.8		1.263	227.4
163	Methylene diiodide	17.8	3.9	5.5		1.267	80.5
33	Butylamine	16.2	4.5	8.0		1.267	99.0
151	Mesityl oxide	16.4	6.1	6.1	0	1.268	115.6
65	1,1-Dichloroethylene	17.0	6.8	4.5		1.271	79.0
201	Propionitrile	15.3	14.3	5.5		1.273	70.9
164	Methyl ethyl ketone	16.0	9.0	5.1	0	1.274	90.1
5	Acetonitrile	15.3	18.0	6.1	0	1.277	52.6
166	Methyl isobutyl carbinol	15.4	3.3	12.3	0	1.279	127.2
103	Ethanethiol	15.7	6.5	7.1		1.280	74.3
168	Methyl methacrylate	17.5	5.5	4.3		1.286	106.5
179	Nitromethane	15.8	18.8	5.1	0	1.286	54.3
44	Chloroform	17.8	3.1	5.7	0	1.293	80.7
76	Diethylene glycol butyl ether acetate	16.0	4.1	8.2		1.295	208.2
38	Butyronitrile	15.3	12.4	5.1	0	1.298	87.3
129	Trichloroethylene	18.0	3.1	5.3	0	1.298	90.2
50	Cyclohexylamine	17.2	3.1	6.5		1.301	113.8

TABLE 9.4 (continued)
Calculated Solubility Sphere for Lignin Solubility

No.	Solvent	D	P	H	SOLUB	RED	V
156	Methyl acrylate	15.3	9.3	5.9		1.302	113.8
106	Ethyl acetate	15.8	5.3	7.2	0	1.306	98.5
208	Propylene glycol monoisobutyl ether	15.1	4.7	9.8		1.313	132.2
206	Propylene glycol monobutyl ether	15.3	4.5	9.2		1.317	132.0
88	Di(2-Methoxyethyl) ether	15.7	6.1	6.5		1.318	142.0
122	Ethylene glycol butyl ether acetate	15.3	4.5	8.8		1.330	171.2
169	1-Methyl naphthalene	20.6	0.8	4.7		1.331	138.8
22	Bromochloromethane	17.3	5.7	3.5		1.336	65.0
36	Butyric acid	14.9	4.1	10.6	0	1.337	110.0
83	Diethyl ketone	15.8	7.6	4.7		1.346	106.4
107	Ethyl acrylate	15.5	7.1	5.5		1.351	108.8
226	Tributyl phosphate	16.3	6.3	4.3		1.356	345.0
74	Diethyl carbonate	16.6	3.1	6.1		1.366	121.0
41	Chlorobenzene	19.0	4.3	2.0	0	1.369	102.1
133	Furan	17.8	1.8	5.3	0	1.372	72.5
95	Dioctyl phthalate	16.6	7.0	3.1		1.372	377.0
69	Di-iso-butyl carbinol	14.9	3.1	10.8		1.374	177.8
152	Methacrylonitrile	15.3	10.8	3.6		1.389	83.9
223	Tetrahydronaphthalene	19.6	2.0	2.9	0	1.392	136.0
32	Butyl acrylate	15.6	6.2	4.9		1.395	143.8
30	Butyl acetate	15.8	3.7	6.3	0	1.403	132.5
215	Stearic acid	16.3	3.3	5.5		1.408	326.0
155	Methyl isoamyl ketone	16.0	5.7	4.1	0	1.411	142.8
112	Ethyl butyl ketone	16.2	5.0	4.1		1.417	139.0
186	Octanoic acid	15.1	3.3	8.2		1.418	159.0
216	Styrene	18.6	1.0	4.1	0	1.421	115.6
51	Cyclohexylchloride	17.3	5.5	2.0	0	1.424	118.6
9	Amyl acetate	15.8	3.3	6.1		1.427	148.0
35	Butyraldehyde	14.7	5.3	7.0		1.428	88.5
31	*sec*-Butyl acetate	15.0	3.7	7.6		1.432	133.6
108	Ethyl amyl ketone	16.2	4.5	4.1		1.434	156.0
142	Isoamyl acetate	15.3	3.1	7.0	0	1.447	148.8
20	Biphenyl	21.4	1.0	2.0		1.450	154.1
67	Dichloromonoflouromethane	15.8	3.1	5.7		1.451	75.4
203	Propyl chloride	16.0	7.8	2.0		1.462	88.1
159	Methyl butyl ketone	15.3	6.1	4.1		1.464	123.6
167	Methyl isobutyl ketone	15.3	6.1	4.1	0	1.464	125.8
158	Methyl amyl acetate	15.2	3.1	6.8		1.465	167.4
143	Isobutyl acetate	15.1	3.7	6.3		1.470	133.5
160	Methyl chloride	15.3	6.1	3.9		1.473	55.4
157	Methylal	15.0	1.8	8.6	0	1.479	169.4
87	Di(isobutyl) ketone	16.0	3.7	4.1	0	1.480	177.1
113	Ethyl chloride	15.7	6.1	2.9		1.485	70.0
233	Tridecyl alcohol	14.3	3.1	9.0		1.486	242.0
228	1,1,1-Trichloroethane	16.8	4.3	2.0	0	1.500	99.3
64	1,1-Dichlorethane	16.5	8.2	0.4		1.502	84.8
42	1-Chlorobutane	16.2	5.5	2.0	0	1.506	104.5
246	*o*-Xylene	17.8	1.0	3.1		1.512	121.2
145	Isobutyl isobutyrate	15.1	2.9	5.9	0	1.516	163.0
245	Xylene	17.6	1.0	3.1	0	1.524	123.3

TABLE 9.4 (continued)
Calculated Solubility Sphere for Lignin Solubility

No.	Solvent	D	P	H	SOLUB	RED	V
190	Oleyl alcohol	14.3	2.6	8.0		1.535	316.0
225	Toluene	18.0	1.4	2.0	0	1.538	106.8
86	Diethyl sulfide	16.8	3.1	2.0	0	1.543	107.4
71	Diethyl amine	14.9	2.3	6.1	0	1.552	103.2
13	Benzene	18.4	0.0	2.0	0	1.582	89.4
175	Naphtha.high-flash	17.9	0.7	1.8		1.585	181.8
39	Carbon disulfide	20.5	0.0	0.6	0	1.586	60.0
189	Oleic acid	14.3	3.1	5.5		1.603	320.0
82	Diethyl ether	14.5	2.9	5.1	0	1.605	104.8
237	Triethylene glycol monooleyl ether	13.3	3.1	8.4		1.614	418.5
109	Ethylbenzene	17.8	0.6	1.4	0	1.615	123.1
170	Methyl oleate	14.5	3.9	3.7		1.628	340.0
97	Dipropylamine	5.3	1.4	4.1	0	1.631	136.9
60	Dibutyl stearate	14.5	3.7	3.5		1.643	382.0
235	Triethylamine	17.8	0.4	1.0		1.645	138.6
240	Trimethylbenzene	17.8	0.4	1.0		1.645	133.6
149	Isopropyl palmitate	14.3	3.9	3.7		1.647	330.0
59	Dibutyl sebacate	13.9	4.5	4.1		1.652	339.0
52	*cis*-Decahydronaphthalene	18.8	0.0	0.0		1.669	156.9
73	*para*-Diethyl benzene	18.0	0.0	0.6		1.673	156.9
150	Mesitylene	18.0	0.0	0.6		1.673	139.8
40	Carbon tetrachloride	17.8	0.0	0.6	0	1.683	97.1
53	*trans*-Decahydronaphthalene	18.0	0.0	0.0		1.704	156.9
43	Chlorodiflouromethane	12.3	6.3	5.7		1.719	72.9
47	Cyclohexane	16.8	0.0	0.2	0	1.761	108.7
161	Methyl cyclohexane	16.0	0.0	1.0		1.774	128.3
101	Eicosane	16.5	0.0	0.0		1.790	359.8
230	Trichlorofluoromethane	15.3	2.0	0.0		1.797	92.8
138	Hexadecane	16.3	0.0	0.0		1.803	294.1
100	Dodecane	16.0	0.0	0.0		1.823	228.6
247	Mineral spirits	15.8	0.1	0.2		1.823	125.0
54	Decane	15.7	0.0	0.0		1.844	195.9
182	Nonane	15.7	0.0	0.0		1.844	179.7
185	Octane	15.5	0.0	0.0		1.858	163.5
231	1,1,2-Trichlorotrifluoroethane	14.7	1.6	0.0		1.860	119.2
137	Heptane	15.3	0.0	0.0		1.873	147.4
140	Hexane	14.9	0.0	0.0	0	1.904	131.6
191	Pentane	14.5	0.0	0.0		1.936	116.2
221	Tetraethylorthosilicate	13.9	0.4	0.6		1.944	224.0
26	Butane	14.1	0.0	0.0		1.969	101.4
241	2,2,2,4-Trimethylpentane	14.1	0.0	0.0		1.969	166.1
147	Isopentane	13.7	0.0	0.0		2.003	117.4
68	1,2-Dichlorotetrafluoroethane	12.6	1.8	0.0		2.042	117.6
63	Dichlorodifluoromethane	12.3	2.0	0.0		2.065	92.3
244	Water	15.5	16.0	42.3		2.081	18.0
194	Perfluoro(dimethylcyclohexane)	12.4	0.0	0.0		2.122	217.4
196	Perfluoromethylcyclohexane	12.4	0.0	0.0		2.122	196.0
195	Perfluoroheptane	12.0	0.0	0.0		2.161	227.3
25	Bromotrifluoromethane	9.6	2.4	0.0		2.340	97.0

```
--M1β4M1β4M1β4G1β4M1β4G1β4M1β4M1β4G–
   2    3        3        2    3        6
  Ac   Ac        β       Ac   Ac        α
  (LIGN)         1       (LIGN)         1
                 M                      Ga
               (CELL)                 (CELL)
```

FIGURE 9.5 Expected generalized sketch of the configuration of cellulose, hemicelluloses, and lignin in wood cell walls. See text or Reference 6 for further details. The sketch is for glucomannan. M is mannose monomer; G is glucose monomer; Ga is galactose monomer; Ac is an acetyl group; (LIGN) is a region similar in HSP to lignin (or acetal etc.); (CELL) is a region similar in HSP to cellulose, being any of cellulose, hemicellulose backbone, or hemicellulose side chain with an alcohol group (M, Ga).

An important effect which may have been overlooked in the solubility of wood and wood components is that there are acid groups present in hemicelluloses, for example, and these can be neutralized by bases. This gives an organic salt with high HSP.[9] Such a salt is hydrophilic and will collect water. This may lead to phase separation, and some destruction of ultrastructure is possible. This is an effect which is known to have caused blistering in coatings.

UREA

Data for the HSP correlation for urea solubility in organic solvents are given in Table 9.1. All of the parameters are rather high, which is characteristic of a low molecular weight solid. The data fit is very good. Perhaps the most interesting thing about this correlation is that it clearly shows that additions of urea to water will improve solubility for a variety of materials including proteins. This is the reason for the improved solubility discussed previously in connection with the destruction of hydrophilic bonding in proteins. The saturated solution of urea and water is also the best physically acting solvent for whole, dried blood which the author could locate in a previous (unpublished) study.

The fact of high HSP for urea/water mixtures has led to its use in many varied types of "products."[7] The saturated solution of urea in water has found particular successes in the following examples.

1. Lithographic stones were previously conditioned to make them more receptive to ink by application of this liquid to change wetting behavior.
2. The saturated solution of urea and water, which swells and softens wood, has been used to give wood flexibility so that it can easily be formed.
3. It has been used by Eskimos to soften seal skins by swelling and softening them. A similar application in Mexico involves "curing" leather. This application probably originated in prehistoric times.
4. It has been used to improve the flow of house paints on cold days or when no other source of liquid has been available (such as on a scaffold), since it is a good solvent miscible in many paints.
5. It is reported to have been used to set hair, since it also softens and swells hair.
6. It was used in the early manufacture of gunpowder as a dispersion medium during grinding because of improved wetting for the powder.
7. Amazonian Indians used this liquid to coagulate latex prior to sale and shipment (solubility). This was practiced particularly during the World War II.

Other unspecified and undocumented uses include those possible because the liquid has the ability to soften human skin, thus allowing easier transport of medicinal chemicals into the body.

WATER

Water has been discussed in detail in Chapter 1. Briefly stated, one can use the HSP for water or a correlation for water solubility to get a general explanation for observed phenomena. Accurate calculation of the HSP for solvent–water mixtures cannot be expected because of the irregularities of water associating with itself, the solvent, and a potential solute. Lindenfors[10] described the association of two molecules of water with one molecule of dimethyl sulfoxide, a solvent frequently mentioned in connection with biological systems. A simplistic approach based on the ratio of δ_H for water as a single molecule vs. that in the correlation(s) for water solubility suggests that $(42.3/16.5)^2$ or about six water molecules are linked by hydrogen bonding into some type of entity. Various structures for assemblies of water molecules have been discussed in the literature. The clusters with six water molecules are among the more probable ones.[11] The data on water solubility used in the HSP correlations are reported by Wallström and Svenningsen.[12]

SURFACE MOBILITY

Surface mobility allows given segments of molecules to orient at surfaces in a direction where their HSP match more closely. The surfaces of hydrophobic polymers (peat moss) can become hydrophilic when contacted with water. One can speculate as to why this occurs. One possibility is that this phenomenon conserves water within the structure. Whenever water is present on an otherwise hydrophobic surface, it will become hydrophilic. This allows the water to enter the structure. When this is accomplished, the surface dries and becomes hydrophobic once more. The hydrophobic surface helps prevent evaporation of water, since water is not particularly soluble in it, and the hydrophilic segments oriented toward the interior of the structure will help bind the water where it is. The basis of the orientation effects described earlier for hemicelluloses is another example of orientation toward regions where HSP matches better. These phenomena are also discussed in Chapter 11. It is also appropriate to repeat that solvent quality has a great deal to do with pigment dispersion stability, in that the adsorbed stabilizing polymer should remain on the pigment surface. A solvent which is too good can remove it. This is discussed in detail in Chapter 3.

The implication of these examples is that solvent quality is very important for the orientation of molecules at interfaces. A change in solvent quality can easily lead to a change in the configuration of molecules at surfaces. It is not surprising that nature has used this to advantage in various ways.

CONCLUSION

Many materials of biological significance have been assigned HSP based on their interaction with a large number of solvents whose HSP are known. Such data have been used to find predictably synergistic solvent mixtures where two nonsolvents dissolve cholesterol when mixed. The ethanol/aliphatic hydrocarbon synergistic mixture is discussed as being of particular interest to the fate of cholesterol in lipid layers. The HSP of chlorophyll and lignin are quite similar, indicating they will be compatible with very much the same kind of surroundings. The physical interrelationships for wood chemicals and wood polymers (lignin, hemicelluloses, and cellulose) are discussed. The side chains on hemicelluloses which contain alcohol groups and the hemicellulose backbone will be most compatible with cellulose and will orient toward this. The hemicellulose side chains without alcohol groups (acetal, acid) are closer in HSP to lignin and will orient in this direction. The acetal side chains actually have lower HSP than will dissolve lignin, for which reason they are expected to lie on the surface of the lignin or perhaps penetrate slightly into the lignin at very special local points where the HSP match is better than the average values seen over the lignin molecule as a whole.

HSP analyses of relative affinities can be applied to a large number of other biological materials and may provide insights into relationships which are not readily obvious or cannot be studied otherwise. The best situation is where the materials in question can be tested directly, otherwise the calculational procedures described in Chapter 1 can be used with some loss of reliability in predictions.

REFERENCES

1. Hansen, C. M., The Three Dimensional Solubility Parameter — Key to Paint Component Affinities I. Solvents, Plasticizers, Polymers, and Resins, *J. Paint Technol.,* 39(505), 104–117, 1967.
2. Hansen, C. M. and Andersen, B. H., The Affinities of Organic Solvents in Biological Systems, *Am. Ind. Hyg. Assoc. J.,* 49(6), 301–308, 1988.
3. Hansen, C. M., The Universality of the Solubility Parameter, *Ind. Eng. Chem. Prod. Res. Dev.,* 8(1), 2–11, 1969.
4. Ursin, C., Hansen, C. M., Van Dyk, J. W., Jensen, P. O., Christensen, I. J., and Ebbehoej, J., Permeability of Commercial Solvents Through Living Human Skin, *Am. Ind. Hyg. Assoc. J.,* 56, 651–660, 1995.
5. Hansen, C. M., 25 Years with the Solubility Parameter (in Danish: 25 År med Opløselighedsparametrene), *Dan. Kemi,* 73(8), 18–22, 1992.
6. Hansen, C. M. and Björkman, A., The Ultrastructure of Wood from a Solubility Parameter Point of View, *Holzforschung*, 52(4), 335–344, 1998.
7. Hansen, C. M., Solvents for Coatings, *Chem. Technol.,* 2(9), 547–553, 1972.
8. Hansen, C. M., Cohesion Energy Parameters Applied to Surface Phenomena, *Handbook of Surface and Colloid Chemistry,* Birdi, K. S., Ed., CRC Press, Boca Raton, FL, 1997, 313–332.
9. Hansen, C. M., Some Aspects of Acid/Base Interactions (in German: Einige Aspekte der Säure/Base-Wechselwirkung), *Farbe und Lack*, 7, 595–598, 1977.
10. Lindenfors, S., Solubility of Cellulose Ethers (in German: Löslichkeit der Celluloseeäther), *Das Papier,* 21, 65–69, 1967.
11. Gregory, J. K., Clary, D. C., Liu, K., Brown, M. G., and Saykally, R. J., The Water Dipole Moment in Water Clusters, *Science*, 275, 814–817, 1997.
12. Wallström, E. and Svenningsen, I., *Handbook of Solvent Properties*, Report T1-84, Scandinavian Paint and Printing Ink Research Institute, Hoersholm, Denmark, 1984.

10 Applications — Safety and Environment

CONTENTS

ABSTRACT

Hansen solubility parameters (HSP) can be used to gain insight into many safety and environmental issues. These include substitution to less dangerous materials, products, and processes, where a listing of solvents having HSP similarity to the one(s) to be substituted provides an overview of the potential choices for improvement. Selection of suitable chemical protective clothing can be improved by considering HSP correlations of breakthrough time. Evaluating risks for inadvertent chemical uptake in recycled plastic can be helped by HSP correlations of chemical resistance and/or permeation phenomena. Similarity of HSP suggests which chemicals are most likely to be rapidly absorbed into given plastic types. These same approaches can be used to evaluate the potential for uptake of chemicals through human skin.

INTRODUCTION

Many organic materials are potential safety hazards. They can also be harmful to the environment. Unfortunately, it is often a matter of experience before the risks are uncovered because of damage being done. Thus, over the years, there have been a series of substitutions with or without the aid of HSP to eliminate or to at least reduce such problems. An example is the current lack of emphasis on the use of ethylene glycol ethers as solvents because of their teratogenic effects, whereas they were used in massive quantities earlier. The problem of replacing ozone-depleting chemicals is a case involving the external environment. Other large-scale substitutions can also be cited, but a list of this type is not the purpose of this chapter. The emphasis here is on the use of sound formulating principles to reduce the potential hazard in terms of reformulation or substitution. When a satisfactory substitution cannot be found, personal protection of one type or another may be required. Here again, HSP can be a help.

Evaluating other forms of environmental risks can be aided by using HSP. An example is the occasional misuse of plastic containers normally used for soft drinks to store chemicals such as herbicides and pesticides. These are likely to diffuse into the plastic container wall itself, making customary washing insufficient. HSP can indicate which chemicals can do this, thus providing information on the means to improve handling of the problem. This type of information can be generated for any polymer type where an HSP correlation of chemical resistance, weight gain, etc. can be generated.

All of the these situations are discussed in more detail in the following.

SUBSTITUTION

Substitution involves the replacement of a potentially dangerous process or chemical with a new process or chemical having less hazardous properties. The hazards can be judged using accepted approaches, for example, labeling requirements, toxicology assessments, biodegradability, and physical properties for the chemical or products. The volatility of products is also a significant factor with lower volatility being preferred due to reduced workplace concentrations and reduced replacement requirements for cleaners and the like which often recirculate in nearly closed systems. On the other hand, the problem should not just be transferred from the air exhaust system to the sewer.

The use of technologies involving water or mechanical methods such as mechanical joints rather than the use of solvents are preferred. Examples of preferred coatings technologies are the use of powders which flow at higher temperatures or polymerization by radiation, both of which use solvent-free base products to provide the coating. Other product types which may be targeted include cutting fluids, cleaners of various types, adhesives, sealers, and fillers.

In general, one primarily wishes to substitute for

- Carcinogens or suspected carcinogens
- Substances with risk phrases for being very toxic, toxic, allergenic, carcinogenic, teratogenic, mutagenic or causing cumulative or irreversible effects
- Substances with moderate or serious aquatic toxicity
- Nonbiodegradable substances
- Substances with high predicted aquatic effects, for example, chemicals which preferentially distribute to a nonaqueous phase to a very high degree

Efforts should be made to develop products with the lowest possible hazard. Those who have read the earlier chapters in this book will immediately recognize HSP as a tool to aid in the substitution and systematic formulation for reduced safety and environmental risks. A key element in this is a listing of solvents where those most resembling the candidate for substitution are at the top of the list. The program described in Chapter 1 can do this by entering the HSP for the solvent to be replaced and requesting a listing with those solvents most similar to it (i.e., the lowest RED numbers as defined in Chapter 1, Equations 1.10) at the top of the list. One must then sort through these potential replacement candidates using other information to arrive at a better alternative.

It is clear that much more data than HSP are required to make the desired substitutions. However, a further discussion of this is beyond the scope of this chapter which emphasizes HSP only. The currently used HSP techniques and correlations can aid in some aspects of substitution, and it is anticipated that future correlations will help in this endeavor. Many cases of substitutions in practice have been listed by Goldschmidt,[1] Olsen,[2] Soerensen and Petersen,[3] and Filskov et al.[4] A long list of references for the whole Danish experience with occupational risks and solutions is given by Soerensen and Petersen.[3]

ALTERNATIVE SYSTEMS

Alternative systems with less solvent or no solvent have been focused on by the coatings and printing ink industries for many years. Examples of such systems are coatings with higher solids, radiation-curable inks and coatings, powder coatings, electrodeposition coatings, and other water-reducible products. It might appear that solvent technology and use of HSP will not be as important as it has been in the past. This is not the case, however, as demonstrated in earlier chapters. For example, HSP principles can be used to aid in improved adhesion, to predict polymer/filler interactions, to improve barrier polymers, and to aid in understanding some biological phenomena.

The use of solvents in alternative coatings systems has been the topic of several previous publications by the author.[5-8] Some general principles of solvent selection have been discussed in Chapter 6 and earlier,[9] as well as elsewhere more recently.[10]

SOLVENT FORMULATION AND PERSONAL PROTECTION FOR LEAST RISK

Solubility parameter principles have been used in formulating alternative, low VOC (volatile organic compound) products. A number of the general formulation principles can be briefly stated for the sake of completeness. These include the following:

1. Solvents with lower viscosity most often lead to polymer solutions with lower viscosity. Such a change allows the use of higher solids at the original viscosity. However, these may evaporate more rapidly and can be expected to have a lower flash point.
2. Solvents with linear and smaller structures diffuse more rapidly than those with branched and larger structures. Inclusion of slower evaporating, more linear solvents can hasten the through-drying of a coating.
3. Two (or more) mixed solvents with lower labeling requirements may be able to replace a single solvent. HSP can be used in this type of endeavor.
4. The surface tension of water-reducible coatings can often be significantly reduced by relatively small additions of ethanol or other alcohol-type solvent. These can, of course, also be used in conjunction with other surface active materials.

Materials with least potential risk are to be used in the Nordic countries wherever possible. The risk must be indicated by the seller/producer in terms of a labeling code. The risk can then be assessed by users or, perhaps more specifically, by primarily professional users. Such labeling is required on paints, printing inks, cleaners, or for any product containing significant amounts of solvent or hazardous chemical. The labeling code dictates the personal protection required for the product depending on the way it is used. Spraying product in a smaller room with limited ventilation requires much more protection than applying paint with a brush outdoors. Tables have been published which give the protection required (gloves, dust mask, fresh air mask, body suit, etc.) for a given set of application conditions for a wide variety of products from paints and printing inks through cleaners.[11] A key element in these tables is the labeling code developed in Denmark according to the MAL (in Danish: Maletekniske Arbejshygieniske Luftbehov) system. For present purposes, this is translated as the "FAN" (fresh air number). Higher MAL/FAN dictate that more extensive personal protection is required.

THE DANISH MAL SYSTEM — THE FAN[12]

As indicated previously, the quality of the working environment must be considered in all cases where organic solvents are being used. The Danish MAL system or other labeling system can be systematically used for this purpose. The Danish MAL reflects the cubic meters of fresh air required for ventilation of 1 liter of product to below the threshold limit value (TLV). This number is modified by a constant, depending on the evaporation rate (or vapor pressure). Higher evaporation rates imply greater hazard, so the multiplier is larger.

The concept behind the MAL system can be better understood in English by translating the MAL number as the FAN. Other numbers in addition to the TLV/GV/OEL (occupational exposure limit) and FAN have been generated to help evaluate risks by inclusion of evaporation rate/vapor pressure considerations. The vapor pressure divided by the TLV is called the vapor hazard ratio (VHR) and the actual calculated vapor composition (using activity coefficients) divided by the TLV has been called SUBFAC. A Danish publication comparing several of these is available.[4] To demonstrate the principle, the simple MAL = FAN has been tabulated in Table 10.1 for several solvents.

Each product-containing solvent is assigned a two-digit number to place it into a potential hazard category. This number is a summation of the hazards possible for the components which are considered potentially hazardous. The first number relates to the potential hazard from the vapors and will vary from 00 through 0, 1, 2, 3, 4, 5, to 6 as the potential hazard increases. The second number varies similarly and relates to the potential hazard from direct contact with the skin, eyes, breathing system, and by ingestion. This second number will not be less than 1 if organic solvents are included in the product in significant amounts. The following is a list of several solvents which are considered less desirable based on this second number being 3 (or higher for higher concentrations in some cases): Toluene and Xylene at >10%, all common ethylene glycol based ethers and their acetates (including diethylene glycol monobutyl ether, for example), terpenes, monomers at rather low concentrations, amines at moderate concentrations, and the most common chlorinated solvents. A "3" in this category places the protection required in a significantly higher category with requirements for gloves as a minimum and frequently fresh air masks as well.

As indicated, a two-digit MAL code defines which safety precautions are required for each of a large number of processing operations and conditions, including interior and exterior painting and gluing, whether or not large surfaces are involved, the quality of ventilation provided, surface preparation, painting of ships, larger construction sites, each of the printing processes, and industrial coating (spray boxes, cabinets, etc.).[11] The protective measure required may be a face guard, eye protection, a dust mask, a gas filter mask, a combination filter mask, a fresh air supplied mask, or a body suit, in order of increasing requirements.

Examples of complete labeling of products and solvents are beyond the scope of this chapter. The purpose of the discussion is to suggest that possible substituting solvents can be listed, such as in Table 10.1, in an attempt to find a substitute with a lower labeling requirement.

Systematic consideration of labeling requirements is becoming a significant parameter in commercial applications of solvents and products containing solvents. This is happening all over the world, both with regard to worker safety as well as to the external environment. Such a procedure has been used to arrive at optimum commercially useful solvent compositions with the lowest possible risk for workers in the serigraphic printing industry as described in Danish patents DK 153797B (1989) and DK 160883 (1991) which correspond to European patents EP 0 205 505 B1 and EP 0 270 654 B1.[13,14] The preferred compositions reduce the MAL number to a minimum and also consider lowest possible internationally required labels as a requirement. The low label requirements of DBE (dibasic esters) have been emphasized in comparison with other solvents.[15] Systematic solvent selection procedures have also been strongly suggested for use in the selection of solvents for restoring older paintings.[16] This is discussed in Chapter 3.

SELECTION OF CHEMICAL PROTECTIVE CLOTHING

HSP correlations for barrier properties of some types of chemical protective clothing are given in Chapter 8, Table 8.1. These correlations are based on data presented by Forsberg and Kieth.[17] Other

TABLE 10.1
Fresh Air Numbers (FAN/MAL) for Selected Solvents from the Danish MAL Labeling System[a]

FAN/MAL	Solvent	FAN/MAL	Solvent
1400	Chloroform	20	*n*-Propanol
1100	Tetrachloromethane	19	Propylene glycol monomethyl ether acetate
880	Benzene	17	Propyl acetate
110	Dichloromethane	15	Propylene glycol monopropyl ether
88	Trichloroethylene	14	Mineral spirits/white spirit
78	*n*-Hexane	14	Butyl acetates
74	Toluene	13	Ethyl acetate
58	C_9 Aromatics	13	Cyclohexane
54	Methanol	13	Benzin (petroleum ether as heptane)
48	Methyl ethyl ketone	12	Heptane
46	Xylene	7	Ethanol
29	2-Propanol	6	Propylene glycol monobutyl ether
28	Propylene glycol monomethyl ether	5	Dipropylene glycol monomethyl ether
26	1,1,1-Trichloroethane	4[b]	DBE (Dibasic esters)[13]
25	$C_{>9}$ Aromatics	0	Ethylene glycol
24	Tetrahydrofurane	0	Propylene glycol
23	Acetone		

[a] These numbers are developed primarily with regard to health hazards from vapors. The second number in the FAN code is added for hazard for skin contact, eye contact, respiratory system contact, and/or ingestion. In addition to these, the European Union requires use of X_i, X_n, C, T, etc. Several of the solvents require such labeling as well. One must also consider R (risk) and S (protective measure) labeling requirements.

[b] Estimated from composition of the mixed solvent.

examples of HSP correlations of barrier properties of protective clothing are discussed in Chapter 8. Earlier publications also include HSP correlations of barrier properties of chemical protective clothing.[18-21]

The procedure for using these correlations requires knowledge of the HSP of the chemicals involved. These may be found in a suitable table or can be calculated according to the procedures outlined in Chapter 1. One then evaluates the RED number for the situation of interest. The RED number is discussed in Chapter 1 (Equation 1.10). If this number is less than 1, the system in not expected to be suitable for use. If the RED number is close to 1.0, there may be some doubt about the recommendation. RED numbers significantly greater than 1.0 can be considered for use. As discussed in Chapter 8, the molecular size of the chemical involved is important in these evaluations.

The major use of such correlations is to evaluate potential barrier types for chemicals where test results are not available. One can usually divide the results into groups of clearly not acceptable, questionable, and worthy of further consideration.

There have been recent attempts to improve on the direct correlation of breakthrough times and permeation rates with HSP by trying to estimate the solubility and diffusion coefficients separately using HSP.[22-25] These efforts have been discussed in Chapters 2 and 8.

UPTAKE OF CONTENTS BY A PLASTIC CONTAINER

Plastic containers have become increasingly popular in recent years. They have many advantages (which will not be discussed here), but there is also one disadvantage which HSP can shed more light on. This is the fact that plastic materials are able to absorb various liquids to some extent.

The extent of absorption clearly depends on the HSP of the plastic used in the container compared with the HSP of the liquid which is in contact with it. Containers in contact with food have been tested well for suitability for this purpose, including barrier properties relative to the contents. This is not the point of the present discussion. A problem exists with the inadvertent storage of hazardous liquids in the plastic container prior to its expected recycling as a container for a food or beverage. Many types of liquids can be temporarily stored in such containers. Whereas the earlier glass or metal containers could not absorb potentially dangerous materials, a plastic container can do this. A simple washing operation cannot be expected to remove all of the absorbed material. Washing only removes what is on the surface or what can diffuse to the surface during the washing process, which presumably takes place at some higher temperature.

HSP concepts can focus attention on the types of chemicals which can absorb into a given type of plastic container. This is useful information in terms of what analyses should be performed prior to recycling. The principles discussed here can possibly contribute in other ways to improve the recycling process based on the increased level of knowledge. There may be other ways to reduce the problem.

SKIN PENETRATION

Human skin is a complicated system. Nevertheless, it has been possible to characterize some aspects of the behavior of human skin by HSP. The HSP found in a correlation of permeation rates of liquids in contact with viable skin[26] are similar to those found for the swelling of psoriasis scales.[27,28] This has been discussed in Chapter 9 in more detail, but also relates to worker safety. The HSP for these correlations are included in Chapter 8, Table 8.1 and Chapter 9, Table 9.1.

A skin penetration warning has been attached to many liquids taken up in the lists of limiting values for workplaces which are published in different countries. It was found earlier based on the HSP correlation with the swelling of psoriasis scales that this practice could be misleading, since HSP predicted many liquids without this warning also swelled psoriasis scales (keratin) and could therefore be expected to penetrate the skin.[27] The lack of a skin penetration warning for these liquids is partly attributable to the fact that this warning is based on experience. The bad experience giving the warning includes a combination of all effects, most notably the combination of dose and toxicity, rather than the potential dose effect only which is indicated by similarity of HSP. Earlier discussions also led to the impression that those involved in this area did not consider the swelling of psoriasis skin as having relevance to the permeation of living skin. The finding that comparable δ_P and δ_H are found from correlating the permeation rates of solvents through living skin is a new input into this discussion. It is recognized that the δ_D parameter is different, but reasons for this are not clear. An improved HSP correlation of the permeation rates of solvents through living skin based on a larger number of solvents than the 13 included in the work of Ursin et al.[26] is perhaps required to give improved predictions in marginal cases, i.e., those near the boundaries of the HSP sphere describing the situation. The size and shape of the penetrating liquid molecules must also be considered.

Predictions of the barrier properties of viable human skin should receive more attention. In addition, there is some discussion of the use of HSP in this respect in Chapter 9.

CONCLUSION

In conclusion, it can be noted that HSP provides a tool to aid in substitution and in systematic formulation of less hazardous products and processes. One can also use HSP to more rapidly arrive at an optimum choice of chemical protective clothing. HSP provides other insights with regard to uptake of undesirable chemicals in the human skin, in packaging materials, and perhaps even in a wide variety of other materials such as those found in nature.

REFERENCES

1. Goldschmidt, G., An Analytical Approach for Reducing Workplace Health Hazards Through Substitution, *Am. Ind. Hyg. Assoc. J.*, 54, 36–43, January 1993.
2. Olsen, E., Substitution: A Method to Fulfill the Working Environment Law for Air Quality (in Danish: Substitution: En Metode til at Overholde Arbejdsmiljoelovens Krav til Luftkvaliteten, *Dan. Kemi,* 67(5), 146–153, 1986.
3. Soerensen, F. and Petersen, H. J. S., Substitution of Hazardous Chemicals and the Danish Experience, *Occup. Hyg.*, 1, 261–278, 1995.
4. Filskov, P., Goldschmidt, G., Hansen, M. K., Höglund, L., Johansen, T., Pedersen, C. L., and Wibroe, L., Substitution in Practice — Experience from BST (in Danish: Substitution i Praksis — Erfaringer fra BST), Arbejsmiljøfondet, Copenhagen, 1989.
5. Hansen, C. M., Solvents in Water-Borne Coatings, *Ind. Eng. Chem. Prod. Res. Dev.,* 16(3), 266–268, 1977.
6. Hansen, C. M., Organic Solvents in High Solids and Water-Reducible Coatings, *Prog. Org. Coat.,* 10(3), 331–352, 1982.
7. Holten-Andersen, J. and Hansen, C. M., Solvent and Water Evaporation from Coatings, *Prog. Org. Coat.,* 11(3), 219–240, 1983.
8. Saarnak, A. and Hansen, C. M., Evaporation from High Solids Coatings (in Swedish: Avdunstningen Från LF-Fårger), *Färg och Lack,* 30(5), 100–105, 1984.
9. Hansen, C. M., Solvents for Coatings, *Chem. Technol.,* 2(9), 547–553, 1972.
10. Wu, D. T., Formulating Solvents to Remove Hazardous Air Polutants, *Polym. Paint Colour J.*, 185, 20–23, December 1995.
11. Anonymous, Directive on Work with Products Having Codes (in Danish: *Bekendtgørelse om arbejde med kodenummererede produkter, Arbejdstilsynets Bekendtgørelse nr.302 af 13. maj 1993*), Danish Directorate for Labor Inspection.
12. Anonymous, Directive on Determination of Codes (in Danish: *Bekendtgørelse om fastsættelse af kodenumre, Arbejdstilsynets bekendtgørelse nr. 301 af 13 maj 1993*), Danish Directorate for Labor Inspection.
13. Madsen, C. H. and Hansen, C. M. EP 0 205 505 B1, 1988. Assigned to CPS Kemi Aps (Now a part of the Autotype Concern).
14. Madsen, C. H. and Hansen, C. M. EP 0 270 654 B1, 1991. Assigned to CPS Kemi Aps (Now a part of the Autotype Concern).
15. Altnau, G., Risikopotentiale von Lösemitteln systematisch bewerten, *Farbe+Lack*, 103(9), 34–37, 1997; Systematic Evaluation of Risk Potentials of Solvents, *Eur. Coat. J.*, 6/98, 454–457, 1998.
16. Hansen, C. M., Conservation and Solubility Parameters, *Nordic Conservation Congress Preprints*, Copenhagen, 1994, 1–13.
17. Forsberg, K. and Kieth, L. H., *Chemical Protective Clothing Performance Index*, 4th ed., Instant Reference Sources, Inc., Austin, TX, 1991.
18. Hansen, C. M., The Systematic Choice of Material in Personal Protection Equipment for Organic Solvents in Safety and Health Aspects of Organic Solvents, *Progress in Chemical and Biological Research* 220, Riihimäki, V. and Ulfvarson, U., Eds., Alan R. Liss, New York, 1986, 297–302.
19. Hansen, C. M. and Hansen, K. M., Which Gloves Should I Put On? (in Danish: Hvilke Handsker Skal Jeg Tage På?), *Färg och Lack*, 33(3), 45–49, 1987.
20. Hansen, C. M. and Hansen, K. M., Solubility Parameter Prediction of the Barrier Properties of Chemical Protective Clothing, *Performance of Protective Clothing: Second Symposium,* ASTM STP 989, Mansdorf, S. Z., Sager, R., and Nielsen, A. P., Eds., American Society for Testing and Materials, Philadelphia, 1988, 197–208.
21. Hansen, C. M., Billing, C. B., and Bentz, A. P., Selection and Use of Molecular Parameters to Predict Permeation Through Fluoropolymer-Based Protective Clothing Materials, *The Performance of Protective Clothing*; Fourth Volume, ASTM STP 1133, McBriarty, J. P. and Henry, N. W., Eds., American Society for Testing and Materials, Philadelphia, 1992, 894–907.
22. Zellers, E. T., Three-Dimensional Solubility Parameters and Chemical Protective Clothing Permeation. I. Modeling the Solubility of Organic Solvents in Viton® Gloves, *J. Appl. Polym. Sci.,* 50, 513–530, 1993.

23. Zellers, E. T. and Zhang, G.-Z., Three-Dimensional Solubility Parameters and Chemical Protective Clothing Permeation. II. Modeling Diffusion Coefficients, Breakthrough Times, and Steady-State Permeation Rates of Organic Solvents in Viton® Gloves, *J. Appl. Polym. Sci.,* 50, 531–540, 1993.
24. Zellers, E. T., Anna, D. H., Sulewski, R., and Wei, X., Critical Analysis of the Graphical Determination of Hansen's Solubility Parameters for Lightly Crosslinked Polymers, *J. Appl. Polym. Sci.,* 62, 2069–2080, 1996.
25. Zellers, E. T., Anna, D. H., Sulewski, R., and Wei, X., Improved Methods for the Determination of Hansen's Solubility Parameters and the Estimation of Solvent Uptake for Lightly Crosslinked Polymers, *J. Appl. Polym. Sci.,* 62, 2081–2096, 1996.
26. Ursin, C., Hansen, C. M., Van Dyk, J. W., Jensen, P. O., Christensen, I. J., and Ebbehoej, J., Permeability of Commercial Solvents through Living Human Skin, *Am. Ind. Hyg. Assoc. J.,* 56, 651–660, 1995.
27. Hansen, C. M., The Absorption of Liquids into the Skin, Report No. T 3-82, Scandinavian Paint and Printing Ink Research Institute, Hoersholm, Denmark, 1982.
28. Hansen, C. M. and Andersen, B. H., The Affinities of Organic Solvents in Biological Systems, *Am. Ind. Hyg. Assoc. J.,* 49(6), 301–308, 1988.

11 The Future

CONTENTS

ABSTRACT

Hansen solubility parameters (HSP) help to quantify the statements "Like Dissolves Like" and "Like Seeks Like." These parameters have found use in many fields of research and practice, primarily because their unique predictive capabilities are based on sound theoretical principles. HSP have extended the original Hildebrand single solubility parameter approach by quantitatively taking into account the molecular permanent dipole–permanent dipole and molecular hydrogen bonding (electron interchange) interactions. HSP and the Prigogine corresponding states theory of polymer solutions are mutually confirming with regard to treatment of specific interactions, as shown in Chapter 2. This is important, since it confirms that the HSP correlations must continue to include a constant not too different from the currently used "4" (or 0.25). This is necessary to differentiate between the atomic (δ_D) and the molecular (specific) interactions (δ_P and δ_H). Neglecting this differentiation will lead to misinterpretations. The geometric mean average for the interaction of unlike molecules is inherently used in the Hildebrand approach and in the HSP approach as well. This same mean must be used in the Prigogine corresponding states theory if agreement is to be found with the HSP correlations presented in this book. Since the agreement is general, the conclusion must be that the geometric mean can be used to average not only dispersion interactions, but also those attributable to permanent dipoles and to hydrogen bonding.

Based on the large number of current uses of HSP, one can easily suppose that there are many more practical uses which remain to be discovered and developed. One need not necessarily extend its theoretical scope to accomplish this. The existing data can be used in a strictly empirical manner if so desired. However, a glimpse has been given of a very general energetic approach to systematically predict and control molecular interactions among many materials of widely different composition. The general predictions possible for these physical interactions have been demonstrated for both bulk phenomena (solubility, swelling, compatibility) and surface phenomena (adsorption, dewetting, spontaneous spreading). In the future, the theory should be explored and used with this general applicability in mind.

Problems and situations clearly needing further attention are discussed, including calculation of HSP for polymers; difficulties in obtaining reliable data; potential experimental problems, and expanding applications into areas including water, gases, organic and inorganic salts, organometallic compounds, fragrances and aromas; and the uptake (and later release) of potentially dangerous chemicals in various types of materials.

INTRODUCTION

There are many matters related to HSP which still need clarification and expansion. Some limitations are clear, but others are not so clear. Since this book is primarily directed toward the practitioner, the following discussions will start with more practical topics.

The first matter of concern is the availability of data. This book attempts to help improve this situation by publishing HSP for a large number of liquids in the Appendix, Table A.1. The book also contains data for many HSP correlations which have never been published before. Many of these are given as examples in the text, and others are included in the Appendix, Table A.2. Other sources are discussed below.

The second matter of concern is how reliable the HSP data are and how accurately the correlations can predict the behavior of untested systems. Qualitative indications of this for the data generated by the author are given in the relevant tables.

A third point which is sometimes irritating is that the scope of the characterizations possible is limited to the cohesive energy spectrum of the test liquids. A situation is often met where only a few solvents having high solubility parameters dissolve a polymer which has still higher HSP. Similarly, only a few solvents may interact intimately with a surface which has very high HSP. These surfaces are clearly wet because of the lower surface tension of all of the liquids, but only a few with high HSP prolong suspension of finer particles, for example. The energy characteristics of such surfaces are apparently higher than those of any liquids which can be used to study them by these techniques. Very high cohesive energies lead to the formation of solids, so there are no pure liquids which can be used to test the very high energy materials. New thinking and new techniques are required to accurately characterize such high energy materials. A full understanding of the behavior of water, organometallic materials, and salt solutions might be helpful in these situations (see the following corresponding sections). The current practice is to extrapolate into the region of very high HSP using Chapter 1, Equation 1.9 which includes the constant "4." It is assumed that this constant is still valid, even for these very high energy characterizations. The given good solvents are often in the boundary region of the HSP spherical characterizations. The solubility of Dextran C (British Drug Houses)[1] is an example of this as shown in Tables 11.1 and 11.2. See Chapters 3 and 5 for further discussion of this problem which is present for both polymers and particulate matter. In a sense, the problem is similar to measuring the surface tension of a surface which has such a high value that even water spontaneously spreads on it. One can only conclude that its surface tension is greater than that of water. In the present case, there is a model to extrapolate HSP to higher values than can be measured directly.

TABLE 11.1
Calculated Solubility Sphere for Dextran C Solubility

D = 24.3 P = 19.9 H = 22.5 RAD. = 17.4 FIT = 0.999 NO = 50

The solvents with their parameters

No.	Solvent	D	P	H	SOLUB	RED	V
4	Acetone	15.5	10.4	7.0	0	1.454	74.0
6	Acetophenone	19.6	8.6	3.7	0	1.371	117.4
10	Aniline	19.4	5.1	10.2	0	1.241	91.5
12	Benzaldehyde	19.4	7.4	5.3	0	1.346	101.5
13	Benzene	18.4	0.0	2.0	0	1.776	89.4
27	1,3-Butanediol	16.6	10.0	21.5	0	1.054	89.9
28	1-Butanol	16.0	5.7	15.8	0	1.313	91.5
30	Butyl acetate	15.8	3.7	6.3	0	1.640	132.5
37	gamma-Butyrolactone	19.0	16.6	7.4	0	1.077	76.8
39	Carbon disulfide	20.5	0.0	0.6	0	1.756	60.0
40	Carbon tetrachloride	17.8	0.0	0.6	0	1.858	97.1
41	Chlorobenzene	19.0	4.3	2.0	0	1.601	102.1
44	Chloroform	17.8	3.1	5.7	0	1.556	80.7
46	*m*-Cresol	18.0	5.1	12.9	0	1.246	104.7
48	Cyclohexanol	17.4	4.1	13.5	0	1.312	106.0
49	Cyclohexanone	17.8	6.3	5.1	0	1.473	104.0
56	Diacetone alcohol	15.8	8.2	10.8	0	1.363	124.2
61	*o*-Dichlorobenzene	19.2	6.3	3.3	0	1.474	112.8
62	2,2-Dichlorodiethyl ether	18.8	9.0	5.7	0	1.313	117.2
75	Diethylene glycol	16.6	12.0	20.7	0	1.000	94.9
82	Diethyl ether	14.5	2.9	5.1	0	1.795	104.8
90	Dimethyl formamide	17.4	13.7	11.3	0	1.082	77.0
94	Dimethyl sulfoxide	18.4	16.4	10.2	1*	1.000	71.3
96	1,4-Dioxane	19.0	1.8	7.4	0	1.485	85.7
98	Dipropylene glycol	16.5	10.6	17.7	0	1.080	130.9
104	Ethanol	15.8	8.8	19.4	0	1.180	58.5
105	Ethanolamine	17.0	15.5	21.2	1	0.880	59.8
106	Ethyl acetate	15.8	5.3	7.2	0	1.559	98.5
120	Ethylene dichloride	19.0	7.4	4.1	0	1.416	79.4
121	Ethylene glycol	17.0	11.0	26.0	1*	1.003	55.8
123	Ethylene glycol monobutyl ether	16.0	5.1	12.3	0	1.406	131.6
124	Ethylene glycol monoethyl ether	16.2	9.2	14.3	0	1.211	97.8
126	Ethylene glycol monomethyl ether	16.2	9.2	16.4	0	1.170	79.1
131	Formamide	17.2	26.2	19.0	1	0.915	39.8
136	Glycerol	17.4	12.1	29.3	1	0.991	73.3
140	Hexane	14.9	0.0	0.0	0	2.037	131.6
148	Isophorone	16.6	8.2	7.4	0	1.410	150.5
153	Methanol	15.1	12.3	22.3	0	1.144	40.7
162	Methylene dichloride	18.2	6.3	6.1	0	1.411	63.9
166	Methyl isobutyl carbinol	15.4	3.3	12.3	0	1.517	127.2
167	Methyl isobutyl ketone	15.3	6.1	4.1	0	1.679	125.8
177	Nitrobenzene	20.0	8.6	4.1	0	1.336	102.7
179	Nitromethane	15.8	18.8	5.1	0	1.399	54.3
181	2-Nitropropane	16.2	12.1	4.1	0	1.479	86.9
204	Propylene carbonate	20.0	18.0	4.1	0	1.172	85.0
205	Propylene glycol	16.8	9.4	23.3	0	1.053	73.6

TABLE 11.1 (continued)
Calculated Solubility Sphere for Dextran C Solubility

No.	Solvent	D	P	H	SOLUB	RED	V
222	Tetrahydrofuran	16.8	5.7	8.0	0	1.450	81.7
223	Tetrahydronaphthalene	19.6	2.0	2.9	0	1.618	136.0
225	Toluene	18.0	1.4	2.0	0	1.744	106.8
229	Trichloroethylene	18.0	3.1	5.3	0	1.560	90.2

Another concern related to reliable HSP values is based on the fact that most chemicals in the intermediate molecular weight range, such as that characteristic of plasticizers, are soluble in almost all of the test liquids, except for, for example, glycerin, water, and hexane. It is impossible to establish the three HSP based on such data. One generally has to rely on group contribution methods or other calculations or comparisons, and there will be some uncertainty involved with this.

Once the necessary reliable HSP data are available, decisions and ideas are needed on how the data should be used. It is here that the existing theory and future extensions of it are most important. In many cases, engineering approximations leading to a systematic course of action have been possible using data which is currently available. One can often arrive at a prediction for expected behavior using the "like seeks like" principle, even though accurate numbers and an appropriate detailed theory may be lacking. It is hoped that this book will aid in the generation of still more HSP data having a uniformly high quality, such that the interactions among still more materials can be predicted. Logical applications for HSP will be found in self-assembling systems, for example. One example is the self-stratifying paints discussed in Chapter 6, and another is the ultrastructure of cell walls in wood discussed in Chapter 9.

HANSEN SOLUBILITY PARAMETER DATA AND DATA QUALITY

The author and others including most solvent suppliers and some paint companies (at least) have databases including HSP data for solvents and HSP correlations for polymer solubility etc. Tables of HSP data for many materials are also included in standard reference works.[2-5] There is still a tendency to regard the contents of such databases as proprietary information for the benefit of the owner and/or his/her customers. Exxon, for example, has indicated a computer program based on HSP where data for over 500 solvents and plasticizers, 450 resins, and 500 pesticides are included.[6,7] The use of these parameters is becoming so commonplace that, in many studies, the δ_D, δ_P, and δ_H parameters appear without any specific reference to where they came from or what they actually represent.

The solvent listing in the Appendix, Table A.1 includes the previously published set of some 240 solvents which have appeared earlier in several sources.[2,4,5,8,9] Some of the values have been revised over the years. There are many additions to this original set of data. The additions are primarily based on the calculational procedures described in Chapter 1. The calculated values have been checked against performance data reported in the literature where this has been possible. An example is the solubility data reported for poly(vinylidine chloride) (PVDC).[10] Appendix, Table A.1 also includes HSP for a number of low molecular weight solids. Low molecular weight solids with relatively low melting points have been treated as if they were liquids for extrapolation of latent heats to 25°C. This seems to be satisfactory, and it is consistent with the treatment of high boiling liquids. See Chapter 1 for details of the calculations.

HSP correlations in addition to those given in connection with examples in the text are included in Appendix, Table A.2. Only data judged (reasonably) reliable are reported. There are limitations on the accuracy of the HSP data derived from Burrell's solvent range studies reported in standard

TABLE 11.2
Calculated Solubility Sphere for Dextran C Solubility

D = 24.3 P = 19.9 H = 22.5 RAD. = 17.4 FIT = 0.999 NO = 50

No.	Solvent	D	P	H	SOLUB	RED	V
217	Succinic anhydride	18.6	19.2	16.6		0.739	66.8
234	Triethanolamine	17.3	22.4	23.3		0.819	133.2
93	Dimethyl sulfone	19.0	19.4	12.3		0.846	75.0
117	Ethylene cyanohydrin	17.2	18.8	17.6		0.866	68.3
213	2-Pyrolidone	19.4	17.4	11.3		0.867	76.4
105	Ethanolamine	17.0	15.5	21.2	1	0.880	59.8
131	Formamide	17.2	26.2	19.0	1	0.915	39.8
70	Diethanolamine	17.2	10.8	21.2		0.972	95.9
14	1,3-Butanediol	18.0	8.4	21.0		0.984	87.5
136	Glycerol	17.4	12.1	29.3	1	0.991	73.3
94	Dimethyl sulfoxide	18.4	16.4	10.2	1	1.000	71.3
75	Diethylene glycol	16.6	12.0	20.7	0	1.000	94.9
121	Ethylene glycol	17.0	11.0	26.0	1*	1.003	55.8
205	Propylene glycol	16.8	9.4	23.3	0	1.053	73.6
27	1,3-Butanediol	16.6	10.0	21.5	0	1.054	89.9
81	Diethylenetriamine	16.7	13.3	14.3		1.063	108.0
236	Triethyleneglycol	16.0	12.5	18.6		1.068	114.0
37	gamma-Butyrolactone	19.0	16.6	7.4	0	1.077	76.8
98	Dipropylene glycol	16.5	10.6	17.7	0	1.080	130.9
90	Dimethyl formamide	17.4	13.7	11.3	0	1.082	77.0
8	Allyl alcohol	16.2	10.8	16.8		1.117	68.4
154	*o*-Methoxyphenol	18.0	8.2	13.3		1.121	109.5
139	Hexamethylphosphoramide	18.5	8.6	11.3		1.132	175.7
118	Ethylenediamine	16.6	8.8	17.0		1.136	67.3
153	Methanol	15.1	12.3	22.3	0	1.144	40.7
135	Furfuryl alcohol	17.4	7.6	15.1		1.144	86.5
243	Trimethylphosphate	16.7	15.9	10.2		1.147	115.8
17	Benzyl alcohol	18.4	6.3	13.7		1.152	103.6
116	Ethylene carbonate	19.4	21.7	5.1		1.152	66.0
197	Phenol	18.0	5.9	14.9		1.167	87.5
126	Ethylene glycol monomethyl ether	16.2	9.2	16.4	0	1.170	79.1
204	Propylene carbonate	20.0	18.0	4.1	0	1.172	85.0
104	Ethanol	15.8	8.8	19.4	0	1.180	58.5
218	1,1,2,2-Tetrabromoethane	22.6	5.1	8.2		1.199	116.8
124	Ethylene glycol monoethyl ether	16.2	9.2	14.3	0	1.211	97.8
89	*N,N*-Dimethyl acetamide	16.8	11.5	10.2		1.215	92.5
45	3-Chloro-1-propanol	17.5	5.7	14.7		1.216	84.2
141	Hexylene glycol	15.7	8.4	17.8		1.219	123.0
172	Methyl-2-pyrrolidone	18.0	12.3	7.2		1.220	96.5
134	Furfural	18.6	14.9	5.1		1.230	83.2
10	Aniline	19.4	5.1	10.2	0	1.241	91.5
46	*m*-Cresol	18.0	5.1	12.9	0	1.246	104.7
199	1-Propanol	16.0	6.8	17.4		1.250	75.2
15	Benzoic acid	18.2	6.9	9.8		1.258	100.0
238	Triethylphosphate	16.7	11.4	9.2		1.259	171.0
214	Quinoline	19.4	7.0	7.6		1.265	118.0
79	Diethylene glycol monoethyl ether	16.1	9.2	12.2		1.272	130.9
3	Acetic anhydride	16.0	11.7	10.2		1.277	94.5

TABLE 11.2 (continued)
Calculated Solubility Sphere for Dextran C Solubility

No.	Solvent	D	P	H	SOLUB	RED	V
232	Tricresyl phosphate	19.0	12.3	4.5		1.277	316.0
132	Formic acid	14.3	11.9	16.6		1.284	37.8
224	Tetramethylurea	16.7	8.2	11.0		1.285	120.4

reference works,[2,11,12] but many correlations based on these data are included for reference anyway. The solvent range chosen for the studies does not completely fill out the possibilities selection of different liquids would have allowed. The problem of estimating a sphere based on limited data which do not experimentally define the whole sphere becomes more acute. This problem is greatest for polymers with high HSP, since not only is there a lack of possible data, but much of the volume of the HSP sphere is located where there are no liquids. The cohesion energies are so high that no liquids are possible and only solids are present. An example of a good HSP correlation from the solvent range studies is that of polyethylene sulfide. This polymer has relatively low HSP, and the solvents in the test series provide nonsolvents at higher HSP than those of the polymer to locate the boundaries with sufficient accuracy. This is shown in Tables 11.3 and 11.4. A comparison of the solvents included in Table 11.3 with those in Table 11.1 shows which ones are lacking in the high HSP range. An example of a poor correlation using solvent range data is that of the solubility of polyvinyl alcohol. Only two of the solvents, ethanol and 2-propanol, dissolve it. This leads to a correlation with the following data: δ_D;δ_P;δ_H;Ro equal to 17.0;9.0;18.0;4.0 with a perfect fit for two good solvents out of 56 in the set of data. The use of these data is not recommended. Ro is clearly too small by comparison with Ro found in HSP correlations for solubility for other water-soluble polymers.

One of the problems with some of the reported correlations in the Appendix, Table A.2 is that the data on which they are based were not generated for this purpose. There are shortcomings in terms of lack of full coverage of the HSP space as well as in the total number of liquids for which there are data. Note that a standard set of test solvents such as that used in Table 11.1 takes full coverage into account. However, some of these liquids must be handled with care for reasons of toxicity. Data for chemical resistance, permeation, and other phenomena related to solubility which can be correlated with HSP are practically never accumulated with an HSP correlation in mind. This does not prevent use of such data as demonstrated elsewhere in this book, but it does place some limitations on the reliability of the predictions obtainable from the correlations. A qualitative indication of the reliability of the correlations is given for this reason.

Reliable HSP data for many polymers of practical importance are not available at this time. It would seem advisable for raw material suppliers to determine the HSP for their relevant products in a reliable manner and to publish these data on their product data sheets or elsewhere. Including them in a possible future edition of this book may also be a possibility.

For the sake of completeness, a couple of warnings are appropriate before proceeding to the next section. As noted in Chapter 1, the three partial solubility parameters tabulated by Hoy[13,14] are not compatible with those of the author. As discussed in the next section, the group contribution procedure presented by van Krevelen and Hoftyzer[15] does not give satisfactory agreement with the procedures given in Chapter 1. Finally, water (or its mixtures) should not be included currently in any HSP correlations without a very careful analysis of the results. The small molecular volume, exceptionally high δ_H parameter, and tendency to self-associate depending on the local environment all lead to the likely result that water will be an outlier for the correlation and result in HSP values which are less reliable, and have lower predictive ability than had water been neglected. Mixtures of organic solvents with water are still more problematic when used as test liquids (see Figure 11.1

TABLE 11.3
Calculated Solubility Sphere for Polyethylenesulfide

D = 17.8 P = 3.8 H = 2.2 RAD. = 4.1 FIT = 0.981 NO = 56

The solvents with their parameters

No.	Solvent	D	P	H	SOLUB	RED	V
5	Acetic acid	14.5	8.0	13.5	0	3.352	57.1
7	Acetone	15.5	10.4	7.0	0	2.285	74.0
10	Acetonitrile	15.3	18.0	6.1	0	3.793	52.6
46	Aniline	19.4	5.1	10.2	0	2.125	91.5
52	Benzene	18.4	0.0	2.0	1	0.973	89.4
92	1-Butanol	16.0	5.7	15.8	0	3.462	91.5
103	*sec*-Butyl acetate	15.0	3.7	7.6	0	1.898	133.6
113	Butyraldehyde	14.7	5.3	7.0	0	1.947	88.5
122	Carbon tetrachloride	17.8	0.0	0.6	0	1.006	97.1
148	Chlorobenzene	19.0	4.3	2.0	1	0.600	102.1
814	*p*-Chlorotoluene	19.1	6.2	2.6	0*	0.869	118.3
169	*m*-Cresol	18.0	5.1	12.9	0	2.631	104.7
181	Cyclohexane	16.8	0.0	0.2	0	1.155	108.7
188	Cyclopentanone	17.9	11.9	5.2	0	2.107	89.1
224	1,2-Dichloro ethylene (*cis*)	17.0	8.0	3.2	1*	1.123	75.5
234	*o*-Dichlorobenzene	19.2	6.3	3.3	1	0.954	112.8
235	2,2-Dichlorodiethyl ether	18.8	9.0	5.7	0	1.605	117.2
236	Dichlorodifluoromethane (Freon 12)	12.3	2.0	0.0	0	2.771	92.3
242	Dichloromonofluoromethane	15.8	3.1	5.7	0	1.308	75.4
252	Diethyl amine	14.9	2.3	6.1	0	1.744	103.2
255	Diethyl ether	14.5	2.9	5.1	0	1.772	104.8
263	Diethylene glycol	16.6	12.0	20.7	0	4.970	94.9
203	Di-isobutyl ketone	16.0	3.7	4.1	0*	0.993	177.1
285	*N,N*-Dimethyl acetamide	16.8	11.5	10.2	0	2.752	92.5
297	Dimethyl formamide	17.4	13.7	11.3	0	3.286	77.0
306	1,4-Dioxane	19.0	1.8	7.4	0	1.480	85.7
325	Ethanol	15.8	8.8	19.4	0	4.476	58.5
328	Ethyl acetate	15.8	5.3	7.2	0	1.604	98.5
345	2-Ethyl hexanol	15.9	3.3	11.8	0	2.521	156.6
363	Ethylene carbonate	19.4	21.7	5.1	0	4.491	66.0
368	Ethylene glycol	17.0	11.0	26.0	0	6.077	55.8
375	Ethylene glycol monobutyl ether	16.0	5.1	12.3	0	2.634	131.6
376	Ethylene glycol monoethyl ether	16.2	9.2	14.3	0	3.325	97.8
404	Furfural	18.6	14.9	5.1	0	2.825	83.2
405	Furfuryl alcohol	17.4	7.6	15.1	0	3.286	86.5
406	Glycerol	17.4	12.1	29.3	0	6.916	73.3
429	Isoamyl acetate	15.3	3.1	7.0	0	1.699	148.8
754	Isoamyl alcohol	15.8	5.2	13.3	0	2.898	109.4
440	Isopropyl acetate	14.9	4.5	8.2	0	2.043	117.1
456	Methanol	15.1	12.3	22.3	0	5.483	40.7
464	Methyl acetate	15.5	7.2	7.6	0	1.919	79.7
481	Methyl ethyl ketone	16.0	9.0	5.1	0	1.697	90.1
498	Methyl *n*-amyl ketone	16.2	5.7	4.1	0	1.019	139.8
532	Nitroethane	16.0	15.5	4.5	0	3.038	71.5
534	Nitromethane	15.8	18.8	5.1	0	3.852	54.3
540	Octane	15.5	0.0	0.0	0	1.551	163.5

TABLE 11.3 (continued)
Calculated Solubility Sphere for Polyethylenesulfide

No.	Solvent	D	P	H	SOLUB	RED	V
542	1-Octanol	17.0	3.3	11.9	0	2.401	157.7
550	Pentane	14.5	0.0	0.0	0	1.933	116.2
552	1-Pentanol	15.9	4.5	13.9	0	3.005	108.6
570	2-Propanol	15.8	6.1	16.4	0	3.642	76.8
577	Propionitrile	15.3	14.3	5.5	0	2.948	70.9
584	Propylene carbonate	20.0	18.0	4.1	0	3.655	85.0
604	Styrene	18.6	1.0	4.1	1	0.913	115.6
611	*t*-Butyl alcohol	15.2	5.1	14.7	0	3.317	95.8
618	Tetrahydronaphthalene	19.6	2.0	2.9	1	0.996	136.0
697	Xylene	17.6	1.0	3.1	1	0.724	123.3

and the following discussion). A goal of future work should be to be able to account for the behavior of water in a reliable manner, such that it can be included in studies leading to HSP correlations. The HSP values for water found from the correlation for total water solubility reported in Chapter 1 (Table 1.3) appears promising for some applications where the HSP values for water as a single molecule are clearly not applicable.

GROUP CONTRIBUTION METHODS

Suggested calculational procedures to arrive at the HSP for solvents are given in Chapter 1. The group contribution methods need expansion with new groups. New group contributions should be checked for reliability of the predictions in some way, which is not always possible within the timeframe of most projects. The group contribution values consistently used by the author are reported in Chapter 1. Values added over the years are appended to the original table which was attributable to Beerbower.[4,17,18] As stated previously, the group contributions tabulated by van Kevelen[15] have not been found reliable. The δ_D parameter, in particular, is not predicted well. The author chose not to use these at an early date, although many other authors have chosen to do so. The use of various predictive methods which arrive at different results has always been a problem. Koenhen and Smolders[19] evaluated various equations for predicting HSP.

Methods for reliable *a priori* calculation of the HSP for polymers are not available. This is a serious shortcoming. The author has tried several times to calculate the HSP for individual polymers using the same group contributions suggested for the liquids, and almost every time has ultimately resorted to experiment. Calculation of the radius of interaction is a particular problem in this respect. This is definitely an area requiring attention. Chapter 2 discusses some of the factors which must be taken into account when calculating the radius of interaction. If one consistently uses the same method of estimating HSP, it can be assumed that some of the inherent errors will not affect relative evaluations. Utracki and co-workers[20] estimated HSP for a number of polymers assuming their δ_D parameters were not different and group contributions for the δ_P and δ_H parameters. This is discussed in Chapter 3.

POLYMERS AS POINTS — SOLVENTS AS SPHERES

One way to possibly improve predicting the behavior of polymers is to consider them as points (or more accurately, spheres with very small radii of interaction which depend on molecular weight) rather than as spheres with large radii, as is presently done. A given solvent is assigned a rather large radius of interaction. This radius is larger for smaller molar volume in this inverted approach.

TABLE 11.4
Calculated Solubility Sphere for Polyethylenesulfide

D = 17.8 P = 3.8 H = 2.2 RAD. = 4.1 FIT = 0.981 NO = 56

No.	Solvent	D	P	H	SOLUB	RED	V
148	Chlorobenzene	19.0	4.3	2.0	1	0.600	102.1
697	Xylene	17.6	1.0	3.1	1	0.724	123.3
814	*p*-Chlorotoluene	19.1	6.2	2.6	0*	0.869	118.3
604	Styrene	18.6	1.0	4.1	1	0.913	115.6
234	*o*-Dichlorobenzene	19.2	6.3	3.3	1	0.954	112.8
52	Benzene	18.4	0.0	2.0	1	0.973	89.4
203	Di-isobutyl ketone	16.0	3.7	4.1	0*	0.993	177.1
618	Tetrahydronaphthalene	19.6	2.0	2.9	1	0.996	136.0
122	Carbon tetrachloride	17.8	0.0	0.6	0	1.006	97.1
498	Methyl *n*-amyl ketone	16.2	5.7	4.1	0	1.019	139.8
224	1,2-Dichloro ethylene (*cis*)	17.0	8.0	3.2	1*	1.123	75.5
181	Cyclohexane	16.8	0.0	0.2	0	1.155	108.7
242	Dichloromonofluoromethane	15.8	3.1	5.7	0	1.308	75.4
306	1,4-Dioxane	19.0	1.8	7.4	0	1.480	85.7
540	Octane	15.5	0.0	0.0	0	1.551	163.5
328	Ethyl acetate	15.8	5.3	7.2	0	1.604	98.5
235	2,2-Dichlorodiethyl ether	18.8	9.0	5.7	0	1.605	117.2
481	Methyl ethyl ketone	16.0	9.0	5.1	0	1.697	90.1
429	Isoamyl acetate	15.3	3.1	7.0	0	1.699	148.8
252	Diethyl amine	14.9	2.3	6.1	0	1.744	103.2
255	Diethyl ether	14.5	2.9	5.1	0	1.772	104.8
103	sec-Butyl acetate	15.0	3.7	7.6	0	1.898	133.6
464	Methyl acetate	15.5	7.2	7.6	0	1.919	79.7
550	Pentane	14.5	0.0	0.0	0	1.933	116.2
113	Butyraldehyde	14.7	5.3	7.0	0	1.947	88.5
440	Isopropyl acetate	14.9	4.5	8.2	0	2.043	117.1
188	Cyclopentanone	17.9	11.9	5.2	0	2.107	89.1
46	Aniline	19.4	5.1	10.2	0	2.125	91.5
7	Acetone	15.5	10.4	7.0	0	2.285	74.0
542	1-Octanol	17.0	3.3	11.9	0	2.401	157.7
345	2-Ethyl hexanol	15.9	3.3	11.8	0	2.521	156.6
169	*m*-Cresol	18.0	5.1	12.9	0	2.631	104.7
375	Ethylene glycol monobutyl ether	16.0	5.1	12.3	0	2.634	131.6
285	*N,N*-Dimethyl acetamide	16.8	11.5	10.2	0	2.752	92.5
236	Dichlorodifluoromethane (Freon 12)	12.3	2.0	0.0	0	2.771	92.3
404	Furfural	18.6	14.9	5.1	0	2.825	83.2
754	Isoamyl alcohol	15.8	5.2	13.3	0	2.898	109.4
577	Propionitrile	15.3	14.3	5.5	0	2.948	70.9
552	1-Pentanol	15.9	4.5	13.9	0	3.005	108.6
532	Nitroethane	16.0	15.5	4.5	0	3.038	71.5
297	Dimethyl formamide	17.4	13.7	11.3	0	3.286	77.0
405	Furfuryl alcohol	17.4	7.6	15.1	0	3.286	86.5
611	*t*-Butyl alcohol	15.2	5.1	14.7	0	3.317	95.8
376	Ethylene glycol monoethyl ether	16.2	9.2	14.3	0	3.325	97.8
5	Acetic acid	14.5	8.0	13.5	0	3.352	57.1
92	1-Butanol	16.0	5.7	15.8	0	3.462	91.5
570	2-Propanol	15.8	6.1	16.4	0	3.642	76.8
584	Propylene carbonate	20.0	18.0	4.1	0	3.655	85.0

TABLE 11.4 (continued)
Calculated Solubility Sphere for Polyethylenesulfide

No.	Solvent	D	P	H	SOLUB	RED	V
10	Acetonitrile	15.3	18.0	6.1	0	3.793	52.6
534	Nitromethane	15.8	18.8	5.1	0	3.852	54.3
325	Ethanol	15.8	8.8	19.4	0	4.476	58.5
363	Ethylene carbonate	19.4	21.7	5.1	0	4.491	66.0
263	Diethylene glycol	16.6	12.0	20.7	0	4.970	94.9
456	Methanol	15.1	12.3	22.3	0	5.483	40.7
368	Ethylene glycol	17.0	11.0	26.0	0	6.077	55.8
406	Glycerol	17.4	12.1	29.3	0	6.916	73.3

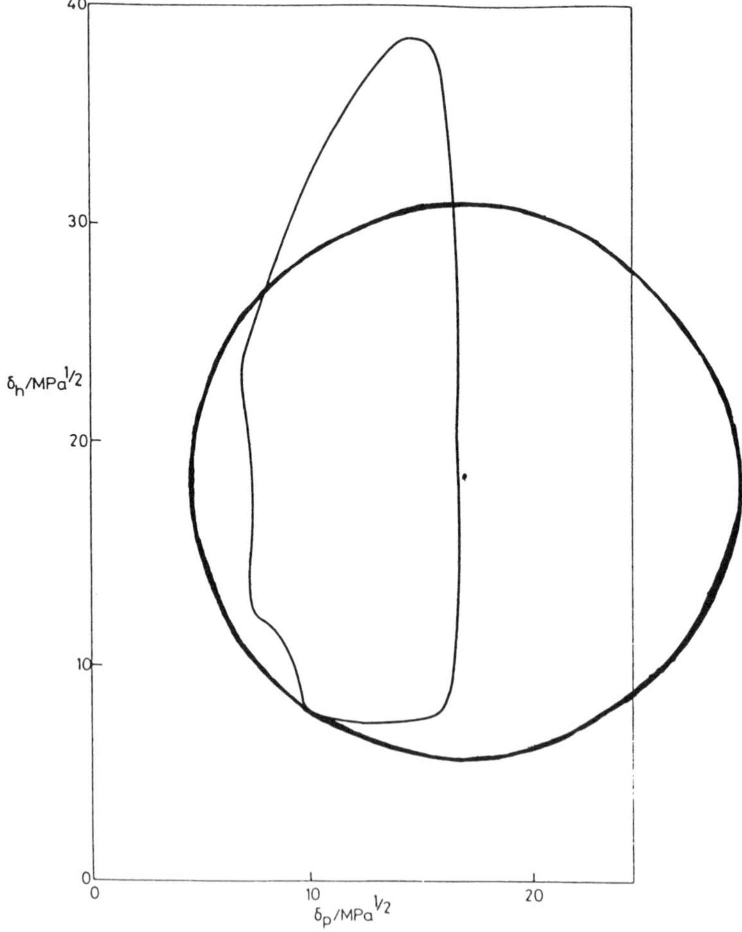

FIGURE 11.1 HSP plot of characterization of Rhodamin FB (C.I. Basic Violet 10) showing potential problems with incorporation of water mixtures as test solvents (see text for discussion). Data from Riedel.[16]

This idea was presented many years ago,[8,21] but it has never been fully explored. The first indications were that there seemed to be no real benefit in terms of improved reliability of predictions for polymer solubility in organic solvents, which was of primary interest, so there was no need to start all over again with this inverted system. On the other hand, there may be advantages in terms of more reliable prediction of polymer–polymer miscibility, for example. This was not explored. The requirement of polymer miscibility will be that the respective points (very small spheres) for the polymers must be very close to each other; comparing distances between small spheres is relatively easy. This type of comparison is sometimes difficult to make in the present approach where the degree of overlapping of rather large spheres is used to estimate polymer–polymer miscibility. No fixed rules of thumb have been established to estimate how much overlap is required for miscibility. However, guidelines for improving polymer–polymer miscibility are easily found in the present approach. These include selection of an improved solvent, reduction of polymer molecular weight, and modification of a polymer's HSP in a desired direction based on the HSP group contributions of its repeating unit or comonomers, for example.

CHARACTERIZING SURFACES

The characterization of surfaces with HSP, or perhaps more correctly cohesion parameters (having exactly the same numerical values), is still in its infancy. This possibility was demonstrated many years ago, however.[22] As shown in Chapters 4 and 5, this type of approach can lead to a new understanding of surface phenomena, which in turn allows systematic study and design of surfaces for desired behavior.

Data on surface characterizations, in addition to that in Chapters 4 and 5, are not provided here. This is primarily because such data are lacking, but also because surface cohesion parameters may not be reflected by nominal bulk composition. The same basic pigment or filler, for example, can have widely different surface cohesion parameters depending on how it has been surface treated. Neither has the effect of adsorbed water been clarified. Likewise, a surface characteristic for a polyvinyl chloride or a polyethylene cannot be expected to be valid for all polymers normally said to be of these compositions. There may also be additives which have different compositions and which may have migrated to the surfaces.

It appears that the relative simplicity of the surface characterizations discussed in this book would lead to their wider use. One current problem is that blindly entering wetting or spontaneous spreading data into the usual computer routine for finding the HSP values often leads to negative numbers for one or more of them. This was discussed in Chapter 4. Currently, the best approach is to compare plots or even to just compare tabulated data for the test solvents to determine where two surfaces differ in affinities. Guides for action can also be found by simple comparison of the HSP of those solvents which show a difference in behavior. A more systematic approach for the use of cohesion parameters to describe surface phenomena would be desirable.

MATERIALS AND PROCESSES SUGGESTED FOR FURTHER ATTENTION

Examples of the use of HSP for many types of materials and phenomena have been presented in earlier chapters. Some special types of materials are singled out here as worthy of still more attention in the near future. These include water, gases, organic and inorganic salts, organometallic materials, and aromatic (fragrances) materials. The uptake of potentially dangerous chemicals in recyclable packaging also needs attention. An additional area of interest may be found in that many commonly used reaction solvents have similar HSP. These include dimethyl sulfoxide, dimethyl formamide, dimethyl acetamide, and sulfolane, for example. It seems unlikely that this is a coincidence, in that the solubility of an activated species is probably determining the reaction rate. Still another area of major interest is the systematic formulation of filled systems using HSP. This is also still in its infancy. Pigments and fillers need to be characterized. Several of these applications are discussed in more detail below. Surface active materials have remained essentially untouched in terms of HSP, although Beerbower started on this many years ago.[8,17,23]

Surface Active Agents

Surface active agents have not been systematically characterized by HSP, although Beerbower has developed some aspects of a theory for given situations.[8,17,23] The statement "like seeks like" indicates that surface active agents should be extensively treated in terms of HSP. Each end of such molecules will require its own set of HSP, as demonstrated by the example of lithium stearate discussed later in the Organometallic Compounds section. Calculating HSP for hydrophobic ends can be done by the methods described in Chapter 1. Calculating HSP for the hydrophilic ends is more problematic for anionic and cationic surfactants, for example, but should be possible for nonionic surfactants. A major question here is "How much of a given molecule belongs to the hydrophilic portion?" This is also an area where much needs to be done. Solubility tests to determine the overall HSP for a surfactant are often plagued by the inaccuracies involved with their being too soluble. One finds that almost all of the test solvents dissolve (or appear to dissolve) such low molecular weight materials. This leads to uncertain HSP values using this procedure.

An example to help illustrate the type of analysis possible is to try to answer the question of where the hydrophobic end of a given surfactant will tend to preferentially reside. An aliphatic end group would have lower affinity for polystyrene, for example, than an aromatic one. Octane will not dissolve polystyrene, whereas toluene will. This is reflected by their cohesion energy parameters. This same reasoning applies to other polymers. A surfactant with a fluorinated end will not dissolve in many polymers where a hydrocarbon end will. The cohesion energy parameters characteristic of fluorocarbons are too low. While these examples are obvious to those skilled in the science of surfaces, they point to the possibility of quantifying affinities of surface active materials in terms of the cohesion energy parameters of their respective end groups. Those familiar with cohesion energy parameters can already discern differences which may improve the chances of success.

Surface Mobility (Self-Assembly)

The rule of thumb that "like seeks like" can be very useful in understanding the structure of complicated systems. That this type of consideration can lead to useful results can be seen in the way that the behavior of wood polymers and the ultrastructure of cell walls in wood was treated in Chapter 8 and in much more detail by Hansen and Björkman.[24] Hemicelluloses appear to function much like surfactants with the backbone and those side chains containing hydroxyl groups favoring placement toward cellulose (or their own kind). Hemicellulose side chains containing acetyl, acid, or ether groups are expected to favor orientation toward lignin regions. In this example, it is interfacial mobility which is in focus, and it can be expected that the orientations may be changed with the transport or presence of other materials such as water through a given local environment. These predictions and inferences appear to agree with what is expected or has been established by independent measurement, but it is too early to say that confirmation has been obtained independently. The treatment of different segments of block copolymers as separate entities is a related endeavor where more quantitative predictions of compatibility should be possible. It is known that additions of a block of C to both A and B improves their chances of compatibility (at some molecular level).

The rotation of some hydrophobic materials to become more hydrophilic when in contact with water is still another example of like seeking like. Peat moss is an example. A drop of water initially pearls on the surface, but shortly thereafter disappears into the interior in a spontaneous manner. The peat moss has become hydrophilic (but returns to the hydrophobic state on drying again). This phenomena was actually employed to develop an electrodeposition coating for an evaporator where film wetting by water was required for good evaporation efficiency.[25] After several hours of exposure of a fresh coating to water, the static contact angles with water disappeared and a coherent water film was obtained.

Many surface phenomena can be understood from the preferences of given segments or materials to seek out regions of similar HSP. Some inferences may be possible from the studies performed on compatible (or nearly compatible) polymers. The HSP data leading to formulation of self-stratifying coatings may also provide useful information[26] (see also Chapter 4). Systematic studies of these effects are badly needed. One such recent study[27] confirmed that the rotation ability (mobility) of aging polymer surfaces could be followed by measuring the (static) receding contact angle of water. Aging can be expected to lead to increased oxygenation and perhaps also to a decrease in average molecular weight. These effects both contribute to the tendency/ability for oxygenated species attached to an otherwise more hydrophobic polymer to rotate into an applied water droplet. When the (static) receding contact angle for water was measured, it fell with exposure time/aging at shorter times, while the (static) advancing contact remained constant. At longer exposure times, when the surface was oxygenated to a greater extent, the advancing contact angle also started to fall.

Surface mobility also has an important role in biological processes, as described in Chapter 9. The orientation of molecules to allow given segments to locate in regions of similar HSP is presumed to be a general phenomenon. Hydrophilic bonding (usually referred to in the present context as intermolecular hydrogen bonding) is responsible for the configurations of proteins in water. The proteins which can be dissolved in mixtures of water and urea or given salts, for example, are no longer "hydrogen bonded" in the conventional usage of the term, since they are now truly dissolved by an effectively good solvent which can also dissolve these segments/bonding sites. The usual solvent, water, does not have the correct set of HSP to truly dissolve these segments of the protein molecules. The urea additions correct for this deficiency, and the protein is said to be denatured in the process. The concept of hydrophilic (hyperphilic?) bonding, which is the opposite of hydrophobic bonding, is discussed in more detail with examples in Chapter 9.

Many of the concepts discussed here are directly applicable for self-assembling systems.

Water

The current treatment of the HSP for water discussed in Chapters 1 and 9 needs confirmation and/or modification. As noted earlier on several occasions, water is very special because of its low molecular volume, its very high δ_H parameter for a liquid, and its tendency to self-associate or to associate with other materials forming special structures. The HSP correlations for the solubility of solvents in water presented in Chapter 1, Table 1.3 have not been tested extensively as of yet, but do seem promising. They are clearly useful to make predictions for the solubility of untested solvents in water, but whether or not these HSP data for water can be used in a larger context remains to be determined. General behavior can be predicted, but can specific behavior be predicted? More research is needed in this area, but, in the meantime, water can be considered as having (at least) duality. Sometimes it acts like a single molecule, and sometimes it acts as a cluster of about six molecules (according to the HSP comparison, at least). There may also be other possibilities. The use of the HSP for water found from the correlation of total water solubility appears to be the most promising set of values to work with at the present time. This is especially true for water in lower energy systems.

It is not yet advisable to include water in a standard set of test liquids for experimental evaluation of the HSP for polymers or other materials because of its tendency to be an outlier. This means a challenge still exists to understand how to be able to incorporate water into a standard set of HSP test liquids without always being concerned about special interpretations for water, and only for water. An example of how this can lead to oddities is discussed in the following.

A characterization problem caused by nonideal mixtures with water is the interpretation of HSP correlations for materials such as the dye Rhodamin FB (C.I. Basic Violet 10).[2,16] Use of mixtures of solvent and water as test solvents led to a very nonspherical (noncircular) cohesion energy parameter plot (see Figure 11.1). The irregular plot can presumably still be used as such, but the characterization of the dye in question is not useful in relation to prediction of interactions with other materials. The plot has given several individuals the impression that there are significant problems with the HSP approach when it is applied to this kind of material. This is not true. A

TABLE 11.5
Calculated Solubility Sphere for Rhodamin FB

No.	Solvent	d_D	d_P	d_H	Solubility	RED	V
4	Acetone	15.5	10.4	7.0	0	1.125	74.0
13	Benzene	18.4	0.0	2.0	0	1.991	89.4
28	1-Butanol	16.0	5.7	15.8	0*	0.999	91.5
30	Butyl acetate	15.8	3.7	6.3	0	1.517	132.5
37	gamma-Butyrolactone	19.0	16.6	7.4	0*	0.988	76.8
47	Cyclohexane	16.8	0.0	0.2	0	2.076	108.7
56	Diacetone alcohol	15.8	8.2	10.8	1*	1.001	124.2
75	Diethylene glycol	16.6	12.0	20.7	1	0.486	94.9
80	Diethylene glycol monomethyl ether	16.2	7.8	12.6	1	0.934	118.0
81	Diethylenetriamine	16.7	13.3	14.3	1	0.487	108.0
90	Dimethyl formamide	17.4	13.7	11.3	1	0.677	77.0
94	Dimethyl sulfoxide	18.4	16.4	10.2	1	0.741	71.3
98	Dipropylene glycol	16.5	10.6	17.7	1	0.570	130.9
104	Ethanol	15.8	8.8	19.4	1	0.732	58.5
121	Ethylene glycol	17.0	11.0	26.0	1	0.815	55.8
126	Ethylene glycol monoethyl ether	16.2	9.2	16.4	1	0.707	79.1
129	2-Ethyl hexanol	15.9	3.3	11.8	0	1.294	156.6
136	Glycerol	17.4	12.1	29.3	1	0.996	73.3
153	Methanol	15.1	12.3	22.3	1	0.589	40.7
165	Methylisoamyl ketone	16.0	5.7	4.1	0	1.530	142.8
172	Methyl-2-pyrrolidone	18.0	12.3	7.2	1*	1.042	96.5
193	1-Pentanol	15.9	4.5	13.9	0	1.138	108.6
199	1-Propanol	16.0	6.8	17.4	0*	0.889	75.2
205	Propylene glycol	16.8	9.4	23.3	1	0.772	73.6
222	Tetrahydrofuran	16.8	5.7	8.0	0	1.295	81.7
225	Toluene	18.0	1.4	2.0	0	1.902	106.8
229	Trichloroethylene	18.0	3.1	5.3	0	1.615	90.2
244	Water	15.5	16.0	42.3	0	1.965	18.0

Note: $\delta_D = 16.7$; $\delta_P = 17.5$; $\delta_H = 18.5$; R = 12.2; FIT = 0.930; NO = 28.
Data from Riedel, G., *Farbe and Lack,* 82(4), 281–287, 1976.
Source: Birdi, K.-S., Ed., *Handbook of Surface and Colloid Chemistry,* CRC Press, Boca Raton, FL, 1997, chap. 10.

computer analysis based on the pure solvent data given by Riedel[16] confirms that a good "spherical" characterization of Rhodamin FB is possible using the same data otherwise used in Figure 11.1.[5] The HSP data for this correlation are given in Table 11.5. The data fit was 0.93 for 28 data points. Figure 11.1 clearly shows that this HSP sphere covers more space than the data, with a significant portion in the high energy region where there are no liquids. The use of Chapter 1, Equation 1.9 (with the constant 4) was used in this correlation, as it has been used in all the other HSP correlations in this book. The HSP correlations for water-soluble polymers and other high energy materials involved similar extrapolations into domains where there are no liquids. This procedure may be subject to revision at some future point in time, but for the present it seems to be the only procedure possible to maintain consistency in the HSP procedures developed. It should be remembered that many (most) liquids with high HSP (water, methanol, glycols) also have low molecular volumes (V). This makes them "better" solvents than expected by comparison with all the other solvents (whose average V is closer to 100 cc/mol). This fact might give the impression that the constant 4 in Chapter 1, Equation 1.9 should be increased. This is discussed more in Chapter 2.

A unifying concept and procedure for the use of water in all testing is needed. The HSP considerations discussed in this book provide help toward reaching this goal.

GASES

HSP can also be used to improve understanding of the solubility behavior of gases. Solubility parameters are usually derived from data at the normal boiling points. HSP derived from these numbers seem to be in good qualitative agreement with expectations (even at 25°C), and in many cases quantitative agreement with physical behavior has also been found. Some examples are given by Barton.[2] Solubility parameter correlations for oxygen[28] (Chapter 8) and nitrogen (Chapter 8) have been used as examples in this book. The δ_P and δ_H parameters for these two gases are zero. HSP for many gases where this is not true are reported in Chapter 8, Table 8.4. A specific example where this is not the case is carbon dioxide. The HSP for carbon dioxide are also in general agreement with its performance as a solvent at higher pressures being comparable to that expected of a higher ketone. The HSP for carbon dioxide and 2-heptanone, for example, are δ_D;δ_P;δ_H equal to 15.3;6.9;4.1 and 16.2;5.7;4.1, respectively, which are not too different.[29] All values are in MPa^{½}.

HSP can be applied to understanding the solvent properties of supercritical gases, although the author is not aware of any published reports of this type of study. The general need to use supercritical carbon dioxide for more oxygenated materials corresponds to its higher HSP, compared with, for example, aliphatic hydrocarbon gases, which are more useful in "nonpolar" systems. Studies of the variation of HSP with pressure and temperature would appear to be useful for many purposes.

In the process of calculating the HSP for gases, it was found necessary to extrapolate the data in Chapter 1, Figure 1.1 to lower molar volumes. Figure 11.2 is derived from this. This figure is worthy of some consideration from a theoretical point of view. The basis of the HSP is a corresponding states calculation for E_D as the energy of vaporization of a corresponding hydrocarbon solvent (same V and structure) at the same reduced temperature. This is, of course, 298.15 K divided by the solvents critical temperature. The reduced temperature at the boiling point is indicated in parenthesis in Figure 11.2. Questions can be raised as to why the noble gases differ from the

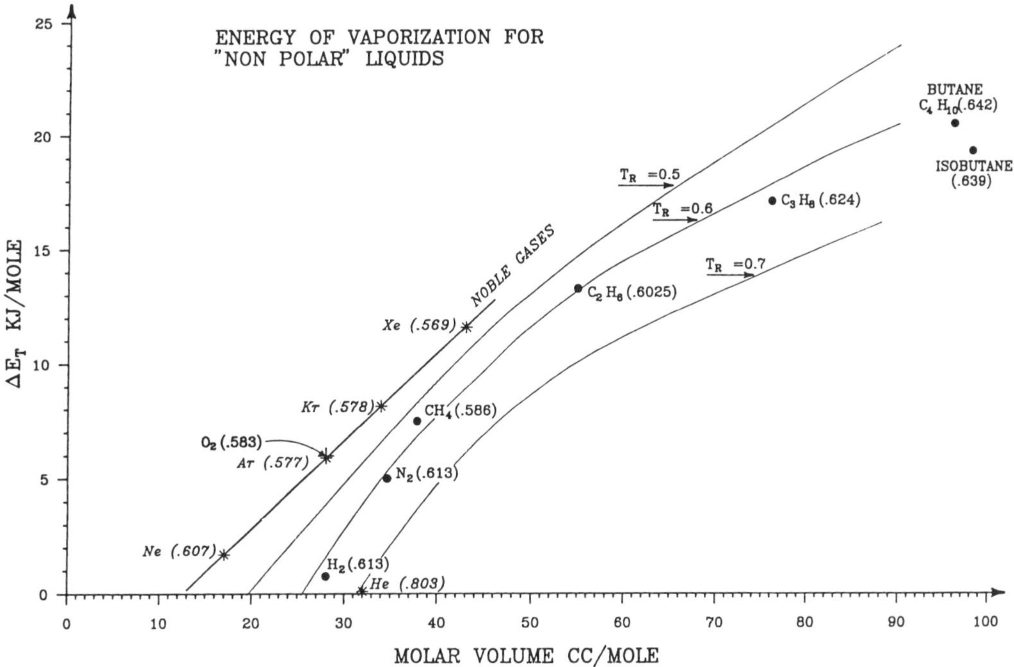

FIGURE 11.2 Cohesion energy for various low molecular weight materials as a function of molecular volume and reduced temperature (given by curves or in parenthesis) (see text for discussion).

hydrocarbon solvents and whether the hydrocarbon solvents were the best choice as reference materials. Also, why is oxygen among the noble gases in cohesion behavior rather than with the other gases? At the time of choice for a reference for the dispersion bonding energies, there was ample data on latent heats for the hydrocarbons and the aliphatic hydrocarbons were considered as having δ_P and δ_H values equal to zero. This may not quite be true, but the corrections would be minor, and the necessary data for a revised reference are lacking. It appears that the currently chartered course of using hydrocarbon solvents as a basis will be maintained. Some additional consideration may be found in further study of the relation of HSP to the corresponding states theory of Prigogine and co-workers as discussed in Chapter 2. The behavior of the hydrocarbon solvents appears to be included within the Prigogine parameter for differences in size (ρ).

Organic Salts

The HSP of several organic salts were compared with the HSP for the organic acids and organic bases from which they were made.[30] The result was that the organic salts always had considerably higher HSP than either of the components making it up. As examples, the salts made from formic acid and acetic acid combined individually with dimethyl ethanolamine had $\delta_D;\delta_P;\delta_H$ equal to 17.2;21.5;22.5 and 16.8;19.8;19.8, respectively, whereas the $\delta_D;\delta_P;\delta_H$ are 14.3;11.9;16.6 for formic acid, 14.5;8.0;13.5 for acetic acid, and 16.1;9.2;15.3 for dimethyl ethanolamine. All of these values are in $MPa^{½}$. This general relationship was also found for other salts formed by combinations of organic acids with a variety of amines. The HSP for the salts were generally close to those mentioned earlier. These values are high enough to make the salt entities insoluble in most polymers. Their affinities for water will be very high, however, both because of high HSP and also because of the charges associated with the salt groups. There was about 10% shrinkage in volume compared with the original volumes of the acids and bases. In some cases, the cohesion energy of the salts is high enough to make them solids rather than liquids. This study showed that organic salts can indeed by characterized by HSP. More work is necessary, however. In particular, the acid groups found in nature, such as in hemicelluloses, deserve more attention (see Chapter 9 and Reference 24).

Inorganic Salts

The solubility of magnesium nitrate [$Mg(NO_3)_2 \cdot 6H_2O$] was evaluated in a standard set of solvents[1] and later correlated more precisely with HSP. The HSP derived from this are $\delta_D;\delta_P;\delta_H$ and Ro equal to 19.5;22.1;21.9 and 13.2, respectively, all in $MPa^{½}$. Nitrates are known to be among the most soluble of salts. Somewhat less soluble than the nitrates are chlorides. These are only partly soluble in a few organic liquids with very high HSP. Group contributions to the HSP from the nitrate group are expected to result in lower HSP, and, in particular, lower δ_D for the nitrate portion of a salt than would be expected from the group contributions from a chloride. This would lead to greater solubility of the nitrates in organic solvents, which is indeed the case. The δ_D parameter seems to be qualitatively capable of describing the behavior of metals to some extent. It may be possible to arrive at an approximate description of inorganic salt solubility in organic media (perhaps water too) using HSP or some modification/extension thereof. The salting in and salting out of various polymers can perhaps provide clues to assign HSP in this connection. Finally, it should be noted that an excellent HSP correlation of the chemical resistance of an inorganic zinc silicate coating is reported in Chapter 7, Table 7.1.

Organometallic Compounds

No systematic studies of the HSP of organometallic compounds have been made. An exception is perhaps that shown in Figure 11.3 where Beerbower[31] used data from Panzer[32] to show that lithium stearate does indeed have two distinct regions of solvent uptake and that an HSP plot can show

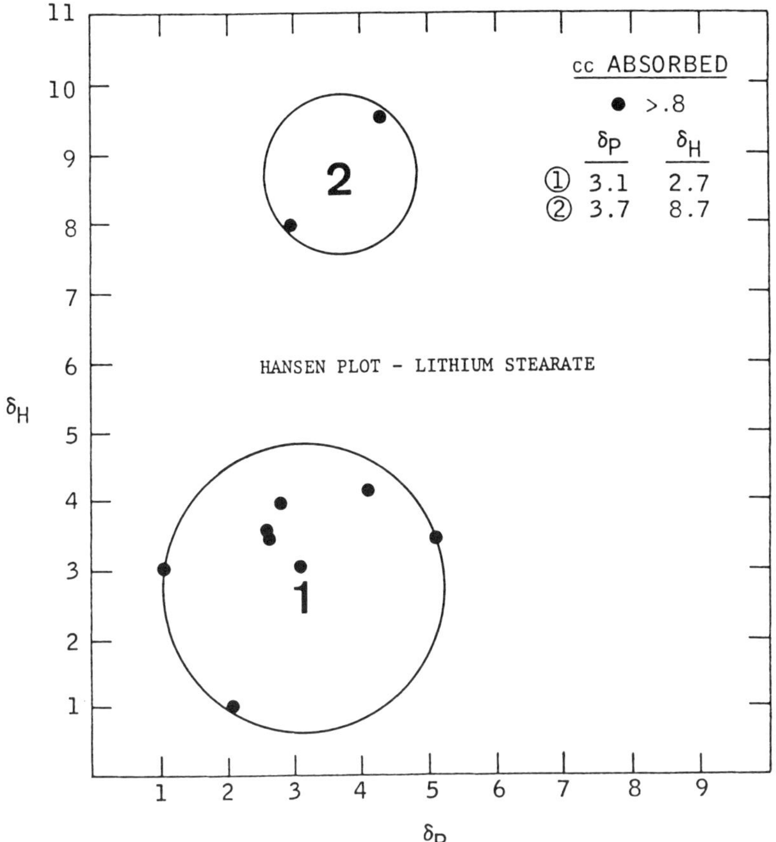

FIGURE 11.3 HSP plot for solvent uptake by lithium stearate.[31,32] Units for δ_P and δ_H are $(cal/cm^3)^{1/2}$. (From Alan Beerbower, personal communication.)

why. This example shows that one can calculate HSP values where the relevant data can be found in the literature and then test these with relevant experiments. Group contributions would be valuable. Metallic bonds differ in nature from those usually discussed in connection with organic compounds. A suspicion is that, at least in practice, the cohesion energy derived from the "metallic" bonding in organometallic compounds can be coupled with the dispersion parameter. There is also a question, for example, of whether metal atoms in the center of more complicated molecules are effectively shielded from any (surface) contact with a solvent. Surface contacts are clearly important, but it appears that the nature of the central atom also has an effect. Finally, it might be noted that Hildebrand and Scott presented a chapter on the solubility parameters of metals.[33] Unfortunately, we do not often deal with pure metals in this context, but rather metal oxides for which no HSP work has been reported, as least not to the author's knowledge.

Aromas and Fragrances

Aromas and fragrances are important in connection with packaging materials, foodstuffs, cosmetics, chewing gum, etc. A recent report[34] discussed HSP in connection with fragrances and aromas. It is clear that HSP exist for these materials, but very little work has been published in the area. One of the examples included in Reference 34 was the development of an artificial nose based on coated oscillating sensors which oscillate more slowly when they gain weight. Matching HSP for the coating and material to be detected leads to increased weight gain and increased sensitivity. Other examples where HSP could be systematically used include counteracting undesirable odors using fragrances having reasonably similar HSP; absorption of odors into plastics, coatings, sealants, etc.; development of packaging with designed HSP to either function as a barrier or as a sink; and an estimate of where in a given food product a given aromatic material is likely to reside. The key to interpretation is, as usual, that similarity of HSP means higher affinity.

Absorption of Chemicals in Recyclable Packaging

HSP correlations exist for chemical resistance, permeation phenomena, and uptake of solvents in many polymers. The recycling of polymeric containers has a potential problem in that the polymers used are able to readily absorb those chemicals whose HSP are not too different from their own. Once a chemical has absorbed into a polymer, and particularly if it is a rigid polymer with a relatively high glass transition temperature, it can be very difficult to get it out again. A relatively slow diffusional process is required to do this. It is suggested that an extensive HSP analysis be done for those polymers where potential misuse or contamination of containers prior to recycling is a possibility. This can point out which chemicals are most likely to present the greatest problems.

Chemical Resistance

Chemical resistance studies have generally been performed with too few liquids and without the necessary spread of HSP to allow the data to be correlated with confidence. In addition, attainment of equilibrium is not usually confirmed. These shortcomings mean that HSP correlations of chemical resistance must be done with great care. This has been discussed in Chapter 7 in more detail. An additional activity which should be done for practical reasons is to assign effective HSP to various test materials such as mustard, ketchup, and other given products which often appear in tests of chemical resistance. Such data will allow greater use of the correlations in that guidelines for potential improvements can be obtained.

THEORETICAL PROBLEMS AWAITING FUTURE RESOLUTION

Polymer Solubility

The Flory chi parameter has been used to describe polymer–solvent interactions for many years.[35,36] If this single parameter is to be complete in this function, it must include both the atomic/dispersion interactions as well as the specific interactions reflected by the δ_P and δ_H parameters. Attempts to calculate chi using HSP are reported in Chapter 2. More understanding is required before chi can be calculated with reasonable accuracy, but intensified efforts seem warranted. Zellers and co-workers have recently made an attempt to use this theory in conjunction with HSP studies.[37-40] A major problem is the reliability of the chi parameter (and also HSP) values in the literature (see Chapter 2 for more details).

The author's current view as expressed in Chapter 2 is that the Flory approach is a special case of the more general Prigogine corresponding states theory. This is in agreement with the view of Patterson discussed in Chapter 2. Furthermore, the very general applications of the HSP approach demonstrated in this book and elsewhere and the apparent agreement in the treatment of specific interactions by both the HSP and Prigogine treatments leads to the need to confirm more substantially that the HSP treatment is indeed a practical expression of the Prigogine theory. The geometric mean must be used in the Prigogine theory to arrive at this similarity of treatment. The author believes that this is a valid conclusion and that still more work related to the similarities of HSP and the Prigogine theory is warranted. In particular, the "structural parameters" in the Prigogine theory have not been studied in this respect.

SURFACE PHENOMENA

Surfaces can clearly be characterized by HSP as demonstrated in Chapters 4 and 5. The work of Beerbower contained in References 17 and 23 has also shown applications of HSP to such varied phenomena as the work of adhesion of liquids on mercury, friction of polyethylene untreated and treated with sulfuric acid, the Rehbinder effect — the crushing of Al_2O_3 under various liquids, and the Joffé effect — effect of liquid immersion on the fracture strength of soda-lime glass. Here again, the successful use of HSP to such applications might not have been anticipated had it been considered as a parameter for use in bulk systems only. A formalized unifying theory linking HSP to both bulk and surface phenomena is still lacking. Presently, the best that can be said is that the generality "like dissolves like" can be quantified in many cases. The extension of this, "like seeks like," also seems to have been demonstrated. It is the surfaces of molecules which interact with each other (also in bulk and solution phenomena), so it is not surprising that cohesion parameters can be applied with success to both solubility and surface phenomena. Much more research needs to be done with these relations. A good starting point is the *Handbook of Surface and Colloid Chemistry,* edited by K.S. Birdi.[41] If we consider chromatographic techniques as depending primarily on surface phenomena, mention should also be made of the extension of the three-parameter HSP approach to a five-parameter approach by Karger and co-workers.[42]

CONCLUSION

HSP have been shown useful in solvent selection; predicting polymer–polymer miscibility; characterizing the surfaces of polymers, fillers, and fibers; correlating permeation phenomena; characterizing organic salts and inorganic salts; etc. No other parameter can be assigned to such a range of materials spanning from gases and liquids, over surfaces, to inorganic salts. These results and the close relation with the Prigogine corresponding states theory of polymer solutions indicate that a still more general theory exists. This theory should quantify the adage "like seeks like," i.e., include surface phenomena as well as bulk phenomena.

Specific areas needing more theoretical work related to HSP in the near future include water, gases, organic salts, inorganic salts, and organometallic compounds. Water remains special because of its low molecular volume and high δ_H. Most materials having HSP in the range of the customary test liquids can be studied using HSP with reasonable success. This is not fully the case for gases, which generally have much lower HSP than the well-studied liquids, and salts, many of which apparently have HSP very much higher than any of the liquids. Extensions of practical applications related to chemical resistance and the uptake of potentially dangerous materials in recyclable polymers are also required.

REFERENCES

1. Hansen, C. M., The Universality of the Solubility Parameter, *Ind. Eng. Chem. Prod. Res. Dev.,* 8(1), 2–11, 1969.
2. Barton, A. F. M., *Handbook of Solubility Parameters and Other Cohesion Parameters*, CRC Press, Boca Raton, FL, 1983.
3. Barton, A. F. M., *Handbook of Polymer-Liquid Interaction Parameters and Solubility Parameters*, CRC Press, Boca Raton, FL, 1990.
4. Hansen, C. M., Solubility Parameters, in *Paint Testing Manual*, Manual 17, Koleske, J. V., Ed., American Society for Testing and Materials, Philadelphia, 1995, 383–404.
5. Hansen, C. M., Cohesion Energy Parameters Applied to Surface Phenomena, *CRC Handbook on Surface Science,* Birdi, K. S., Ed., CRC Press, Boca Raton, FL, 1997, chap. 10.
6. Anonymous, Brochure: Co-Act — A Dynamic Program for Solvent Selection, Exxon Chemical International Inc., 1989.

7. Dante, M. F., Bittar, A. D., and Caillault, J. J., Program Calculates Solvent Properties and Solubility Parameters, *Mod. Paint Coat.,* 79(9), 46–51, 1989.
8. Hansen, C. M. and Beerbower, A., Solubility Parameters, in *Kirk-Othmer Encyclopedia of Chemical Technology,* Suppl. Vol., 2nd ed., Standen, A., Ed., Interscience, New York, 1971, 889–910.
9. Hansen, C. M. and Andersen, B. H., The Affinities of Organic Solvents in Biological Systems, *J. Am. Ind. Hyg. Assoc.,* 49(6), 301–308, 1988.
10. Wessling, R. A., The Solubility of Poly(Vinylidene Chloride), *J. Appl. Polym. Sci.,* 14, 1531–1545, 1970.
11. Grulke, E. A., Solubility Parameter Values, in *Polymer Handbook,* 3rd ed., Brandrup, J. and Immergut, E. H., Eds., John Wiley-Interscience, New York, 1989, VII/519–559.
12. Burrell, H., Solubility of Polymers, in *Kirk Othmer Encycopedia of Polymer Science and Technology,* Vol. 12, 2nd ed., John Wiley & Sons, New York, 1970, 618–626.
13. Hoy, K. L., New Values of the Solubility Parameters from Vapor Pressure Data, *J. Paint Technol.,* 42(541), 76–118, 1970.
14. Hoy, K. L., *Tables of Solubility Parameters*, Union Carbide Corp., Research and Development Dept., South Charleston, WV, 1985 (1st ed. 1969).
15. van Krevelen, D. W. and Hoftyzer, P. J., *Properties of Polymers: Their Estimation and Correlation with Chemical Structure,* 2nd ed., Elsevier, Amsterdam, 1976.
16. Riedel, G., The Solubility of Colorants in Organic Solvents (in German: Löslichkeit von Farbstoffen in organischen Lösungsmitteln), *Farbe und Lack,* 82(4), 281–287, 1976.
17. Beerbower, A., Environmental Capability of Liquids, in *Interdisciplinary Approach to Liquid Lubricant Technology,* NASA Publication SP-318, 1973, 365–431.
18. Fedors, R. F., A Method for Estimating Both the Solubility Parameters and Molar Volumes of Liquids, *Polym. Eng. Sci.,* 14(2), 147–154, 472, 1974.
19. Koenhen, D. N. and Smolders, C. A., The Determination of Solubility Parameters of Solvents and Polymers by Means of Correlation with Other Physical Quantities, *J. Appl. Polym. Sci.,* 19, 1163–1179, 1975.
20. Utracki, L., Personal communication, 1995.
21. Kähler, T. and Knudsen, S. L., Student Report, Technical University of Denmark, 1967.
22. Hansen, C. M., Characterization of Surfaces by Spreading Liquids, *J. Paint Technol.,* 42(550), 660–664, 1970.
23. Beerbower, A., Boundary Lubrication — Scientific and Technical Applications Forecast, AD747336, Office of the Chief of Research and Development, Department of the Army, Washington, D.C., 1972.
24. Hansen, C. M. and Björkman, A., The Ultrastructure of Wood from a Solubility Parameter Point of View, *Holzforschung*, 52(4), 335–344, 1998.
25. Hansen, C. M. and Knudtson, J., Spreading and Wetting of an Electrodeposition Coating (in German: Ausbreitung und Benetzung bei einem Elektrotauchlack), *Farbe und Lack*, 2, 115–118, 1973.
26. *Progress in Organic Coatings,* Special issue devoted to Self-Stratifying Coatings, 28(3), July 1996.
27. Thulstrup, D., Characterization of Surfaces by Static Contact Angle Measurements (lecture in Danish: Karakterisering af Overflader ved Statisk Kontaktvinkelmåling), Surface Properties and Modification of Plastics (Conference Notes), Ingeniørforening i Danmark, November 13, 1997, Copenhagen. See also DSM Materialenyt 2:97, Dansk Selskab for Materialeproevning og -forskning, Ingeniørforening in Danmark, Copenhagen, 1997, 63–69.
28. König, U. and Schuch, H., Molecular Composition and Permeability of Plastics, (in German: Konstitution und Permeabilität von Kunststoffen) *Kunststoffe*, 67(1), 27–31, 1977.
29. Hansen, C. M., 25 Years with Solubility Parameters (in Danish: 25 År med Opløselighedsparametrene), *Dan. Kemi,* 73(8), 18–22, 1992.
30. Hansen, C. M., Some Aspects of Acid/Base Interactions (in German: Einige Aspekte der Säure/Base-Wechselwirkung), *Farbe und Lack*, 83(7), 595–598, 1977.
31. Beerbower, A., Personal communication — Predicting Wetting in the Oil-Brine-Rock System, Unedited manuscript for presentation at the National A.I.Ch.E. Meeting, Tulsa, OK, March 10–13, 1974.
32. Panzer, J., Components of Solid Surface Free Energy from Wetting Measurements, *J. Colloid Interface Sci.,* 44(1), 142–161, July 1973.
33. Hildebrand, J. and Scott, R. L., *The Solubility of Nonelectrolytes,* 3rd ed., Reinhold, New York, 1950.

34. Hansen, C. M., Solubility Parameters for Aromas and Scents (in Danish: Aromastoffers Opløselighedsparametre), *Plus Process*, 11(9), 16–17, 1997.
35. Flory, P. J., *Principles of Polymer Chemistry*, Cornell University Press, New York, 1953.
36. Eichinger, B. E. and Flory, P. J., Thermodynamics of Polymer Solutions, *Trans. Faraday Soc.*, 64(1), 2035–2052, 1968; 64(2), 2053–2060, 1968; 64(3), 2061–2065, 1968; 64(4), 2066–2072, 1968.
37. Zellers, E. T., Three-Dimensional Solubility Parameters and Chemical Protective Clothing Permeation. I. Modelling the Solubility of Organic Solvents in Viton® Gloves, *J. Appl. Polym. Sci.*, 50, 513–530, 1993.
38. Zellers, E. T., Three-Dimensional Solubility Parameters and Chemical Protective Clothing. II. Modelling Diffusion Coefficients, Breakthrough Times, and Steady-State Permeation Rates of Organic Solvents in Viton® Gloves, *J. Appl. Polym. Sci.*, 50, 531–540, 1993.
39. Zellers, E. T., Anna, D. H., Sulewski, R., and Wei, X., Critical Analysis of the Graphical Determination of Hansen's Solubility Parameters for Lightly Crosslinked Polymers, *J. Appl. Polym. Sci.*, 62, 2069–2080, 1996.
40. Zellers, E. T., Anna, D. H., Sulewski, R., and Wei, X., Improved Methods for the Determination of Hansen's Solubility Parameters and the Estimation of Solvent Uptake for Lightly Crosslinked Polymers, *J. Appl. Polym. Sci.*, 62, 2081–2096, 1996.
41. Birdi, K. S., *Handbook of Surface and Colloid Chemistry*, CRC Press, Boca Raton, FL, 1997.
42. Karger, B. L., Snyder, L. R., and Eon, C., Expanded Solubility Parameter Treatment for Classification and Use of Chromatographic Solvents and Adsorbents, *Anal. Chem.*, 50(14), 2126–2136, 1978.

Appendix

CONTENTS

TABLE A.1
Hansen Solubility Parameters for Selected Solvents (Solvents are in Alphabetical Order)

No.	Solvent	Dispersion	Polar	Hydrogen Bonding	Molar Volume
1	ACETALDEHYDE	14.7	8.0	11.3	57.1
2	ACETALDOXIME	16.3	4.0	20.2	61.2
3	ACETAMIDE	17.3	18.7	22.4	60.8
4	ACETANILIDE	20.6	13.3	12.4	110.9
5	ACETIC ACID	14.5	8.0	13.5	57.1
6	ACETIC ANHYDRIDE	16.0	11.7	10.2	94.5
7	ACETONE	15.5	10.4	7.0	74.0
8	ACETONECYANHYDRIN	16.6	12.2	15.5	94.0
9	ACETONEMETHYLOXIME	14.7	4.6	4.6	96.7
10	ACETONITRILE	15.3	18.0	6.1	52.6
11	ACETOPHENONE	19.6	8.6	3.7	117.4
12	ACETOXIME	16.3	3.7	10.9	80.2
765	1-ACETOXY-1,3-BUTADIENE	16.1	4.4	8.3	118.4
13	*N*-ACETYL CAPROLACTAM	18.9	8.7	4.8	155.0
14	*N*-ACETYL MORPHOLINE	18.3	5.3	7.8	115.6
15	*N*-ACETYL PIPERIDINE	18.5	10.0	6.5	125.8
16	*N*-ACETYL PYRROLIDONE	17.8	13.1	8.3	127.0
785	2-ACETYL THIOPHENE	19.1	12.2	9.3	108.0
17	ACETYLACETONE	16.1	11.2	6.2	103.0
18	ACETYLBROMIDE	16.7	10.6	5.2	74.0
19	ACETYLCHLORIDE	16.2	11.2	5.8	71.4
20	ACETYLENE (ETHYNE)	14.4	4.2	11.9	42.1
21	ACETYLFLUORIDE	14.7	14.0	5.7	62.0
22	ACROLEIN	15.0	7.2	7.8	66.7
23	ACRYLAMIDE	15.8	12.1	12.8	63.4
24	ACRYLIC ACID	17.7	6.4	14.9	68.5
25	ACRYLONITRILE	16.0	12.8	6.8	67.1
26	ACRYLYLCHLORIDE	16.2	11.6	5.4	81.3
27	ALLYL ACETATE	15.7	4.5	8.0	108.5
28	ALLYL ACETIC ACID	16.7	4.7	11.3	102.1
769	ALLYL ACETOACETATE	15.9	6.9	8.6	137.8
29	ALLYL ACETONITRILE (4-PENTENENITRILE)	16.3	11.2	5.0	98.5
30	ALLYL ALCOHOL	16.2	10.8	16.8	68.4
31	ALLYL AMINE	15.5	5.7	10.6	74.9
32	ALLYL BROMIDE (3-BROMOPRENE)	16.5	7.3	4.9	86.5
33	ALLYL CHLORIDE	17.0	6.2	2.3	82.3
34	ALLYL CYANIDE	16.0	14.3	5.6	80.5
35	ALLYL ETHYLETHER	15.0	4.8	5.1	112.6
36	ALLYL FLUORIDE	14.9	6.4	1.0	71.5
37	ALLYL FORMATE	15.7	5.4	8.8	91.0
38	ALLYL IODIDE	17.4	6.4	3.3	90.8
39	ALLYL ISOCYANIDE	16.1	13.0	5.4	84.5
40	ALLYL ISOTHIOCYANATE	17.0	11.3	8.5	98.0
41	ALLYL MERCAPTAN	17.5	5.2	9.3	80.2
42	ALLYL MERCAPTAN	16.4	6.2	7.9	80.2
43	ALLYL METHYLETHER	15.0	4.3	5.9	93.7
44	AMMONIA	13.7	15.7	17.8	20.8
45	AMYL ACETATE	15.8	3.3	6.1	148.0

TABLE A.1 (continued)
Hansen Solubility Parameters for Selected Solvents (Solvents are in Alphabetical Order)

No.	Solvent	Dispersion	Polar	Hydrogen Bonding	Molar Volume
699	ANETHOLE (*TRANS*)	19.0	4.3	8.7	150.0
46	ANILINE	19.4	5.1	10.2	91.5
47	*p*-ANISIDINE (METHOXY ANILINE)	19.9	6.5	11.3	113.3
48	ANISOLE	17.8	4.1	6.7	119.1
49	AZIDOETHANE	15.9	8.9	12.9	79.0
50	3-AZIDOPROPENE	16.8	7.7	13.4	83.0
51	BENZALDEHYDE	19.4	7.4	5.3	101.5
52	BENZENE	18.4	0.0	2.0	89.4
53	1,3-BENZENEDIOL	18.0	8.4	21.0	87.5
54	BENZOIC ACID	18.2	6.9	9.8	100.0
55	BENZONITRILE	17.4	9.0	3.3	102.6
56	BENZOPHENONE	13.6	8.6	5.7	164.2
57	BENZOYL CHLORIDE	20.7	8.2	4.5	116.0
58	BENZYL ALCOHOL	18.4	6.3	13.7	103.6
59	BENZYL BUTYL PHTHALATE	19.0	11.2	3.1	306.0
60	BENZYL CHLORIDE	18.8	7.1	2.6	115.0
61	BENZYL METHACRYLATE	16.8	4.1	4.1	167.8
62	*N*-BENZYL PYRROLIDONE	18.2	6.1	5.6	160.0
63	BIPHENYL	21.4	1.0	2.0	154.1
64	BORINECARBONYL	17.0	10.2	6.9	41.8
65	2-BROMO ALLYL ALCOHOL	17.1	9.9	16.2	84.5
66	2-BROMO PROPENE	16.1	6.0	4.9	88.8
67	1-BROMO PROPENE (*CIS*)	16.3	6.4	5.0	84.7
68	4-BROMO-1-BUTANE	16.5	6.0	4.5	102.0
69	4-BROMO-1,2-BUTADIENE	17.0	6.5	4.7	93.3
854	5-BROMO-2-NITROBENZOTRIFLUORIDE	20.0	6.0	4.9	150.1
815	1-BROMO-4-ETHOXY BENZENE	19.5	7.7	5.3	140.0
70	BROMOACETYLENE	15.7	9.9	5.6	67.7
800	*o*-BROMOANISOLE	19.8	8.4	6.7	124.5
71	BROMOBENZENE	20.5	5.5	4.1	105.3
806	*p*-BROMOBENZONITRILE	20.4	9.3	5.8	113.8
805	*p*-BROMOBENZOYL CHLORIDE	20.2	6.5	5.5	137.2
729	2-BROMOBUTANE	16.3	7.7	4.4	109.5
802	*o*-BROMOCHLOROBENZENE	20.3	7.7	4.6	116.9
72	BROMOCHLOROMETHANE	17.3	5.7	3.5	65.0
73	BROMOETHYLENE	16.1	6.3	2.3	71.6
74	BROMOFORM	21.4	4.1	6.1	87.5
748	BROMOMETHYL METHYL ETHER	16.9	8.5	7.0	81.6
75	1-BROMONAPHTALENE	20.3	3.1	4.1	140.0
803	*p*-BROMONITROBENZENE	20.9	9.9	5.9	103.7
76	BROMOPRENE	16.9	6.4	4.7	95.2
77	3-BROMOPROPYNE	18.1	6.5	5.3	75.3
710	*o*-BROMOSTYRENE	19.5	5.2	5.3	130.0
78	2-BROMOTHIOPHENE	20.1	5.2	4.6	96.8
713	*o*-BROMOTOLUENE	19.3	5.0	4.2	119.0
714	*p*-BROMOTOLUENE	19.3	6.8	4.1	122.9
743	BROMOTRICHLORO METHANE #1	18.3	0.8	0.0	99.2
744	BROMOTRICHLORO METHANE #2-GC	18.3	8.1	6.0	99.2

TABLE A.1 (continued)
Hansen Solubility Parameters for Selected Solvents (Solvents are in Alphabetical Order)

No.	Solvent	Dispersion	Polar	Hydrogen Bonding	Molar Volume
745	BROMOTRICHLORO METHANE #3	18.3	8.1	0.0	99.2
79	BROMOTRIFLUOROMETHANE (FREON 1381)	9.6	2.4	0.0	97.0
80	1,2-BUTADIENE	14.7	1.7	6.2	82.3
81	1,3-BUTADIENE	14.8	2.8	5.6	83.2
82	1,3-BUTADIENE-1-CHLORO	15.6	8.5	2.0	92.2
83	2,3-BUTADIENE-1-OL	16.2	6.6	16.8	76.5
84	1,3-BUTADIENE-1,2-DI-CHLORO	17.0	10.7	2.8	102.5
85	BUTADIENE-4-CYANO	16.2	11.7	5.2	93.7
86	BUTADIENEDIOXIDE	18.3	14.4	6.2	77.8
87	BUTADIONE	15.7	5.1	6.8	87.8
88	1,4-BUTANDIOL DIACRYLATE	16.8	9.1	4.2	194.4
89	BUTANE	14.1	0.0	0.0	101.4
90	1,3-BUTANEDIOL	16.6	10.0	21.5	89.9
730	1,4-BUTANEDIOL	16.6	15.3	21.7	88.9
91	1-BUTANETHIOL	16.3	5.3	4.5	107.8
92	1-BUTANOL	16.0	5.7	15.8	91.5
93	2-BUTANOL	15.8	5.7	14.5	92.0
94	1-BUTENE	13.2	1.3	3.9	94.3
95	2-BUTENE (*CIS*)	14.7	1.3	4.1	90.3
96	2-BUTENE (*TRANS*)	14.6	0.0	2.9	92.9
97	3-BUTENENITRILE	16.2	14.3	5.6	80.5
98	1-BUTENYL METHYL ETHER (*CIS*)	15.1	5.3	5.2	111.9
99	BUTENYL METHYL ETHER (*CIS*)	15.2	3.4	5.0	118.8
100	2-BUTENYL METHYL ETHER (*TRANS*)	15.3	4.4	4.3	110.4
101	BUTOXY ETHOXY PROPANOL	15.5	6.5	10.2	185.9
728	3-BUTOXYBUTANOL	15.9	5.5	10.6	166.3
102	*n*-BUTYL ACETATE	15.8	3.7	6.3	132.5
103	*SEC*-BUTYL ACETATE	15.0	3.7	7.6	133.6
768	*n*-BUTYL ACETO ACETATE	16.6	5.8	7.3	164.3
104	*n*-BUTYL ACRYLATE	15.6	6.2	4.9	143.8
611	*t*-BUTYL ALCOHOL	15.2	5.1	14.7	95.8
105	*n*-BUTYL AMINE	16.2	4.5	8.0	99.0
718	BUTYL BENZOATE	18.3	2.9	5.5	178.0
774	*n*-BUTYL BUTYRATE	15.6	2.9	5.6	166.7
722	*n*-BUTYL CYCLOHEXANE	16.2	0.0	0.6	176.7
723	*n*-BUTYL CYCLOPENTANE	16.4	0.0	1.0	162.0
106	BUTYL ISOPROPENYL ETHER	14.8	5.3	5.0	145.7
107	BUTYL LACTATE	15.8	6.5	10.2	149.0
108	*n*-BUTYL METHACRYLATE	15.6	6.4	6.6	159.4
717	2-BUTYL OCTANOL	16.1	3.6	9.3	224.2
109	*N*-*n*-BUTYL PYRROLIDONE	17.5	9.9	5.8	148.0
764	*n*-BUTYL SALICYLATE	17.9	4.8	11.7	181.7
724	3-*n*-BUTYL TOLUENE	17.4	0.1	1.0	173.7
747	BUTYL-6-METHYL-3-CYCLOHEXENE CARBOLYATE	16.1	2.5	5.7	209.1
725	*n*-BUTYLBENZENE	17.4	0.1	1.1	157.0
110	2,3-BUTYLENE CARBONATE	18.0	16.8	3.1	105.5
111	BUTYLENEOXIDE	16.3	6.2	5.9	87.5
719	*o*-*n*-BUTYLTOLUENE	17.6	0.1	1.0	171.3

TABLE A.1 (continued)
Hansen Solubility Parameters for Selected Solvents (Solvents are in Alphabetical Order)

No.	Solvent	Dispersion	Polar	Hydrogen Bonding	Molar Volume
720	*p-n*-BUTYLTOLUENE	17.4	0.1	1.0	174.2
112	2-BUTYNEDINITRILE	15.2	16.2	8.0	78.4
113	BUTYRALDEHYDE	14.7	5.3	7.0	88.5
114	BUTYRIC ACID	14.9	4.1	10.6	110.0
767	BUTYRIC ANHYDRIDE	16.0	6.3	7.7	164.4
115	GAMMA-BUTYROLACTONE	19.0	16.6	7.4	76.8
116	BUTYRONITRILE	15.3	12.4	5.1	87.3
117	BUTYRYLCHLORIDE	16.8	9.4	4.8	103.6
118	CAPROLACTONE (EPSILON)	19.7	15.0	7.4	110.8
119	CARBON DIOXIDE	15.3	6.9	4.1	38.0
120	CARBON DISULFIDE	20.5	0.0	0.6	60.0
121	CARBON DISULFIDE-2	19.9	5.8	0.6	60.0
122	CARBON TETRACHLORIDE	17.8	0.0	0.6	97.1
123	CARBONYL SULFIDE	17.4	3.7	0.0	51.0
124	CARBONYLCYANIDE	15.0	6.3	8.0	71.2
125	CHLORAL	17.2	7.4	7.6	97.5
126	CHLORINE	17.3	10.0	0.0	46.0
127	1-CHLORO-1-FLUORO ETHYLENE	16.0	6.9	4.0	67.6
128	CHLORO ACETALDEHYDE	16.2	16.1	9.0	60.4
129	CHLORO ACETIC ACID	17.7	10.4	12.3	68.6
797	*p*-CHLORO ACETOPHENONE	19.6	7.6	4.0	129.7
130	3-CHLORO ALLYL ALCOHOL	17.2	10.3	16.5	78.6
131	2-CHLORO ALLYL ALCOHOL	17.1	10.2	16.4	79.6
132	1-CHLORO METHYL ACRYLATE	15.9	7.3	8.5	101.4
133	2-CHLORO PROPENE (ISOPROPENYL CHLORIDE)	15.5	6.7	2.2	84.9
135	1-CHLORO VINYL ETHYL ETHER	16.8	6.5	5.7	104.5
136	1-CHLORO-1-NITROETHANE	16.8	13.5	4.7	85.1
137	3-CHLORO-1-PROPANOL	17.5	5.7	14.7	84.2
138	4-CHLORO-1,2-BUTADIENE	16.6	8.0	6.7	89.5
139	1-CHLORO-2-BUTENE	16.2	7.7	2.0	97.4
817	1-CHLORO-2-ETHOXY BENZENE	19.2	8.1	4.4	138.7
140	3-CHLORO-2-METHYL PROPENE	16.2	5.6	2.0	98.8
141	1-CHLORO-2-METHYL PROPENE	16.1	7.1	4.2	95.6
813	6-CHLORO-2-NITROTOLUENE	20.3	9.6	3.8	132.0
819	4-CHLORO-2-NITROTOLUENE	19.9	11.8	3.8	132.0
816	1-CHLORO-4-ETHOXY BENZENE	19.3	6.3	4.4	139.2
142	CHLOROACETONE	16.8	9.6	5.5	80.5
143	CHLOROACETONITRILE	17.4	13.6	2.0	63.3
144	CHLOROACETYLCHLORIDE	17.5	9.2	5.5	79.5
145	CHLOROACETYLENE	16.2	2.1	2.5	63.7
766	2-CHLOROALLYLIDENE DIACETATE	16.7	7.3	8.8	160.2
146	4-CHLOROANISOLE	19.6	7.8	6.7	122.5
147	4-CHLOROBENZALDEHYDE	19.9	7.2	5.6	113.4
148	CHLOROBENZENE	19.0	4.3	2.0	102.1
811	4-CHLOROBENZONITRILE	19.5	8.0	4.1	125.1
853	4-CHLOROBENZOTRICHLORIDE	20.3	5.5	3.5	153.8
812	*p*-CHLOROBENZOYL CHLORIDE	19.9	6.7	5.1	127.1
149	4-CHLOROBENZYL ALCOHOL	19.6	7.5	13.0	117.7

TABLE A.1 (continued)
Hansen Solubility Parameters for Selected Solvents (Solvents are in Alphabetical Order)

No.	Solvent	Dispersion	Polar	Hydrogen Bonding	Molar Volume
150	3-CHLOROBENZYLCHLORIDE	19.9	9.3	2.6	117.7
151	1,2-CHLOROBROMOETHYLENE	17.2	6.6	2.3	78.7
152	1-CHLOROBUTANE	16.2	5.5	2.0	104.5
750	2-CHLOROBUTANE	15.8	7.6	2.0	106.8
780	2-CHLOROCYCLOHEXANONE	18.5	13.0	5.1	113.9
153	CHLOROCYCLOPROPANE	17.6	7.2	2.2	84.9
154	CHLORODIFLUOROMETHANE (FREON 22)	12.3	6.3	5.7	72.9
155	*N*-CHLORODIMETHYLAMINE	16.0	7.8	7.9	87.4
772	2-CHLOROETHYL ACETATE	16.7	9.6	8.8	107.5
756	2-CHLOROETHYL ETHYL ETHER	16.3	7.9	4.6	109.5
834	2-CHLOROETHYL ETHYL SULFIDE	17.2	5.0	6.1	116.9
808	*o*-CHLOROFLUOROBENZENE	19.4	8.7	2.0	104.9
156	CHLOROFORM	17.8	3.1	5.7	80.7
157	*BIS*(CHLOROMETHYL) ETHER	17.2	4.9	6.6	86.6
158	CHLOROMETHYLSULFIDE	16.6	6.4	2.0	95.0
809	*p*-CHLORONITROBENZENE	20.4	9.6	4.2	103.7
159	CHLORONITROMETHANE	17.4	13.5	5.5	65.1
160	2-CHLOROPHENOL	20.3	5.5	13.9	102.3
161	CHLORPRENE	16.1	5.4	2.1	93.2
162	2-CHLOROPROPENAL	17.1	12.9	8.1	75.5
163	CHLOROPROPENE	15.3	6.9	2.2	82.3
164	2-CHLOROPROPENOIC ACID	19.1	9.4	12.4	86.6
165	3-CHLOROPROPIONALDEHYDE	17.0	13.3	8.2	73.0
166	CHLOROPROPIONITRILE	17.3	15.9	6.1	77.4
167	3-CHLOROPROPYNE	16.7	7.4	2.3	72.4
711	*p*-CHLOROSTYRENE	18.7	4.3	3.9	128.3
712	*o*-CHLOROSTYRENE	18.7	4.7	3.9	126.8
168	4-CHLOROTHIOPHENOL	20.8	8.6	10.6	100.0
787	*o*-CHLOROTHIOPHENOL	20.2	7.0	10.0	113.4
814	*p*-CHLOROTOLUENE	19.1	6.2	2.6	118.3
134	CHLOROTRIFLUOROETHYLENE (CTFE)	15.3	6.3	0.0	75.6
169	*m*-CRESOL	18.0	5.1	12.9	104.7
170	CROTONALDEHYDE	16.2	14.9	7.4	82.5
171	CROTONIC ACID	16.8	8.7	12.0	84.6
172	CROTONLACTONE	19.0	19.8	9.6	76.4
173	*TRANS*-CROTONONITRILE	16.4	18.8	5.5	81.4
174	CYANAMID (CARBAMONITRILE)	15.5	17.6	16.8	32.8
175	CYANOGEN	15.1	11.8	0.0	54.6
176	CYANOGEN BROMIDE	18.3	15.2	0.0	52.6
177	CYANOGEN CHLORIDE	15.6	14.5	0.0	51.8
178	CYCLOBUTANONE	18.3	11.4	5.2	73.4
179	CYCLODECANONE	16.8	8.0	4.1	161.0
180	CYCLOHEPTANONE	17.2	10.6	4.8	118.2
181	CYCLOHEXANE	16.8	0.0	0.2	108.7
182	CYCLOHEXANOL	17.4	4.1	13.5	106.0
183	CYCLOHEXANONE	17.8	6.3	5.1	104.0
184	CYCLOHEXYLAMINE	17.2	3.1	6.5	113.8
185	CYCLOHEXYLCHLORIDE	17.3	5.5	2.0	118.6

TABLE A.1 (continued)
Hansen Solubility Parameters for Selected Solvents (Solvents are in Alphabetical Order)

No.	Solvent	Dispersion	Polar	Hydrogen Bonding	Molar Volume
186	CYCLOOCTANONE	17.0	9.6	4.5	131.7
187	CYCLOPENTANE	16.4	0.0	1.8	94.9
188	CYCLOPENTANONE	17.9	11.9	5.2	89.1
189	CYCLOPENTENE	16.7	3.8	1.7	89.0
735	2-CYCLOPENTENYL ALCOHOL	18.1	7.6	15.6	86.2
190	CYCLOPRENE	17.2	2.4	2.0	50.0
191	CYCLOPROPANE	17.6	0.0	0.0	58.3
192	CYCLOPROPYLMETHYLKETONE	17.0	11.1	4.6	93.6
193	CYCLOPROPYLNITRILE	18.6	16.2	5.7	75.4
194	*CIS*-DECAHYDRONAPHTHALENE	18.8	0.0	0.0	156.9
195	*TRANS*-DECAHYDRONAPHTHALENE	18.0	0.0	0.0	156.9
196	DECANE	15.7	0.0	0.0	195.9
197	1-DECANOL	17.5	2.6	10.0	191.8
734	2-DECANOL	15.8	3.9	10.0	192.8
726	1-DECENE	15.8	1.0	2.2	190.6
759	DI-(2-CHLOROETHOXY) METHANE	17.1	10.2	7.1	141.1
751	DIBUTYL ETHER	15.3	3.4	3.3	170.4
773	DIBUTYL FUMARATE	16.7	3.0	6.7	232.7
845	DI-ISOBUTYL SULFOXIDE	16.3	10.5	6.1	195.0
198	DI-ISOPROPYL SULFOXIDE	17.0	11.5	7.4	159.9
219	DI-*n*-BUTYL ETHER	15.2	3.4	4.2	170.3
200	DI-*n*-BUTYL SULFOXIDE	16.4	10.5	6.1	195.0
201	DI-*n*-PROPYL SULFOXIDE	17.0	13.0	7.4	159.9
199	DI-*p*-TOLYL SULFOXIDE	20.3	11.4	3.1	209.0
778	DIPHENYL ETHER	19.6	3.2	5.8	160.4
779	DIPHENYL SULFONE	21.1	14.4	3.4	174.3
202	DI-(2-METHOXYETHYL) ETHER	15.7	6.1	6.5	142.0
204	DI-(2-CHLORO-ISOPROPYL) ETHER	19.0	8.2	5.1	146.0
205	DI-2-ETHYL HEXYL AMINE	15.6	0.8	3.2	301.5
736	DI-2-ETHYL HEXYL ETHER	15.9	2.6	2.5	300.5
206	1,3-DICHLORO-2-BUTENE	16.9	7.8	2.7	107.7
207	1,4-DICHLORO-2-BUTYNE	17.8	7.6	2.0	97.8
209	DIACETONE ALCOHOL	15.8	8.2	10.8	124.2
210	DIALLYL AMINE	15.6	4.5	6.7	124.1
211	DIALLYL ETHER	15.3	4.4	5.3	118.8
749	1,1-DIALLYLOXYETHANE	15.4	5.1	4.8	163.3
212	DIAZOMETHANE	14.7	6.1	11.3	78.1
213	DIBENZYL ETHER	17.3	3.7	7.3	192.7
214	1,1-DIBROMO ETHYLENE	15.1	4.8	7.0	85.3
215	DIBROMO METHANE	17.8	6.4	7.0	69.8
804	*o*-DIBROMOBENZENE	20.7	6.5	5.3	120.0
216	1,1-DIBROMOETHANE	18.5	8.4	8.8	91.4
217	1,2-DIBROMOETHYLENE	18.0	4.9	3.0	82.7
218	2,3-DIBROMOPRENE	17.7	11.8	6.4	103.4
220	*N,N*-DIBUTYL FORMAMIDE	15.5	8.9	6.2	182.0
221	DIBUTYL PHTHALATE	17.8	8.6	4.1	266.0
222	DIBUTYL SEBACATE	13.9	4.5	4.1	339.0
223	DIBUTYL STEARATE	14.5	3.7	3.5	382.0

TABLE A.1 (continued)
Hansen Solubility Parameters for Selected Solvents (Solvents are in Alphabetical Order)

No.	Solvent	Dispersion	Polar	Hydrogen Bonding	Molar Volume
700	3,4-DICHLORO ααα-TRIFLUOROTOLUENE	20.0	4.7	2.4	145.5
229	2,3-DICHLORO-1,3-BUTADIENE	17.1	2.3	2.8	104.0
825	1,3-DICHLORO-2-FLUOROBENZENE	19.4	9.1	2.7	117.1
739	1,3-DICHLORO-2-PROPANOL	17.5	9.9	14.6	95.5
824	2,4-DICHLORO-3-FLUORONITROBENZENE	19.9	7.2	4.2	140.0
828	2,4-DICHLORO-5-NITROBENZOTRIFLUORIDE	19.9	7.6	3.7	162.0
230	DICHLOROACETALDEHYDE	16.7	9.1	7.5	94.0
231	1,1-DICHLOROACETONE	17.1	7.6	5.4	97.3
232	DICHLOROACETONITRILE	17.4	9.4	6.4	80.3
233	2,6-DICHLOROANISOLE	19.8	8.4	6.5	137.1
798	2,6-DICHLOROANISOLE	19.9	8.4	6.5	137.1
846	2,4-DICHLOROBENZALDEHYDE	19.7	8.8	5.4	135.7
234	*o*-DICHLOROBENZENE	19.2	6.3	3.3	112.8
715	*p*-DICHLOROBENZENE	19.7	5.6	2.7	118.6
716	*m*-DICHLOROBENZENE	19.7	5.1	2.7	114.8
826	2,5-DICHLOROBENZOTRIFLUORIDE	20.0	4.7	2.4	145.5
235	2,2-DICHLORODIETHYL ETHER	18.8	9.0	5.7	117.2
236	DICHLORODIFLUOROMETHANE (FREON 12)	12.3	2.0	0.0	92.3
237	1,1-DICHLOROETHANE	16.5	7.8	3.0	84.2
238	*N,N*-DICHLOROETHYL AMINE	16.8	7.6	7.7	98.3
239	1,1-DICHLOROETHYLENE	16.4	5.2	2.4	79.9
224	1,2-DICHLOROETHYLENE (*CIS*)	17.0	8.0	3.2	75.5
240	1,2-DICHLOROETHYLENE (*CIS*)	17.0	8.0	3.2	75.5
225	*N,N*-DICHLOROMETHYL AMINE	16.5	7.5	8.0	90.8
241	DICHLOROMETHYL METHYL ETHER	17.1	12.9	6.5	90.5
242	DICHLOROMONOFLUOROMETHANE (FREON 21)	15.8	3.1	5.7	75.4
243	2,3-DICHLORONITROBENZENE	19.7	12.6	4.4	132.5
801	3,4-DICHLORONITROBENZENE	20.1	7.2	4.1	130.6
822	2,4-DICHLORONITROBENZENE	20.4	8.7	4.2	133.4
833	1,5-DICHLOROPENTANE	19.0	7.8	1.5	127.5
244	2,5-DICHLOROPHENOL	20.0	6.3	12.1	119.0
852	2,6-DICHLOROPHENOL	20.1	7.5	10.9	119.0
848	3,4-DICHLOROPHENYL ACETONITRILE	20.5	10.8	4.4	148.8
245	1,1-DICHLOROPROPANE	16.1	7.8	3.5	98.3
738	2,3-DICHLOROPROPANOL	17.5	9.2	14.6	95.2
226	2,3-DICHLOROPROPENE	16.2	7.8	3.0	91.6
246	1,1-DICHLOROPROPENE	16.9	6.7	2.9	93.5
227	1,2-DICHLOROPROPENE (*CIS*)	17.0	8.5	2.9	93.9
247	1,2-DICHLOROTETRAFLUOROETHANE (FREON 114)	12.6	1.8	0.0	117.6
248	3,4-DICHLOROTOLUENE	19.8	9.8	2.5	128.7
228	1,2-DICHLOROVINYL ETHYL ETHER	16.9	10.5	6.0	117.8
249	DIETHANOLAMINE	17.2	10.8	21.2	95.9
777	1,1-DIETHOXY BUTANE	15.4	4.9	4.6	177.6
792	DIETHOXY DISULFIDE	15.1	8.3	7.4	141.5
250	1,1-DIETHOXY ETHANE	15.0	3.4	4.0	143.9
791	1,1-DIETHOXY ETHANOL (ACETAL)	15.2	5.4	5.3	142.2
251	*N,N*-DIETHYL ACETAMIDE	16.4	11.3	7.5	124.5
252	DIETHYL AMINE	14.9	2.3	6.1	103.2

TABLE A.1 (continued)
Hansen Solubility Parameters for Selected Solvents (Solvents are in Alphabetical Order)

No.	Solvent	Dispersion	Polar	Hydrogen Bonding	Molar Volume
253	*p*-DIETHYL BENZENE	18.0	0.0	0.6	156.9
721	1,2-DIETHYL BENZENE	17.7	0.1	1.0	153.5
254	DIETHYL CARBONATE	16.6	3.1	6.1	121.0
255	DIETHYL ETHER	14.5	2.9	5.1	104.8
256	*N,N*-DIETHYL FORMAMIDE	16.4	11.4	9.2	111.4
257	DIETHYL KETONE	15.8	7.6	4.7	106.4
258	DIETHYL PHTHALATE	17.6	9.6	4.5	198.0
259	DIETHYL SULFATE	15.7	14.7	7.1	131.5
260	DIETHYL SULFIDE	16.8	3.1	2.0	107.4
261	2-(DIETHYLAMINO) ETHANOL	14.9	5.8	12.0	133.2
262	DIETHYLDISULFIDE	16.7	6.7	5.7	123.1
263	DIETHYLENE GLYCOL	16.6	12.0	20.7	94.9
264	DIETHYLENE GLYCOL BUTYL ETHER ACETATE	16.0	4.1	8.2	208.2
755	DIETHYLENE GLYCOL DIVINYL ETHER	16.0	7.3	7.9	164.1
757	DIETHYLENE GLYCOL DIBUTYL ETHER	15.8	4.7	4.4	248.1
762	DIETHYLENE GLYCOL DIETHYL ETHER	15.8	5.9	5.6	179.8
265	DIETHYLENE GLYCOL HEXYL ETHER	16.0	6.0	10.0	204.3
266	DIETHYLENE GLYCOL METHYL *T*-BUTYL ETHER	16.0	7.2	7.2	193.9
267	DIETHYLENE GLYCOL MONOBUTYL ETHER	16.0	7.0	10.6	170.6
268	DIETHYLENE GLYCOL MONOETHYL ETHER	16.1	9.2	12.2	130.9
269	DIETHYLENE GLYCOL MONOETHYL ETHER ACETATE	16.2	5.1	9.2	175.5
270	DIETHYLENE GLYCOL MONOMETHYL ETHER	16.2	7.8	12.6	118.0
271	DIETHYLENE GLYCOL MONOPROPYL ETHER	16.0	7.2	11.3	153.9
272	DIETHYLENETRIAMINE	16.7	13.3	14.3	108.0
810	*o*-DIFLUOROBENZENE	18.0	9.0	1.0	98.5
829	2,6-DIFLUOROBENZONITRILE	18.8	11.2	3.2	111.6
830	3,5-DIFLUOROBENZONITRILE	18.8	11.2	3.2	111.6
273	1,1-DIFLUOROETHANE	14.9	10.2	3.0	69.5
274	1,1-DIFLUOROETHYLENE	15.0	6.8	3.6	58.2
823	2,4-DIFLUORONITROBENZENE	19.4	11.0	3.7	109.6
275	DIHEXYL PHTHALATE	17.0	7.6	3.6	332.3
276	DIHYDROGEN DISULFIDE	17.3	6.3	10.7	49.6
277	DIHYDROPYRAN	17.5	5.5	5.7	91.2
208	DI-ISOBUTYL CARBINOL	14.9	3.1	10.8	177.8
203	DI-ISOBUTYL KETONE	16.0	3.7	4.1	177.1
279	DI-ISODECYL PHTHALATE	16.6	6.2	2.6	464.2
280	DI-ISOHEPTYL PHTHALATE	16.8	7.2	3.4	364.9
281	DI-ISONONYL ADIPATE	16.2	1.8	3.8	433.7
282	DI-ISONONYL PHTHALATE	16.6	6.6	2.9	432.4
840	DI-ISOPROPYL METHYL PHOSPHONATE	16.4	10.0	5.7	184.4
842	DI-ISOPROPYL PHOSPHONOFLUORIDATE	15.7	10.2	5.9	174.5
283	DIKETENE	16.2	15.1	8.1	75.9
763	1,3-DIMETHOXY BUTANE	15.6	5.5	5.2	140.0
284	1,1-DIMETHOXY ETHANE	15.1	4.9	4.9	106.7
285	*N,N*-DIMETHYL ACETAMIDE	16.8	11.5	10.2	92.5
286	DIMETHYL ACETYLENE	15.1	3.4	7.6	78.9
287	DIMETHYL AMINE	15.3	4.8	11.2	66.2
288	DIMETHYL AMINE-DIMER	15.3	4.8	7.9	132.4

TABLE A.1 (continued)
Hansen Solubility Parameters for Selected Solvents (Solvents are in Alphabetical Order)

No.	Solvent	Dispersion	Polar	Hydrogen Bonding	Molar Volume
289	*N,N*-DIMETHYL BUTYRAMIDE	16.4	10.6	7.4	127.8
290	DIMETHYL CARBONATE	15.5	3.9	9.7	84.2
291	DIMETHYL CELLOSOLVE	15.4	6.0	6.0	104.5
292	DIMETHYL DIETHYLENE GLYCOL	15.8	6.1	9.2	142.0
293	DIMETHYL DIKETONE	15.7	5.3	11.7	88.2
294	DIMETHYL DISULFIDE	17.3	7.8	6.5	88.6
295	DIMETHYL ETHANOLAMINE	16.1	9.2	15.3	101.1
296	DIMETHYL ETHER	15.2	6.1	5.7	63.2
297	DIMETHYL FORMAMIDE	17.4	13.7	11.3	77.0
298	1,1-DIMETHYL HYDRAZINE	15.3	5.9	11.0	76.0
299	DIMETHYL KETENE	15.2	7.4	4.8	87.6
839	DIMETHYL METHYL PHOSPHONATE	16.7	13.1	7.5	106.9
849	2,6-DIMETHYL PHENOL	19.1	4.9	12.9	116.3
850	3,4-DIMETHYL PHENOL	19.2	6.0	13.4	121.0
300	DIMETHYL PHTHALATE	18.6	10.8	4.9	163.0
782	2,5-DIMETHYL PYRROLE	18.3	7.6	6.8	101.8
301	DIMETHYL SULFIDE	16.1	6.4	7.4	73.2
302	DIMETHYL SULFONE	19.0	19.4	12.3	75.0
303	DIMETHYL SULFOXIDE	18.4	16.4	10.2	71.3
304	2,3-DIMETHYL-1-BUTENE	14.9	1.2	2.8	125.2
305	DIOCTYL PHTHALATE	16.6	7.0	3.1	377.0
306	1,4-DIOXANE	19.0	1.8	7.4	85.7
307	1,3-DIOXOLANE	18.1	6.6	9.3	69.9
308	DIPROPYL AMINE	15.3	1.4	4.1	136.9
737	DIPROPYL KETONE	15.8	5.7	4.9	140.8
309	DIPROPYLENE GLYCOL	16.5	10.6	17.7	130.9
310	DIPROPYLENE GLYCOL METHYL ETHER	15.5	5.7	11.2	157.4
311	DIPROPYLENE GLYCOL MONOMETHYL ETHER ACETATE	16.3	4.9	8.0	195.7
312	2,3-DITHIABUTANE	17.3	7.8	6.5	88.6
313	DITRIDECYL PHTHALATE	16.6	5.4	1.9	558.3
314	*p*-DIVINYL BENZENE	18.6	1.0	7.0	142.8
315	DIVINYL SULFIDE	16.5	4.6	5.6	93.6
316	DODECANE	16.0	0.0	0.0	228.6
317	EICOSANE	16.5	0.0	0.0	359.8
318	EPICHLOROHYDRIN	18.9	7.6	6.6	78.4
319	1,2-EPOXY PROPENE	16.5	8.6	6.7	70.0
320	3,4-EPOXY-1-BUTENE	16.6	7.7	7.4	80.7
321	EPSILON-CAPROLACTAM	19.4	13.8	3.9	110.7
322	1,2-ETHANE DITHIOL	17.9	7.2	8.7	83.9
323	ETHANESULFONYLCHLORIDE	17.7	14.9	6.8	94.7
324	ETHANETHIOL (ETHYL MERCAPTAN)	15.7	6.5	7.1	74.3
325	ETHANOL	15.8	8.8	19.4	58.5
326	ETHANOLAMINE	17.0	15.5	21.2	59.8
794	4-ETHOXY ACETOPHENONE	18.8	10.3	6.4	162.6
776	1-ETHOXY ETHOXY-2-PROPANOL	15.9	5.7	11.7	156.0
761	3-ETHOXY PROPIONALDEHYDE	16.0	8.8	7.4	112.1
327	ETHOXYETHYL PROPIONATE	16.2	3.3	8.8	155.5
328	ETHYL ACETATE	15.8	5.3	7.2	98.5

TABLE A.1 (continued)
Hansen Solubility Parameters for Selected Solvents (Solvents are in Alphabetical Order)

No.	Solvent	Dispersion	Polar	Hydrogen Bonding	Molar Volume
329	ETHYL ACETYLENE	15.1	3.4	5.0	81.5
330	ETHYL ACRYLATE	15.5	7.1	5.5	108.8
331	ETHYL AMINE	15.0	5.6	10.7	65.6
332	ETHYL AMYL KETONE	16.2	4.5	4.1	156.0
333	ETHYL BENZENE	17.8	0.6	1.4	123.1
334	ETHYL BROMIDE	16.5	8.4	2.3	74.6
335	ETHYL BUTYL KETONE	16.2	5.0	4.1	139.0
336	ETHYL CARBAMATE	16.8	10.1	13.0	91.2
337	ETHYL CARBYLAMINE	15.6	15.2	5.8	74.4
338	ETHYL CHLORIDE	15.7	6.1	2.9	70.0
339	ETHYL CHLOROFORMATE	15.5	10.0	6.7	95.6
340	ETHYL CINNAMATE	18.4	8.2	4.1	166.8
341	2-ETHYL CROTON ALDEHYDE	16.1	8.0	5.5	115.2
342	ETHYL CYANOACRYLATE	15.2	10.3	9.0	117.1
343	ETHYL ETHYNYLETHER	15.4	7.9	5.9	87.6
344	ETHYL FORMATE	15.5	8.4	8.4	80.2
346	2-ETHYL HEXYL ACETATE	15.8	2.9	5.1	196.0
347	2-ETHYL HEXYL ACRYLATE	14.8	4.7	3.4	208.2
348	ETHYL HYPOCHLORITE	15.7	8.6	6.5	79.5
349	ETHYL IODIDE	17.3	7.9	7.2	81.2
350	ETHYL ISOCYANATE	15.4	12.0	2.5	78.7
351	ETHYL ISOPROPENYL ETHER	14.8	3.5	5.1	113.3
352	ETHYL ISOTHIOCYANATE	17.2	14.7	9.0	87.1
353	ETHYL LACTATE	16.0	7.6	12.5	115.0
354	ETHYL METHACRYLATE	15.8	7.2	7.5	125.8
355	ETHYL METHYL SULFIDE	17.1	4.8	2.5	91.2
356	ETHYL THIOCYANATE	15.4	13.4	5.4	87.1
357	1-ETHYL VINYL ETHYL ETHER	15.3	4.2	4.6	125.6
358	ETHYL VINYLETHER	14.9	4.9	5.6	94.9
359	ETHYL VINYLKETONE	15.8	11.3	4.5	99.3
360	2-ETHYL-1-BUTANOL	15.8	4.3	13.5	123.2
727	2-ETHYL-1-BUTENE	14.9	1.7	3.5	123.1
361	ETHYL-1-PROPYNYLETHER	15.9	6.4	5.4	101.6
362	2-ETHYL-1,3-BUTADIENE	15.3	1.6	3.8	115.2
345	2-ETHYL-HEXANOL	15.9	3.3	11.8	156.6
363	ETHYLENE CARBONATE	19.4	21.7	5.1	66.0
364	ETHYLENE CHLOROHYDRIN	16.9	8.8	17.2	67.1
365	ETHYLENE CYANOHYDRIN	17.2	18.8	17.6	68.3
366	ETHYLENE DIBROMIDE	19.2	3.5	8.6	87.0
367	ETHYLENE DICHLORIDE	19.0	7.4	4.1	79.4
368	ETHYLENE GLYCOL	17.0	11.0	26.0	55.8
369	ETHYLENE GLYCOL BUTYL ETHER ACETATE	15.3	4.5	8.8	171.2
752	ETHYLENE GLYCOL BUTYL ETHYL ETHER	15.3	4.9	4.6	175.5
753	ETHYLENE GLYCOL BUTYL METHYL ETHER	15.5	5.2	4.9	157.2
370	ETHYLENE GLYCOL DI-*T*-BUTYL ETHER	14.7	4.1	8.2	210.0
371	ETHYLENE GLYCOL DIACETATE	16.2	4.7	9.8	132.8
758	ETHYLENE GLYCOL DIBUTYL ETHER	15.7	4.5	4.2	209.5
760	ETHYLENE GLYCOL DIETHYL ETHER	15.4	5.4	5.2	141.6

TABLE A.1 (continued)
Hansen Solubility Parameters for Selected Solvents (Solvents are in Alphabetical Order)

No.	Solvent	Dispersion	Polar	Hydrogen Bonding	Molar Volume
372	ETHYLENE GLYCOL METHYL *T*-BUTYL ETHER	15.3	5.1	8.2	157.4
373	ETHYLENE GLYCOL MONO-2-ETHYL HEXYL ETHER	16.0	4.1	5.1	194.7
731	ETHYLENE GLYCOL MONOBENZYL ETHER	17.8	5.9	12.2	143.0
771	ETHYLENE GLYCOL MONOETHYL ETHER ACRYLATE	15.9	5.1	9.3	147.7
374	ETHYLENE GLYCOL MONO-*T*-BUTYL ETHER	15.3	6.1	10.8	131.0
375	ETHYLENE GLYCOL MONOBUTYL ETHER	16.0	5.1	12.3	131.6
376	ETHYLENE GLYCOL MONOETHYL ETHER	16.2	9.2	14.3	97.8
377	ETHYLENE GLYCOL MONOETHYL ETHER ACETATE	15.9	4.7	10.6	136.1
378	ETHYLENE GLYCOL MONOISOBUTYL ETHER	15.2	4.9	9.6	132.5
379	ETHYLENE GLYCOL MONOISOPROPYL ETHER	16.0	8.2	13.1	115.8
380	ETHYLENE GLYCOL MONOMETHYL ETHER	16.2	9.2	16.4	79.1
381	ETHYLENE GLYCOL MONOMETHYL ETHER ACETATE	15.9	5.5	11.6	121.6
789	ETHYLENE GLYCOL SULFITE	20.0	15.9	5.1	75.1
382	ETHYLENE METHYL SULFONATE	16.9	9.3	9.6	97.9
383	ETHYLENE OXIDE	15.6	10.0	11.0	49.9
384	EHTYLENE SULFIDE	19.3	9.0	6.5	59.5
385	ETHYLENEDIAMINE	16.6	8.8	17.0	67.3
386	ETHYLENEIMINE	18.6	9.8	7.7	51.8
387	ETHYLIDENE ACETONE	16.2	12.1	4.5	99.0
388	ETHYNYL PROPYL ETHER	15.4	3.8	5.4	104.1
389	ETHYNYLMETHYLETHER	15.8	8.1	6.5	70.1
389	ETHYNYLMETHYLETHER	15.8	8.1	6.5	70.1
390	1-FLUORO ACRYLIC ACID	16.0	8.7	13.0	90.0
391	1-FLUORO ACRYLONITRILE	14.1	15.4	5.7	88.8
827	4-FLUORO-3-NITROBENZOFLUORIDE	19.4	7.2	3.7	108.4
799	*p*-FLUOROANISOLE	18.7	7.3	6.7	113.2
392	FLUOROBENZENE	18.7	6.1	2.0	94.7
393	FLUOROETHYLENE	15.2	7.0	1.0	57.5
394	FLUOROMETHANE	13.4	10.6	9.5	40.7
395	FLUOROPRENE	14.2	5.8	1.0	85.5
820	4-FLUOROPROPYLPHENONE	19.6	7.1	3.5	138.8
396	FORMALDEHYDE	12.8	14.4	15.4	36.8
397	FORMAMIDE	17.2	26.2	19.0	39.8
398	FORMIC ACID	14.3	11.9	16.6	37.8
399	FORMYL FLUORIDE	11.2	13.4	7.9	32.0
741	FORMYL FLUORIDE	15.0	10.1	8.6	56.5
400	*N*-FORMYL HEXAMETHYLENE IMINE	18.5	10.4	7.6	127.0
401	*N*-FORMYL PIPERIDINE	18.7	10.6	7.8	111.5
402	FUMARONITRILE	16.7	13.6	7.8	83.0
403	FURAN	17.8	1.8	5.3	72.5
404	FURFURAL	18.6	14.9	5.1	83.2
405	FURFURYL ALCOHOL	17.4	7.6	15.1	86.5
406	GLYCEROL	17.4	12.1	29.3	73.3
407	GLYCIDYL METHACRYLATE	16.3	8.5	5.7	136.4
408	GLYOXAL-ETHANDIAL	15.0	17.0	13.3	50.9
409	HEPTANE	15.3	0.0	0.0	147.4
410	1-HEPTENE	15.0	1.1	2.6	141.9
411	*n*-HEPTYL ACETATE	15.8	2.9	5.5	181.1

TABLE A.1 (continued)
Hansen Solubility Parameters for Selected Solvents (Solvents are in Alphabetical Order)

No.	Solvent	Dispersion	Polar	Hydrogen Bonding	Molar Volume
706	HEXACHLOROACETONE-1 (D ONLY)	18.3	0.0	0.0	151.9
707	HEXACHLOROACETONE-2 (G.C.)	18.3	6.6	6.4	151.9
412	HEXADECANE	16.3	0.0	0.0	294.1
413	HEXAFLUORO 1,3-BUTADIENE	13.8	0.0	0.0	104.3
414	HEXAFLUORO ISOPROPANOL	17.2	4.5	14.7	105.3
415	HEXAFLUOROHEXANOL	15.1	4.4	9.9	123.3
416	HEXAMETHYLPHOSPHORAMIDE	18.5	8.6	11.3	175.7
417	HEXANE	14.9	0.0	0.0	131.6
418	1-HEXENE	14.7	1.1	0.0	126.1
419	HEXYL ACETATE	15.8	2.9	5.9	165.0
420	HEXYLENE GLYCOL	15.7	8.4	17.8	123.0
421	HEXYLENE GLYCOL DIACETATE	15.3	4.5	7.2	204.3
422	HYDRAZINE	14.2	8.3	8.9	32.1
423	HYDROGEN CYANIDE	12.3	17.6	9.0	39.3
424	HYDROGEN SULFIDE	17.9	6.0	10.2	36.1
425	HYDROXYETHYL ACRYLATE	16.0	13.2	13.4	114.9
831	INDOLE	19.8	7.5	6.5	110.1
426	4-IODO-1,2-BUTADIENE	17.4	6.3	6.2	105.1
427	IODOBENZENE	19.5	6.0	6.1	112.0
428	IODOPRENE	17.2	2.5	6.2	104.2
429	ISOAMYL ACETATE	15.3	3.1	7.0	148.8
754	ISOAMYL ALCOHOL (3-METHYL-1-BUTANOL)	15.8	5.2	13.3	109.4
430	ISOBUTYL ACETATE	15.1	3.7	6.3	133.5
770	ISOBUTYL ACRYLATE	15.5	6.2	5.0	145.0
431	ISOBUTYL ALCOHOL	15.1	5.7	15.9	92.8
432	ISOBUTYL ISOBUTYRATE	15.1	2.9	5.9	163.0
786	ISOBUTYL SULFOXIDE	16.3	10.5	6.1	195.0
433	ISOBUTYLENE	14.5	2.0	1.5	89.4
434	ISOBUTYLENEOXIDE	16.1	4.8	5.8	90.0
435	ISOCYANIC ACID	15.8	10.5	13.6	37.7
436	ISOOCTYL ALCOHOL	14.4	7.3	12.9	156.6
437	ISOPENTANE	13.7	0.0	0.0	117.4
438	ISOPHORONE	16.6	8.2	7.4	150.5
439	ISOPRENE (2-METHYL-1,3-BUTADIENE)	14.7	1.4	4.1	100.9
440	ISOPROPYL ACETATE	14.9	4.5	8.2	117.1
441	ISOPROPYL AMINE (2-PROPAN AMINE)	14.8	4.4	6.6	86.8
442	ISOPROPYL CHLORIDE (2-CHLORO PROPANE)	15.9	8.3	2.1	91.7
443	ISOPROPYL ETHER	13.7	3.9	2.3	140.9
444	ISOPROPYL PALMITATE	14.3	3.9	3.7	330.0
445	ISOVALERALDEHYDE	14.7	9.5	5.0	106.0
446	ISOXAZOLE	18.8	13.4	11.2	64.1
447	KETENE	15.4	7.3	5.8	53.0
709	LACTIC ACID (DL)	17.0	8.3	28.4	72.1
448	LAURYL METHACRYLATE	14.4	2.2	5.1	293.1
449	MALONONITRILE	17.7	18.4	6.7	55.5
450	MESITYL OXIDE	16.4	6.1	6.1	115.6
451	MESITYLENE	18.0	0.0	0.6	139.8
452	METHACRYLALDEHYDE	15.7	11.1	7.4	83.3

TABLE A.1 (continued)
Hansen Solubility Parameters for Selected Solvents (Solvents are in Alphabetical Order)

No.	Solvent	Dispersion	Polar	Hydrogen Bonding	Molar Volume
453	METHACRYLAMIDE	15.8	11.0	11.6	76.0
454	METHACRYLIC ACID	15.8	2.8	10.2	84.8
455	METHACRYLONITRILE	15.8	15.1	5.4	83.9
456	METHANOL	15.1	12.3	22.3	40.7
793	4-METHOXY ACETOPHENONE	18.9	11.2	7.0	137.8
781	4-METHOXY BENZONITRILE	19.4	16.7	5.4	121.0
457	1-METHOXY-1,3-BUTADIENE	15.5	8.3	5.4	101.8
458	METHOXYHEXANONE (PENTOXONE)	15.3	6.0	5.9	143.5
459	*o*-METHOXYPHENOL (GUAIACOL)	18.0	8.2	13.3	109.5
460	3-METHOXYPROPIONITRILE	16.6	14.4	7.8	91.1
461	2-METHYL (*CIS*) ACRYLIC ACID	16.8	5.2	12.4	83.9
463	METHYL 1-PROPENYL ETHER	15.0	4.3	5.7	93.5
464	METHYL ACETATE	15.5	7.2	7.6	79.7
465	METHYL ACETYLENE	15.1	3.8	9.2	59.6
467	METHYL ACRYLATE	15.3	6.7	9.4	90.3
468	3-METHYL ALLYL ALCOHOL	16.0	6.0	15.5	84.4
469	METHYL ALLYL CYANIDE	16.4	11.3	5.1	97.7
470	METHYL AMINE	13.0	7.3	17.3	44.4
471	METHYL AMYL ACETATE	15.2	3.1	6.8	167.4
472	METHYL BENZOATE	17.0	8.2	4.7	124.9
473	METHYL BROMIDE	17.0	8.8	2.6	56.8
474	METHYL BUTYL KETONE	15.3	6.1	4.1	123.6
475	METHYL CHLORIDE	15.3	6.1	3.9	55.4
476	METHYL CHLOROFORMATE	16.3	9.5	8.5	77.3
477	METHYL CYCLOHEXANE	16.0	0.0	1.0	128.3
478	3-METHYL CYCLOHEXANONE	17.7	6.3	4.7	122.5
479	2-METHYL CYCLOHEXANONE	17.6	6.3	4.7	121.3
480	METHYL ETHYL ETHER	14.7	4.9	6.2	84.1
481	METHYL ETHYL KETONE	16.0	9.0	5.1	90.1
482	METHYL ETHYL KETOXIME	14.7	4.9	7.8	94.8
483	METHYL FORMATE	15.3	8.4	10.2	62.2
484	METHYL GLYOXAL	15.5	16.1	9.7	68.9
485	METHYL HYDRAZINE	16.2	8.7	14.8	52.7
486	METHYL HYDROPEROXIDE	15.0	15.0	30.0	24.1
487	1-METHYL IMIDAZOLE	19.7	15.6	11.2	79.5
488	METHYL IODIDE	17.5	7.7	5.3	62.3
489	METHYL ISOAMYL KETONE	16.0	5.7	4.1	142.8
490	METHYL ISOBUTYL CARBINOL	15.4	3.3	12.3	127.2
491	METHYL ISOBUTYL KETONE	15.3	6.1	4.1	125.8
492	METHYL ISOCYANATE	15.6	7.3	2.5	61.8
493	METHYL ISOPROPENYL KETONE	15.9	12.1	4.5	99.3
494	METHYL ISOTHIOCYANATE	17.3	16.2	10.1	68.4
495	3-METHYL ISOXAZOLE	19.4	14.8	11.8	57.7
496	METHYL MERCAPTAN	16.6	7.7	8.6	54.1
497	METHYL METHACRYLATE	15.8	6.5	5.4	106.1
498	METHYL *n*-AMYL KETONE	16.2	5.7	4.1	139.8
499	METHYL *n*-PROPYL KETONE	16.0	7.6	4.7	106.7

TABLE A.1 (continued)
Hansen Solubility Parameters for Selected Solvents (Solvents are in Alphabetical Order)

No.	Solvent	Dispersion	Polar	Hydrogen Bonding	Molar Volume
500	1-METHYL NAPHTHALENE	20.6	0.8	4.7	138.8
501	METHYL NITRATE	15.8	14.0	4.8	63.8
502	METHYL OLEATE	14.5	3.9	3.7	340.0
841	METHYL PHOSPHONIC DIFLUORIDE	14.0	14.0	8.4	73.9
503	METHYL PROPIONATE	15.5	6.5	7.7	96.8
504	METHYL SALICYLATE	18.1	8.0	13.9	129.6
740	METHYL SILANE	15.5	3.3	0.0	71.0
505	METHYL SULFIDE	16.1	6.4	7.4	73.2
506	METHYL SULFOLANE	19.4	17.4	5.3	112.7
507	METHYL THIOCYANATE	17.3	15.0	6.0	68.5
508	METHYL VINYL ETHER	14.9	5.3	6.3	75.2
509	METHYL VINYL KETONE	15.6	12.5	5.0	81.2
510	1-METHYL VINYL METHYL ETHER	14.8	4.2	5.6	96.1
511	METHYL VINYL SULFIDE	16.4	4.9	6.0	82.1
512	METHYL VINYL SULFONE	16.8	19.6	4.8	87.6
462	2-METHYL-1-BUTANOL	16.0	5.1	14.3	109.5
513	3-METHYL-1-BUTENE	14.0	1.4	3.8	112.9
514	2-METHYL-1-BUTENE	14.2	1.8	2.3	108.7
515	2-METHYL-1-CHLORO ACROLEIN	17.1	10.6	7.3	91.7
516	2-METHYL-1-PROPANOL	15.1	5.7	15.9	92.8
517	METHYL-1-PROPYNYL ETHER	15.7	6.3	5.9	85.4
518	3-METHYL-1,2-BUTADIENE	15.1	2.5	4.5	99.7
519	2-METHYL-1,3-DIOXOLANE	17.3	4.8	5.8	89.8
732	2-METHYL-2-BUTANOL	15.3	6.1	13.3	109.6
520	2-METHYL-2-BUTENE	14.3	2.0	3.9	106.7
521	*N*-METHYL-2-PYRROLIDONE	18.0	12.3	7.2	96.5
796	METHYL-*P*-TOLUATE	19.0	6.5	3.8	140.4
522	METHYL-*T*-BUTYL ETHER	14.8	4.3	5.0	119.8
523	METHYLAL	15.0	1.8	8.6	169.4
524	METHYLENE DICHLORIDE	18.2	6.3	6.1	63.9
525	METHYLENE DIIODIDE	17.8	3.9	5.5	80.5
526	MORPHOLINE	18.8	4.9	9.2	87.1
844	*N,N,N,N*-TETRAMETHYLTHIOUREA	17.3	6.0	10.5	132.2
529	NAPHTHA.HIGH-FLASH	17.9	0.7	1.8	181.8
530	NAPHTHALENE	19.2	2.0	5.9	111.5
784	*p*-NITRO TOLUENE	20.1	9.6	3.9	98.5
531	NITROBENZENE	20.0	8.6	4.1	102.7
532	NITROETHANE	16.0	15.5	4.5	71.5
533	NITROETHYLENE	16.3	16.6	5.0	59.9
534	NITROMETHANE	15.8	18.8	5.1	54.3
535	1-NITROPROPANE	16.6	12.3	5.5	88.4
536	2-NITROPROPANE	16.2	12.1	4.1	86.9
537	NONANE	15.7	0.0	0.0	179.7
538	NONYL PHENOL	16.5	4.1	9.2	231.0
539	NONYL PHENOXY ETHANOL	16.7	10.2	8.4	275.0
807	α,α-DICHLORO TOLUENE	19.9	6.6	2.4	134.2
540	OCTANE	15.5	0.0	0.0	163.5

TABLE A.1 (continued)
Hansen Solubility Parameters for Selected Solvents (Solvents are in Alphabetical Order)

No.	Solvent	Dispersion	Polar	Hydrogen Bonding	Molar Volume
541	OCTANOIC ACID	15.1	3.3	8.2	159.0
542	1-OCTANOL	17.0	3.3	11.9	157.7
543	2-OCTANOL	16.1	4.9	11.0	159.1
544	1-OCTENE	15.3	1.0	2.4	158.0
545	OLEIC ACID	16.2	3.1	5.5	320.0
546	OLEYL ALCOHOL	14.3	2.6	8.0	316.0
547	OXALYLCHLORIDE	16.1	3.8	7.5	85.8
818	PENTACHLOROCYCLOPROPANE	18.5	10.5	3.7	128.5
788	PENTACHLOROPHENOL	21.5	6.9	12.8	134.7
548	1,3-PENTADIENE (*TRANS*)	14.7	2.5	5.0	101.7
821	PENTAFLUOROBENZOPHENONE	19.3	8.1	5.4	185.5
549	PENTAMETHYLENE SULFIDE	18.5	6.3	8.9	103.6
550	PENTANE	14.5	0.0	0.0	116.2
551	2,4-PENTANEDIONE	17.1	9.0	4.1	103.1
552	1-PENTANOL	15.9	4.5	13.9	108.6
733	2-PENTANOL	15.6	6.4	13.3	109.6
553	4-PENTENAL	15.5	8.1	6.8	98.7
554	1-PENTENE	13.9	1.4	3.8	110.4
555	PERFLUORO ETHYLENE (TETRAFLUORO ETHYLENE)	15.1	0.0	0.0	65.8
556	PERFLUORO (DIMETHYLCYCLOHEXANE)	12.4	0.0	0.0	217.4
557	PERFLUOROHEPTANE	12.0	0.0	0.0	227.3
558	PERFLUOROMETHYLCYCLOHEXANE	12.4	0.0	0.0	196.0
559	PHENOL	18.0	5.9	14.9	87.5
560	2-PHENOXY ETHANOL	17.8	5.7	14.3	124.7
561	BIS-(*M*-PHENOXYPHENYL) ETHER	19.6	3.1	5.1	373.0
851	PHENYL ACETONITRILE	19.5	12.3	3.8	114.9
562	PHOSGENE	16.4	5.3	5.3	71.7
563	PHOSPHORUS TRICHLORIDE	18.4	3.6	0.0	87.3
564	PHTHALIC ANHYDRIDE	20.6	20.1	10.1	96.8
565	PINE OIL	15.6	3.0	9.8	155.0
566	1,2-PROPADIENE (ALLENE)	15.3	3.0	6.8	60.1
567	2-PROPANETHIOL	16.3	6.8	6.5	94.1
568	1-PROPANETHIOL	16.1	5.8	5.7	90.5
569	1-PROPANOL	16.0	6.8	17.4	75.2
570	2-PROPANOL	15.8	6.1	16.4	76.8
571	PROPARGYLALDEHYDE	16.2	11.9	8.7	60.0
572	β-PROPIOLACTONE	19.7	18.2	10.3	65.5
573	PROPIONALDEHYDE	15.1	6.7	10.0	73.4
574	PROPIONALDEHYDE-2,3-EPOXY	17.5	12.4	10.2	72.1
575	PROPIONAMIDE	16.7	9.8	11.5	78.9
576	PROPIONIC ACID	14.7	5.3	12.4	75.0
577	PROPIONITRILE	15.3	14.3	5.5	70.9
578	PROPIONYLCHLORIDE	16.1	10.3	5.3	86.9
579	*n*-PROPYL ACETATE	15.3	4.3	7.6	115.3
580	PROPYL AMINE	16.9	4.9	8.6	83.0
581	PROPYL CHLORIDE	16.0	7.8	2.0	88.1
582	PROPYL METHACRYLATE	15.5	6.3	6.6	158.8
583	PROPYLENE	15.1	1.6	1.5	68.8

TABLE A.1 (continued)
Hansen Solubility Parameters for Selected Solvents (Solvents are in Alphabetical Order)

No.	Solvent	Dispersion	Polar	Hydrogen Bonding	Molar Volume
584	PROPYLENE CARBONATE	20.0	18.0	4.1	85.0
746	PROPYLENE CHLOROHYDRIN	16.8	9.8	15.3	85.4
585	PROPYLENE GLYCOL	16.8	9.4	23.3	73.6
586	PROPYLENE GLYCOL MONO-*T*-BUTYL ETHER	15.3	6.1	10.8	151.6
587	PROPYLENE GLYCOL MONOBUTYL ETHER	15.3	4.5	9.2	132.0
588	PROPYLENE GLYCOL MONOETHYL ETHER	15.7	6.5	10.5	115.6
589	PROPYLENE GLYCOL MONOETHYL ETHER ACETATE	15.6	4.3	9.0	155.1
590	PROPYLENE GLYCOL MONOISOBUTYL ETHER	15.1	4.7	9.8	132.2
857	PROPYLENE GLYCOL MONOISOPROPYL ETHER	15.5	6.1	11.0	134.6
591	PROPYLENE GLYCOL MONOMETHYL ETHER	15.6	6.3	11.6	93.8
592	PROPYLENE GLYCOL MONOMETHYL ETHER ACETATE	15.6	5.6	9.8	137.1
593	PROPYLENE GLYCOL MONOPHENYL ETHER	17.4	5.3	11.5	143.2
594	PROPYLENE GLYCOL MONOPROPYL ETHER	15.8	7.0	9.2	130.3
595	PROPYLENE OXIDE	15.2	8.6	6.7	67.6
596	PROPYNONITRILE	15.5	17.0	6.3	62.5
597	PYRIDAZINE	20.2	17.4	11.7	72.6
598	PYRIDINE	19.0	8.8	5.9	80.9
599	2-PYROLIDONE	19.4	17.4	11.3	76.4
600	PYRROLE	19.2	7.4	6.7	69.2
601	PYRUVONITRILE	15.9	18.9	8.0	70.9
602	QUINOLINE	19.4	7.0	7.6	118.0
704	SALICYLALDEHYDE	19.4	10.7	14.7	104.6
832	SKATOLE	20.0	7.1	6.2	122.6
603	STEARIC ACID	16.3	3.3	5.5	326.0
604	STYRENE	18.6	1.0	4.1	115.6
605	SUCCINALDEHYDE (BUTANEDIAL)	16.8	9.8	10.5	81.2
606	SUCCINIC ANHYDRIDE	18.6	19.2	16.6	66.8
607	SUCCINONITRILE	17.9	16.2	7.9	81.2
608	SULFOLANE	18.4	16.6	7.4	95.3
609	SULFUR DICYANIDE	18.1	13.5	0.0	60.0
610	SULFUR DIOXIDE	15.8	8.4	10.0	44.0
790	TETRAMETHYLENE SULFONE	20.3	18.2	10.9	95.7
612	1,1,2,2-TETRABROMOETHANE	22.6	5.1	8.2	116.8
708	1,2,4,5-TETRACHLOROBENZENE (G.C.)	21.2	10.7	3.4	116.2
613	1,1,2,2-TETRACHLOROPROPANE	17.9	6.7	3.3	123.7
783	2,2,6,6-TETRACHLOROCYCLOHEXANONE	19.5	14.0	6.3	138.8
614	1,1,2,2-TETRACHLOROETHANE	18.8	5.1	5.3	105.2
615	TETRACHLOROETHYLENE	18.3	5.7	0.0	101.2
616	TETRAETHYLORTHOSILICATE	13.9	4.3	0.6	224.0
617	TETRAHYDROFURAN	16.8	5.7	8.0	81.7
618	TETRAHYDRONAPHTHALENE	19.6	2.0	2.9	136.0
619	TETRAHYDROPYRAN	16.4	6.3	6.0	97.8
620	TETRAHYDROTHIAPYRAN	18.5	6.3	8.9	103.6
621	TETRAHYDROTHIOPHENE	18.9	7.5	5.8	88.3
622	TETRAMETHYLENE SULFIDE	18.6	6.7	9.1	88.3
623	TETRAMETHYLENE SULFOXIDE	18.2	11.0	9.1	90.0
624	TETRAMETHYLUREA	16.7	8.2	11.0	120.4
625	2-THIABUTANE	16.2	5.9	5.3	90.4

TABLE A.1 (continued)
Hansen Solubility Parameters for Selected Solvents (Solvents are in Alphabetical Order)

No.	Solvent	Dispersion	Polar	Hydrogen Bonding	Molar Volume
626	THIACYCLOPROPANE	19.3	9.1	5.0	58.0
627	THIAZOLE	20.5	18.8	10.8	42.6
628	THIOACETAMIDE	17.5	20.6	20.2	75.0
629	THIOACETIC ACID	17.0	6.7	8.9	71.5
630	GAMMA-THIOBUTYROLACTONE	19.0	6.9	6.2	86.6
631	THIOCYANIC ACID	16.8	8.9	10.9	51.7
632	THIONYL CHLORIDE	16.9	6.2	5.9	79.0
633	THIOPHENE	18.9	2.4	7.8	79.0
703	THIOPHENOL	20.0	4.5	10.3	102.4
634	THIOUREA	20.0	21.7	14.8	72.8
635	1,4-THIOXANE	19.0	6.6	7.7	93.5
636	TIGALDEHYDE	16.2	12.9	6.8	96.6
637	TOLUENE	18.0	1.4	2.0	106.8
638	TOLYLENE DIISOCYANATE	19.3	7.9	6.1	143.5
639	1,2,3-TRIAZOLE	20.7	8.8	15.0	58.2
640	TRIBROMO ETHYLENE	15.3	9.4	8.0	97.8
641	TRIBUTYL PHOSPHATE	16.3	6.3	4.3	345.0
642	3,3,3-TRICHLORO PROPENE	17.7	15.5	3.4	106.2
643	1,1,2-TRICHLORO PROPENE	17.7	15.7	3.4	104.8
644	1,2,3-TRICHLORO PROPENE	17.8	15.7	3.4	105.0
861	TRICHLOROACETIC ACID	18.3	5.8	11.4	100.2
645	TRICHLOROACETONITRILE	16.4	7.4	6.1	100.0
701	1,2,4-TRICHLOROBENZENE	20.2	6.0	3.2	125.5
646	TRICHLOROBIPHENYL	19.2	5.3	4.1	187.0
647	1,1,1-TRICHLOROETHANE	16.8	4.3	2.0	99.3
648	1,1,2-TRICHLOROETHANE	18.2	5.3	6.8	92.9
649	TRICHLOROETHYLENE	18.0	3.1	5.3	90.2
650	TRICHLOROFLUOROMETHANE (FREON 11)	15.3	2.0	0.0	92.8
702	2,4,6-TRICHLOROPHENOL	20.3	5.1	10.8	132.5
651	2,4,5-TRICHLOROTHIOPHENOL	21.0	4.5	9.1	145.0
652	1,1,2-TRICHLOROTRIFLUOROETHANE (FREON 113)	14.7	1.6	0.0	119.2
653	TRICRESYL PHOSPHATE	19.0	12.3	4.5	316.0
654	TRIDECYL ALCOHOL	14.3	3.1	9.0	242.0
655	TRIETHANOLAMINE	17.3	22.4	23.3	133.2
656	TRIETHYLAMINE	17.8	0.4	1.0	138.6
657	TRIETHYLENE GLYCOL	16.0	12.5	18.6	114.0
856	TRIETHYLENE GLYCOL MONOMETHYL ETHER	16.2	7.6	12.5	160.0
658	TRIETHYLENE GLYCOL MONOOLEYL ETHER	13.3	3.1	8.4	418.5
659	TRIETHYLPHOSPHATE	16.7	11.4	9.2	171.0
795	2,3,4-TRIFLUORO NITROBENZENE	19.5	7.7	3.5	122.0
660	TRIFLUOROACETIC ACID	15.6	9.9	11.6	74.2
661	1,1,1-TRIFLUOROETHANE	14.6	10.7	0.0	64.6
662	TRIFLUOROMETHANE (FREON 23)	14.4	8.9	6.5	46.1
847	4-(TRIFLUOROMETHYL) ACETOPHENONE	18.8	6.1	3.5	151.7
663	TRIISONONYL TRIMELLILATE	16.6	5.7	2.2	602.9
664	TRIISOOCTYL TRIMELLITATE	16.6	6.0	2.5	553.1
665	TRIMETHYL AMINE	14.6	3.4	1.8	90.3
666	2,2,4-TRIMETHYL-1,3-PENTANEDIOL MONOISOBUTYRATE	15.1	6.1	9.8	227.4

TABLE A.1 (continued)
Hansen Solubility Parameters for Selected Solvents (Solvents are in Alphabetical Order)

No.	Solvent	Dispersion	Polar	Hydrogen Bonding	Molar Volume
667	TRIMETHYLBENZENE	17.8	0.4	1.0	133.6
668	TRIMETHYLENEOXIDE	19.0	9.8	7.2	58.0
669	TRIMETHYLENESULFIDE	18.8	7.8	9.4	72.8
670	2,2,4-TRIMETHYLPENTANE	14.1	0.0	0.0	166.1
671	TRIMETHYLPHOSPHATE	16.7	15.9	10.2	115.8
855	1,3,5-TRIOXANE	18.7	9.2	8.6	77.0
672	TRIPROPYLENE GLYCOL MONOMETHYL ETHER	15.3	5.5	10.4	214.0
860	UREA-R = 19.4	20.9	18.7	26.4	45.8
673	VALERONITRILE	15.3	11.0	4.8	103.8
674	VINYL 2-CHLORO ETHYL ETHER	16.3	6.7	5.8	101.7
675	VINYL 2-METHOXY ETHYL ETHER	15.9	6.7	6.8	96.2
676	VINYL ACETATE	16.0	7.2	5.9	92.6
677	VINYL ACETIC ACID	16.8	5.2	12.3	85.3
678	VINYL ACETYLENE	15.1	1.7	12.0	74.3
679	VINYL ALLYL ETHER	14.9	6.5	5.3	105.4
742	VINYL AMINE	15.7	7.2	11.8	51.8
680	VINYL BROMIDE	15.9	6.3	5.4	71.6
681	VINYL BUTYL SULFIDE	16.0	5.0	5.4	136.7
682	VINYL BUTYRATE	15.6	3.9	6.9	126.5
683	VINYL CHLORIDE	16.0	6.5	2.4	68.7
684	VINYL CROTONATE	15.9	5.0	9.0	118.8
685	VINYL ETHER	14.8	4.2	5.8	90.7
686	VINYL ETHYL SULFIDE	16.4	5.8	6.3	101.3
687	VINYL FORMATE	15.3	6.5	9.7	74.7
688	VINYL IODIDE (IODOETHENE)	17.1	5.5	7.3	75.6
689	VINYL PROPIONATE	15.6	8.0	4.7	110.1
690	VINYL PROPYL ETHER	14.9	3.5	5.2	113.0
843	4-VINYL PYRIDINE	18.1	7.2	6.8	107.3
691	VINYL PYRROLIDONE	16.4	9.3	5.9	106.9
692	VINYL SILANE	15.5	2.6	4.0	89.4
693	VINYL TRIFLUORO ACETATE	13.9	4.3	7.6	116.4
694	VINYL TRIMETHYL SILANE	14.5	1.0	2.5	145.3
695	VINYLENECARBONATE	17.3	18.1	9.6	86.0
696	WATER	15.5	16.0	42.3	18.0
859	WATER — 1% IN R = 18.1	15.1	20.4	16.5	18.0
858	WATER — 18	15.5	16.0	42.3	18.0
697	XYLENE	17.6	1.0	3.1	123.3
698	*o*-XYLENE	17.8	1.0	3.1	121.2

TABLE A.2
Hansen Solubility Parameters for Selected Correlations

No.	Polymer	Dispersion	Polar	Hydrogen Bonding	Interaction Radius
1	CELLIT BP-300	16.60	12.00	6.70	10.20
2	CELLIDORA A	18.20	12.40	10.80	7.40
3	ETHOCEL HE 10	17.90	4.30	3.90	5.90
4	ETHOCEL STD 20	20.10	6.90	5.90	9.90
5	ARALD DY025	14.00	7.40	9.40	13.70
6	EPIKOTE 828	23.10	14.60	5.00	20.50
7	1001	20.00	10.32	10.11	10.02
8	1004	17.40	10.50	9.00	7.90
9	1007	21.00	11.10	13.40	11.70
10	1009	19.30	9.37	10.95	8.26
11	PKHH	23.40	7.20	14.80	14.90
12	VERSAMID 100	23.80	5.30	16.20	16.10
13	VERSAMID 115	20.30	6.60	14.10	9.60
14	VERSAMID 125	24.90	3.10	18.70	20.30
15	VERSAMID 140	26.90	2.40	18.50	24.00
16	DESMOPHEN 651	17.70	10.60	11.60	9.50
17	DESMOPHEN 800	19.10	12.20	9.90	8.00
18	DESMOPHEN 850	21.54	14.94	12.28	16.78
19	DESMOPHEN 1100	16.00	13.10	9.20	11.40
20	DESMOPHEN 1150	20.60	7.80	11.60	13.10
21	DESMOPHEN 1200	19.40	7.40	6.00	9.80
22	DESMOPHEN 1700	17.90	9.60	5.90	8.20
23	DESMOLAC 4200	18.70	9.60	9.90	8.20
24	M-NAL SM510N	19.90	8.10	6.00	9.80
25	SUP BECK 1001	23.26	6.55	8.35	19.85
26	PHENODUR 373U	19.74	11.62	14.59	12.69
27	P-LYTE S-100	16.47	0.37	2.84	8.59
28	PICCOPALE 110	17.55	1.19	3.60	6.55
29	P-RONE 450L	19.42	5.48	5.77	9.62
30	POLYSAR 5630	17.55	3.35	2.70	6.55
31	HYCAR 1052	18.62	8.78	4.17	9.62
32	BUNA HULS B10	17.53	2.25	3.42	6.55
33	C-FLEX IR305	16.57	1.41	–0.82	9.62
34	LUTONAL IC/1203	14.20	2.50	4.60	12.40
35	LUTONAL I60	16.90	2.50	4.00	7.20
36	PVB ETHER	17.40	4.30	8.40	7.40
37	LIGNIN	20.17	14.61	15.04	11.66
38	MODAFLOW	16.10	3.70	7.90	8.90
39	VIPLA KR	18.40	6.60	8.00	3.00
40	CERECLOR 70	20.00	8.30	6.80	9.80
41	CHLOROPAR 40	17.00	7.60	7.90	11.90
42	PERGUT S 5	17.40	9.50	3.80	10.00
43	ALLOPREN R10	17.40	4.30	3.90	6.10
44	PARLON P 10	20.26	6.32	5.40	10.64
45	HYPALON 20	18.10	3.40	4.90	3.60
46	HYPALON 30	18.20	4.70	2.00	5.00
47	ALPEX	19.90	0.00	0.00	9.40
48	NITRO CEL H23	15.41	14.73	8.84	11.46

TABLE A.2 (continued)
Hansen Solubility Parameters for Selected Correlations

No.	Polymer	Dispersion	Polar	Hydrogen Bonding	Interaction Radius
49	CELLOLYN 102	21.73	0.94	8.53	15.75
50	PENTALYN 255	17.55	9.37	14.32	10.64
51	PENTALYN 830	20.03	5.81	10.93	11.66
52	ESTER GUM BL	19.64	4.73	7.77	10.64
53	VERSAMID 930	17.43	–1.92	14.89	9.62
54	VERSAMID 961	18.90	9.60	11.10	6.20
55	VERSAMID 965	20.15	6.04	12.90	9.20
56	DES-DUR L	17.50	11.30	5.90	8.50
57	DES-DUR N	17.60	10.00	3.70	9.30
58	SUPRAS F5100	19.70	12.90	12.80	11.40
59	MOW-AL B30H	18.60	12.90	10.30	8.30
60	MOW-AL B60H	20.20	11.20	13.30	11.20
61	BUTVAR B76	18.60	4.36	13.03	10.64
62	LUCITE 2042	17.60	9.66	3.97	10.64
63	LUCITE 2044	16.20	6.80	5.70	9.10
64	P-GUM MB319	18.60	10.80	4.10	11.50
65	P-GUM M527	18.40	9.40	6.50	10.70
66	PMMA	18.64	10.52	7.51	8.59
67	MOWILITH 50	20.93	11.27	9.66	13.71
68	P-STYRENE LG	22.28	5.75	4.30	12.68
69	LAROFLEX MP45	18.40	8.40	5.80	9.00
70	VILIT MB 30	20.00	8.30	6.70	9.40
71	VILIT MC 31	20.00	8.30	6.70	9.40
72	VILIT MC 39	18.40	7.60	6.70	6.80
73	V-LITE VAGD	17.10	10.40	6.50	7.50
74	V-LITE VAGH	16.50	10.90	6.40	7.70
75	V-LITE VMCA	17.70	11.10	6.90	8.70
76	V-LITE VMCC	17.60	11.10	6.80	8.80
77	V-LITE VMCH	17.60	11.10	6.40	8.60
78	V-LITE VYHH	17.40	10.20	5.90	7.80
79	V-LITE VYLF	18.10	10.30	4.20	8.30
80	ALF AC 366	18.60	10.00	5.00	10.40
81	ALF AM 756	23.00	2.20	4.20	16.90
82	ALF AN 896	22.90	15.20	7.60	18.10
83	ALF AN 950	22.60	13.80	8.10	17.10
84	ALF AT 316	20.50	9.30	9.10	12.40
85	LAF AT 576	19.20	5.30	6.30	11.90
86	ALK F261HS	23.60	1.00	7.60	19.00
87	ALK F41	20.60	4.60	5.50	12.60
88	DURO T354	17.30	4.20	7.90	9.30
89	DYNAPOL L812	22.60	13.10	5.80	16.80
90	DYNAPOL L850	20.00	6.20	7.00	9.50
91	PLEXAL C-34	18.50	9.21	4.91	10.64
92	SOALK 1935-EGAX	18.00	11.60	8.50	9.00
93	VEST-IT BL908	18.80	12.00	6.00	11.50
94	VEST-IT BL915	17.70	13.00	7.60	11.50
95	BE 370	20.70	6.10	12.70	14.80
96	BEETLE 681	22.20	–0.40	10.10	18.40

TABLE A.2 (continued)
Hansen Solubility Parameters for Selected Correlations

No.	Polymer	Dispersion	Polar	Hydrogen Bonding	Interaction Radius
97	CYMEL 300	19.35	12.83	12.87	9.82
98	CYMEL 325	25.50	15.20	9.50	22.20
99	DYNOMIN MM9	18.80	14.00	12.30	10.50
100	DYNOMIN UM15	19.90	15.80	13.40	11.70
101	SOAMIN M60	15.90	8.10	6.50	10.60
102	SYNR A 560	22.10	5.00	11.30	15.50
103	PLASTOPALH	20.81	8.29	14.96	12.69
104	UFORM MX-61	22.70	2.80	5.40	16.20
106	PARALOID P400	19.20	9.60	9.30	12.20
107	P-LOID P410	19.60	9.10	6.80	12.20
108	P QR 954	18.40	9.80	10.00	12.40
109	BAYSILONUD125	19.40	9.90	10.10	6.90
110	TEFLON (SL2-)	17.10	8.10	1.30	4.70
111	DEN 438	20.30	15.40	5.30	15.10
112	DEN 444	19.50	11.60	9.30	10.00
113	ZINK SILICATE (CR)	23.50	17.50	16.80	15.60
114	2-COMP EPOXY (CR)	18.40	9.40	10.10	7.00
115	POLYVINYLIDINE FLUORIDE	17.00	12.10	10.20	4.10
116	COAL TAR PITCH	18.70	7.50	8.90	5.80
117	NR 20 MIN	17.50	7.30	6.50	5.10
118	NR 1 HR	16.60	9.10	4.40	10.00
119	NR 4 HR	19.00	12.60	3.80	13.30
120	BR 20 MIN	16.50	1.00	5.10	5.00
121	BR 1 h	15.80	–2.10	4.00	8.20
122	BR 4 h (2)	17.60	2.10	2.10	7.00
123	NAT 20 min	14.50	7.30	4.50	11.00
124	NAT 1 h	15.60	3.40	9.10	14.00
125	NAT 4 h	19.40	13.20	7.70	19.00
126	PVC 20 min	16.10	7.10	5.90	9.30
127	PVC 1 h	14.90	11.10	3.80	13.20
128	PVC 4 h	24.40	4.90	9.90	22.70
129	PVA 20 min	11.20	12.40	13.00	12.10
130	PVA 1 h	15.30	13.20	13.50	8.80
131	PVA 4 h	17.20	13.60	15.40	10.90
132	PE 20 min	16.90	3.30	4.10	8.10
133	PE 1 h	17.10	3.10	5.20	8.20
134	PE 4 h	24.10	14.90	0.30	24.30
135	VIT 20 min	10.90	14.50	3.10	14.10
136	VIT 1 h	16.50	8.10	8.30	6.60
137	VIT 4 h	13.60	15.40	8.60	14.40
138	NEO 20 min	17.60	2.50	5.90	6.20
139	NEO 1 h	19.00	8.00	0.00	13.20
140	NEO 4 h	14.60	13.90	2.30	15.90
141	CH 5100	16.60	5.40	4.00	3.80
142	CH 5200	16.60	6.00	4.80	3.70
143	PVDC (110)	17.60	9.10	7.80	3.90
144	PVDC (130)	20.40	10.00	10.20	7.60
145	PES C=1	18.70	10.50	7.60	9.10

TABLE A.2 (continued)
Hansen Solubility Parameters for Selected Correlations

No.	Polymer	Dispersion	Polar	Hydrogen Bonding	Interaction Radius
146	PES L B+C=1	17.70	9.70	6.40	9.30
147	PPS CR 93	18.80	4.80	6.80	2.80
148	PPS TS60%12MO	18.70	5.30	3.70	6.70
149	PA6 CR	17.00	3.40	10.60	5.10
150	PA66 SOL	17.40	9.80	14.60	5.10
151	PA11 CR	17.00	4.40	10.60	5.10
152	POMH + POMC	17.10	3.10	10.70	5.20
153	PETP CR	18.20	6.40	6.60	5.00
154	PTFE L80 CR	16.20	1.80	3.40	3.90
155	PMMA CR	16.00	5.00	12.00	13.00
156	PE?	16.80	5.40	2.40	4.70
157	PPO CR	17.90	3.10	8.50	8.60
158	PUR CR	18.10	9.30	4.50	9.70
159	ABS CR	16.30	2.70	7.10	7.80
160	PSU CR	16.00	6.00	6.60	9.00
161	VINYL SIL BEERB	16.40	3.70	4.50	10.00
162	CELLOPHAN SW	16.10	18.50	14.50	9.30
163	EVOH SOL	20.50	10.50	12.30	7.30
164	SARANEX 4 h	17.70	18.30	0.70	18.40
165	4H 35 DEG	19.40	13.40	18.00	8.60
166	4H- 58 35D	20.50	11.30	10.30	6.70
167	PVALC	15.00	17.20	17.80	10.20
168	ACETAL CELANESE	21.10	9.30	5.90	11.40
169	ACETALHOMO-DUO	19.00	5.00	8.00	5.00
170	CEL ACET	16.90	16.30	3.70	13.70
171	CAB	17.20	13.80	2.80	12.60
172	CAP	9.80	13.60	11.40	15.20
173	CTFE	14.10	2.70	5.50	6.60
174	FEP	19.00	4.00	3.00	4.00
175	FURAN	19.00	6.00	8.00	5.00
176	FURF ALC	19.90	3.90	5.10	3.80
177	PFA (?)	16.70	7.70	–0.50	8.10
178	PHENOLIC	21.60	5.20	18.80	15.40
179	PETG	18.00	3.00	4.00	6.00
180	HDPE	18.00	0.00	2.00	2.00
181	PP	18.00	0.00	1.00	6.00
182	ECTFE HT	19.50	7.30	1.70	5.10
183	PAN	21.70	14.10	9.10	10.90
184	PEI1200PSI	17.20	6.40	5.20	3.60
185	PEI 2400 PSI	17.40	4.60	9.00	7.20
186	PEI 600 PSI	17.30	5.30	4.70	3.30
187	ESTER GUM	16.90	4.50	6.50	9.20
188	ALKYD 45 SOYA	17.50	2.30	7.70	10.00
189	SILDC-1107?	19.60	3.40	10.80	9.80
190	PVETHYLETHER?	15.10	3.10	11.90	12.90
191	PBUTYLACRYLATE	16.20	9.00	3.00	10.10
192	PBMA?	15.90	5.50	5.90	8.50
193	SILICONE DC 23?	16.40	0.00	7.80	5.50

TABLE A.2 (continued)
Hansen Solubility Parameters for Selected Correlations

No.	Polymer	Dispersion	Polar	Hydrogen Bonding	Interaction Radius
194	PE	16.00	0.80	2.80	3.20
195	GILSONITE	17.10	2.10	3.90	4.90
196	PVBUTETH	17.40	3.40	7.80	8.40
197	NAT RUBBER	16.00	4.00	6.00	1.30
198	HYP 20 CHLSULPE	17.40	3.20	4.00	4.80
199	ETHCEL N22?	22.70	0.50	16.50	20.10
200	CHLRUBBER	17.90	6.30	5.10	7.60
201	DAMMAR GUM	18.40	4.20	7.80	8.30
202	VERSAMID 100?	18.80	3.00	9.20	7.80
203	PS	18.50	4.50	2.90	5.30
204	PVAC	17.60	2.20	4.00	4.10
205	PVC	17.60	7.80	3.40	8.20
206	PHENOLICS	19.80	7.20	10.80	12.80
207	BUNA N BDEACN	17.80	3.20	3.40	3.70
208	PMMA	18.10	10.50	5.10	9.50
209	PEO 4000	21.50	10.90	13.10	15.90
210	PESULFIDE	17.80	3.80	2.20	4.10
211	PC	18.10	5.90	6.90	5.50
212	PLIOLITE P1230	18.10	4.70	3.70	3.90
213	MYLAR PET	18.00	6.20	6.20	5.00
214	VCVA COPOLY	17.30	8.70	6.10	7.80
215	PUR	17.90	6.90	3.70	2.70
216	SAN	16.60	9.80	7.60	4.80
217	VINSOL ROSIN	17.40	10.00	13.00	10.50
218	EPON 1001	17.00	9.60	7.80	7.10
219	SHELLAC	19.70	10.10	15.10	10.70
220	PMACN	17.20	14.40	7.60	3.80
221	CELLACET	18.30	16.50	11.90	8.80
222	CELLNIT	16.90	13.50	10.30	9.90
223	PVOH	17.00	9.00	18.00	4.00
224	NYLON 66	16.00	11.00	24.00	3.00
225	ACRYLOID B-44	19.40	11.20	4.40	10.50
226	ACRYLOID B-66	18.00	9.00	3.00	9.00
227	ACRYLOID B-72	19.20	11.20	1.80	11.00
228	ACRYLOID B-82	19.10	9.10	3.30	9.00
229	R+H PBA	16.00	8.00	8.00	12.00
230	R+H PIBMA	20.70	4.10	10.70	11.50
231	R+H PNBMA	16.00	6.20	6.60	9.50
232	R+H PEMA	19.00	9.00	8.00	11.00
233	R+H PMAA	25.60	11.20	19.60	20.30
234	R+H PMMA	19.10	11.30	4.10	10.30
235	BMA/AN 80/20	17.50	9.90	4.10	9.50
236	ISOB MALANH/CYCLOL 75/25	16.80	–0.40	7.20	8.50
237	MAA/EA/ST 15/38/47	17.60	5.20	7.00	4.50
238	MAA/MA/VA 15/27.5/57.5	25.50	15.70	18.10	21.50
239	MAA/MA/VA 15/17.5/67.5	25.50	15.70	18.10	21.50
240	MMA/CYCLOL 58/42	18.70	9.90	8.70	8.80
241	MMA/EA 50/50	17.50	9.90	4.10	9.50

TABLE A.2 (continued)
Hansen Solubility Parameters for Selected Correlations

No.	Polymer	Dispersion	Polar	Hydrogen Bonding	Interaction Radius
242	MMA/EA 25/75	19.00	9.00	15.00	12.00
243	MMA/EA/AGE 40/40/20	17.60	9.80	5.60	9.70
244	MMA/EA/AA	15.90	15.90	11.50	11.10
245	MMA/EA/AN 55/30/15?	16.70	10.90	8.50	8.50
246	MMA/EA/AN 40/40/20	20.40	13.20	11.00	12.30
247	MMA/EA/BAMA 40/40/20	17.90	8.50	11.70	12.90
248	MMA/EA/CYCLOL	17.60	9.80	6.40	9.80
249	MMA/EA/MAA 40/40/20	19.00	9.00	15.00	12.00
250	MMA/EA/MAM 45/45/10 ?	19.50	11.10	8.70	11.20
251	MMA/EA/VBE 40/40/20	17.80	10.00	6.60	9.80
252	ACID DEGMP	15.30	13.30	14.90	15.60
253	CARB DEG PTH	19.40	13.40	11.60	11.10
254	CRYPLEX 1473-5	19.20	9.40	5.60	8.90
255	DEG ISOPH	19.20	17.20	14.60	11.80
256	DEG PHTH	21.00	15.20	13.20	13.70
257	DPG PHTH	20.10	11.50	6.70	11.60
258	DOW ADIP TEREP	17.80	10.40	6.80	9.30
259	DOW X-2635 MALEATE	17.80	5.60	6.80	4.50
260	VITEL PE LINEAR	14.90	10.10	2.90	6.10
261	VITEL PE101-X	21.30	6.30	4.70	7.30
262	HYD BIS A FUM ISPH	17.00	4.40	6.20	5.00
263	HYD BIS A PG FUM ISPH	18.70	8.90	5.50	8.40
264	PENTA BENZ MAL	19.40	12.20	10.20	10.80
265	SOL MYLAR 49001	19.00	5.00	4.00	5.00
266	SOL MYLAR 49002	19.00	5.00	5.00	5.00
267	TEG EG MAL TEREP	18.80	11.40	9.20	10.20
268	TEG MALEATE	18.10	13.90	12.10	9.70
269	VAREZ 123	17.30	10.90	11.90	10.70
270	AMOCO 18-290	19.30	3.70	7.90	7.80
271	BUTON100 BUTAD-STY	17.00	4.00	3.00	7.30
272	BUTON 300	17.30	3.70	7.30	7.00
273	KOPPERS KTPL-A	19.30	3.70	7.90	7.80
274	RUBBER MOD PS	20.00	5.00	1.00	7.00
275	STY MAL ANH	23.40	13.80	15.20	16.50
276	LYTRON 820	21.10	13.10	14.50	14.40
277	MARBON 9200	19.00	4.00	4.00	6.00
278	PARAPOL S-50	17.90	3.90	4.90	3.90
279	PARAPOL S-60	17.90	3.90	4.90	3.90
280	PICCOFLEX 120	17.40	7.80	3.80	7.70
281	SHELL POLYALDEHYDE EX39	19.60	10.00	3.60	9.40
282	SHELL POLYALDEHYDE EX40	19.60	10.00	3.60	9.40
283	SHELL X-450	19.30	9.50	11.10	11.10
284	SMA 1430A	18.80	11.40	16.40	14.10
285	SAN 85/15	19.10	9.50	3.10	8.70
286	STY/BUTENOL 85/15	17.40	7.80	3.80	7.70
287	STY/CYCLOL 82/18	18.20	5.60	7.20	5.70
288	STY/2EHA/AA 81/11/8	17.70	4.90	5.90	5.90
289	STY/MAA 90/10	18.70	6.30	7.30	6.70

TABLE A.2 (continued)
Hansen Solubility Parameters for Selected Correlations

No.	Polymer	Dispersion	Polar	Hydrogen Bonding	Interaction Radius
290	STY/MA 85/15	18.00	9.00	3.00	9.00
291	STY/HALF ESTER MA 60/40	18.90	10.90	10.70	9.70
292	STY/PROP HALF E MA 57/43	18.00	9.80	8.40	10.10
293	STY/VBE 85/15	17.40	7.80	3.80	7.70
294	STYRON 440M-27 MOD PS	20.00	5.00	1.00	7.00
295	STYRON 475M-27	20.00	5.00	1.00	7.00
296	STYRON 480-27	20.00	6.00	4.00	5.30
297	ACRYLOID K120N	17.60	10.00	3.80	9.50
298	DODA 6225	19.00	2.00	1.00	3.00
299	DODA 3457	19.00	2.00	1.00	3.00
300	ELVAX 250	19.00	2.00	1.00	3.00
301	ELVAX 150	18.70	2.30	0.70	6.00
302	ELVAX EOD 3602-1	17.70	3.30	2.70	5.40
303	EXON 470 PVC	17.40	7.80	3.80	7.70
304	EXON 471	17.90	8.70	2.50	9.00
305	EXON 473	17.40	7.80	3.80	7.70
306	GEON 121	19.50	6.70	11.10	8.00
307	POLYCYCLOL a	19.00	9.00	15.00	12.00
308	PVBE	16.70	3.70	8.30	8.60
309	PVEE	16.00	4.00	12.00	14.00
310	FORMVAR 7/70E PVFORMAL	22.20	12.60	14.20	14.00
311	FORMVAR 15/95E	22.20	12.60	14.20	14.00
312	PVIBE	16.00	1.00	8.00	10.00
313	SARAN F-120 VCL2/AN?	28.80	16.80	0.80	23.70
314	SARAN F-220 ?	28.80	16.80	0.80	23.70
315	SINCLAIR 3840A	18.40	4.00	9.60	7.30
316	VA/EHA/MA 63/33/4	17.70	6.30	7.70	5.30
317	VA/EHA/CYC/MAA/76/12/8/4	21.20	12.40	13.00	12.60
318	VA/EA/CY 70/20/10	20.00	12.00	11.00	15.00
319	VBE/AN/MAA 46/27/27	18.90	11.70	11.10	9.60
320	VBE/MA/MAC 46/27/27	19.40	13.00	13.80	12.30
321	VDC/AA 75/25 ?	20.40	11.00	0.80	11.70
322	VINYLITE AYAA PVAC	22.90	18.30	7.70	20.40
323	VINYLITE VAGH	17.00	7.80	6.80	7.10
324	VINYLITE VMCH	18.30	9.70	7.70	8.50
325	VINYLITE VXCC	18.00	9.40	4.60	8.40
326	VINYLITE VYHH	19.00	11.00	5.00	10.00
327	VINYLITE VYLF	18.00	9.40	4.60	8.40
328	VINYLITE XYHL PVBUTYRAL	19.00	9.00	15.00	12.00
329	VINYLITE XYSG PVBUTYRAL	19.00	9.00	15.00	12.00
330	VYSET 69	17.90	3.50	7.50	5.90
331	ACRYLAMIDE MONOMER	16.90	18.10	19.90	17.00
332	BAKELITE SULFONE P-47	20.00	3.00	6.00	3.00
333	BECKOLIN 27 MODIF OIL	11.40	0.00	3.00	18.10
334	PEO 4000	22.20	11.20	13.20	17.10
335	CHLORINATED RUBBER	18.00	6.00	5.00	7.00
336	CONOCO H-35 HYDROCARBON M	11.40	0.00	3.00	18.10
337	DAMMAR GUM DEWAXED	19.00	2.00	9.00	9.00

TABLE A.2 (continued)
Hansen Solubility Parameters for Selected Correlations

No.	Polymer	Dispersion	Polar	Hydrogen Bonding	Interaction Radius
338	EPOCRYL E-11 ?	17.30	12.90	12.10	8.50
339	ESTANE X-7 ?? DIOX ONLY	19.00	1.80	7.40	1.00
340	HEXADECYL MONOESTER TRIME	19.00	11.60	14.00	11.90
341	HYDR SPERM OIL WX135	20.00	4.00	2.00	5.00
342	HYPALON 20 CHL SULF PE	17.80	3.20	4.40	4.10
343	HYPALON 30	17.80	3.40	3.20	5.10
344	KETONE RESIN S588	18.00	10.80	13.20	12.20
345	SANTOLITE MHP ARYLSULFONA	18.40	12.00	8.40	10.60
346	p-TOLSULFONAMIDE-FORMALDEH	24.60	18.60	16.40	20.90
347	VYHH-NIF REPT	17.40	9.90	6.70	7.50
348	PVF DMF ONLY	17.40	13.70	11.30	2.00
349	PES SOL MUP	19.60	10.80	9.20	6.20
350	LARD 37°C	15.90	1.16	5.41	12.03
351	LARD 23°C	17.69	2.66	4.36	7.98
352	1% IN WATER –AMINES	15.07	20.44	16.50	18.12
353	1% IN WATER +AMINES	14.96	18.33	15.15	16.22
354	BLOOD SERUM	23.20	22.73	30.60	20.48
355	SUCROSE	21.67	26.26	29.62	20.44
356	UREA	20.90	18.70	26.40	19.42
357	PSORIASIS SCALES	24.64	11.94	12.92	19.04
358	LIGNIN	20.61	13.88	15.25	11.83
359	CHOLESTEROL	20.40	2.80	9.40	12.60
360	CHLOROPHYLL	20.20	15.60	18.20	11.10
361	CELLULOSE-PAPER STRENGTH	25.40	18.60	24.80	21.70
362	PSU ULTRASON S	19.70	8.30	8.30	8.00
363	BAREX 210 CR	20.10	9.10	12.70	10.90
364	BAREX 210 CR-STYRENE	17.70	8.90	10.90	6.40
365	PARALOID B72	17.60	7.40	5.60	9.40
366	ESTIMATE DRIED OIL	16.00	6.00	7.00	5.00
367	DAMMAR DEWAXED	19.00	2.00	9.00	9.00
368	LDPE PERM > 80	16.50	4.50	0.50	6.00
369	LDPE PERM > 0.8	15.30	5.30	2.50	10.10
370	TOLONATE HDT (RH-POULENC)	19.00	11.00	3.00	12.00
371	TOLONATE HDB (RH-POULENC)	19.00	11.00	2.00	11.30
372	R ACM (ACRYLIC)	16.80	11.80	11.60	17.00
373	R BUTYL	18.00	0.00	3.00	9.00
374	R ECO	21.30	8.10	6.10	12.00
375	R CSM	28.00	14.00	3.40	28.30
376	R EBONITE 0.722	18.70	6.10	2.70	6.60
377	R ETHYLENE/PROPYLENE	16.60	0.00	5.20	9.10
378	R EPDM	18.60	–3.40	4.40	10.70
379	R FQ FL/SI	15.90	20.10	6.90	16.80
380	R FKM (VITON) 0.76	11.60	23.00	5.00	21.60
381	R NR NAT RUB	20.80	1.80	3.60	14.00
382	R NBR	19.80	17.80	3.20	19.00
383	R CR CHLOROPRENE	24.60	8.60	6.40	20.40
384	R AU ESTER PU	17.90	13.30	10.70	17.10
385	R PEU ETHER PU	17.90	13.30	10.70	17.10

TABLE A.2 (continued)
Hansen Solubility Parameters for Selected Correlations

No.	Polymer	Dispersion	Polar	Hydrogen Bonding	Interaction Radius
386	R T SULPHIDE	25.30	17.30	6.70	23.60
387	R Q SILICONE .748	13.80	5.00	1.20	14.30
388	R SBR	17.20	6.00	4.60	9.80
389	R TFP TETFLPROP .744	16.60	6.80	0.60	7.90
390	R ABS	17.60	8.60	6.40	10.90
391	R CELLULOSE ACETATE	14.90	7.10	11.10	12.40
392	R CHLORINATED PVC	17.50	6.50	5.50	6.30
393	R DIALLYLPHTHALATE	22.20	12.20	8.60	15.80
394	R POM ACETAL	17.20	9.00	9.80	5.30
395	R PA12	18.50	8.10	9.10	6.30
396	R PA66	18.20	8.80	10.80	5.20
397	R POLYAMIDEIMIDE	18.50	5.70	8.70	4.20
398	R POLYBUTYLENETEREPH	18.00	5.60	8.40	4.50
399	R POLYCARBONATE	19.10	10.90	5.10	12.10
400	R HDPE/LDPE	17.50	4.30	8.30	3.90
401	R PET	19.10	6.30	9.10	4.80
402	R POLYIMIDES	24.30	19.50	22.90	21.60
403	R PMMA	19.30	16.70	4.70	17.40
404	R TPX	18.80	1.40	6.40	7.90
405	R POLYPHENYLENEOXIDE	16.90	8.90	2.70	11.70
406	R POLYSULPHONE	19.80	11.20	6.20	11.30
407	R POLYPROPYLENE	17.20	5.60	–0.40	4.50
408	R EPOXY COLD CURING	16.80	10.80	8.80	8.20
409	R EPOXY HOT CURING	18.30	12.30	9.70	7.30
410	R HET RESIN	17.50	11.30	8.30	8.60
411	R ISOPHTHALIC	19.80	17.40	4.20	18.00
412	R TEREPHTHALIC	19.80	17.40	4.20	18.00
413	EPIKOTE 828 60	16.60	14.00	2.80	14.90
414	EPIKOTE 828 30	16.30	16.40	1.90	16.70
415	EPIKOTE 1001 60	15.80	16.30	3.30	16.40
416	EPIKOTE 1001 40	16.30	13.10	6.30	10.90
417	EPIKOTE 1001 20	18.80	13.60	8.90	12.00
418	EPIKOTE 1001 10	18.10	11.40	9.00	9.10
419	EPIKOTE 1004 60	17.70	10.10	7.60	9.80
420	EPIKOTE 1004 30	18.50	9.30	8.00	9.60
421	EPIKOTE 1007 30	18.60	10.60	8.10	8.80
422	EPIKOTE 1009 60	17.00	9.60	8.50	7.60
423	EPIKOTE 1009 30	19.80	10.60	10.30	9.70
424	EPIKOTE 1009 10	19.00	9.10	10.70	8.00
425	PIBMA 10	17.00	4.60	7.60	9.50
426	PIBMA 30	17.10	5.90	0.70	7.30
427	PMMA 10	17.80	10.40	2.90	9.60
428	PMMA 30	17.20	7.20	3.50	4.80
429	PBMA 10	20.60	3.50	7.20	12.80
430	PBMA 30	18.10	5.70	0.00	8.50
431	PMMA 10	17.60	10.10	5.80	9.40
432	PMMA 30	17.50	5.50	3.80	4.50
433	PEMA 10	16.50	8.70	5.00	10.40
434	PEMA 30	16.90	7.80	0.50	7.30

TABLE A.2 (continued)
Hansen Solubility Parameters for Selected Correlations

No.	Polymer	Dispersion	Polar	Hydrogen Bonding	Interaction Radius
435	CRODA AC500 TSA 10	17.80	6.40	4.70	10.70
436	CRODA AC500 TSA 30	21.20	1.40	10.70	12.30
437	CRODA AC550 TSA 10	16.30	10.60	7.40	12.90
438	CRODA AC550 TSA 30	16.30	10.60	7.40	12.90
439	LUMFLON LF200 10	18.50	5.40	6.90	9.90
440	LUMFLON LF200 30	20.10	4.40	3.20	8.50
441	LUMFLON LF916 10	17.50	6.80	10.50	12.50
442	LUMFLON LF916 30	18.10	3.90	8.30	8.80
443	PLASTOKYD S27	20.10	5.70	5.30	20.00
444	PLASTOKYD SC140	25.20	9.20	3.70	20.00
445	PLASTOKYD SC400	23.70	0.50	10.30	20.00
446	PLASTOKYD AC4X	23.90	7.80	8.80	19.90
447	ALLOPRENE R10 10	19.50	9.20	6.90	7.50
448	ALLOPRENE R10 30	17.90	5.60	6.70	5.80
449	ALLOPRENE R10 60	19.60	6.50	5.80	9.10
450	HYPALON 20	20.30	3.20	0.70	11.30
451	POLYISOPRENE SW	17.00	4.00	4.00	7.30
452	BROMOBUTYL RUBBER S	17.60	1.70	2.00	6.00
453	BROMOBUTYL RUBBER L	17.00	3.40	2.00	6.00

DOCUMENTATION FOR HANSEN SOLUBILITY PARAMETERS IN POLYMERS DATABASE

Polymers 1–109:

These polymers are listed in Reference 1 with suppliers. This report from the Scandinavian Paint and Printing Ink Research Institute updates an earlier one from 1982. The institute no longer exists. See also Reference 2.

Polymer 110:

This is an intermediate value for the permeation of chemicals through Challenge® materials.[3] See also Table 8.1 and Figure 8.2. Improved values are found below in Polymers 141 and 142. This correlation was based on few data to help locate additional solvents for testing. Results from tests with these then resulted in the more recent correlations.

Polymers 111 and 112:

These are correlations of true solubilities for the DOW epoxy Novolacs 438 and 444.

Polymers 113 and 114:

These are correlations of the chemical resistance of coatings based on inorganic zinc silicate and a two-component epoxy produced by Hempel´s Marine Paints. Data taken from resistance tables.

Polymer 115:

The data are solubilities determined for PVDF with the correlation being previously published in Reference 4.

Polymer 116:

Data for coal tar pitch generated for the solubility of the solids not dissolved in some cases where the solution was darkened with only partial solution.

Polymers 117–140:

Permeation correlations for protective clothing described in detail in Reference 5. See also Chapter 8, Table 8.1.

Polymers 141 and 142:

Final permeation correlations for Challenge 5100 and 5200 materials. Data from Reference 3 where there is considerable discussion. See also Chapter 8, Table 8.1 and Figure 8.2.

Polymers 143 and 144:

These correlations are based on which solvents dissolve PVDC at elevated temperatures and use data from Wessling.[6] These were additionally used to check new calculations for solubility parameters of the solvents where these were lacking.

Polymers 145–148:

These chemical resistance data for PES (ICI-Victrex®) and PPS (Philips-Ryton®) were based on supplier data sheets and are reported in Reference 7.

Polymers 149–160:

These correlations for many common plastic types are based on the resistance tables reported in the *PLASTGUIDE* (1989) published by the Danish company Dukadan, which no longer exists. A single correlation for the solubility of PA6,6 is based on its solubility only with data from Reference 8.

Polymer 161:

Beerbower presented several sets of data and correlations of swelling and solubility (and other phenomena). Not sure where this one for polyvinyl silane came from originally.

Polymers 162 and 163:

These correlations for swelling of cellophane and solubility of ethylene vinyl alcohol copolymer are based on data generated at NIF (Scandinavian Paint and Printing Ink Research Institute).

Polymers 164–167:

These are supplementary breakthrough time correlations for Saranex®, Safety 4® 4H, and polyvinyl alcohol protective gloves. See also Chapter 8.

Polymers 168–181:

These correlations for common polymer types are based on data in resistance tables in the *Modern Plastics Encyclopedia* in the 1984/1985 issue.[9] Such data are not always sufficiently encompassing to allow good correlations.

Polymer 182:

Correlation based on high temperature solvents for ECTFE.

Polymer 183:

Data for this correlation of solubility of polyacrylonitrile were taken from the *Polymer Handbook*,[10] Table of Solvents and Nonsolvents, p. VII/385–VII/386. See also Chapter 3, Table 3.3.

Polymers 184–186:

Data for the tensile strength of Polyethylene imide (GE-Ultem®) were generated by GE as published in the *Modern Plastics Encyclopedia* 1984/1985.[9]

Polymers 187–224:

The *Handbook of Solubility Parameters and Other Cohesion Parameters*[11] as well as the *Polymer Handbook*[12] included so-called "solvent range" data. Solvents were divided into groups of poor, moderate, and strong hydrogen bonding, and many experiments were run. The correlations show that not all the data were well taken, but a reasonable indication is possible. The full Hansen solubility parameter system is not covered very well by these limited solubility data. These polymers are included in Reference 11 (Table 1, p. 280).

Polymers 225–346:

These entries have the same problem as those in polymers 187–224 in that the data are sometimes questionable and not sufficient enough to do what has been done, i.e., convert solvent range data to Hansen solubility parameter spheres. These entries cover the acrylics, polyesters, polystyrenes, vinyls, and miscellaneous categories. Some categories are not yet included. Data from Reference 11 (Table 2, pp. 281–289).

Polymer 347:

These values for VYHH® (Union Carbide) were taken from Reference 1.

Polymer 348:

This questionable correlation for PVF includes only one solvent as being good.[13]

Polymer 349:

Data on PES true solubility taken by author. See Chapter 3.

Polymers 350–358:

These entries are not all polymers, but mostly biological materials with the source of data being Reference 14.

Polymer 359:

The solubility of cholesterol, data collected by the author. See Chapter 9.

Polymer 360:

Solubility data generated by high school students as part of a project included in Reference 4. Source of chlorophyll was crushed leaves.

Polymer 361:

Correlation on strength of paper immersed in different solvent reported in Reference 4. Data were taken from Reference 15.

Polymer 362:

Solubility of Ultrason® PES has been reported by BASF in their product data. These data were combined with supplementary solubility data for this correlation. Also reported in Reference 16. See Chapter 3.

Polymers 363 and 364:

Chemical resistance of BAREX® 210 from data in BP Chemicals datasheet. Styrene is an outlier in the first, while its removal from consideration gives a perfect fit.

Polymers 365–367:

These data were generated in connection with a lecture to the Nordic Conservation Congress in Copenhagen.[17] All give perfect fits, partly because of too few data, but the correlations can be useful. Paraloid® B72 and Dammar are used as protective lacquers.

Polymers 368 and 369:

These correlations divide the permeation coefficients given in Reference 18 into >80 and >0.8, respectively. The units are $[(c^3 \times cm)/(cm^2 \times s \times Pa)] \times 10^{-13}$. The fits are good. See Chapter 8.

Polymers 370 and 371:

These are correlations of experimental solubility data for the Rhône-Poulenc reactive isocyanates Tolonate® HDT (which gave the same result as Tolonate HDT-LV) and Tolonate HDB (which gave the same results as Tolonate HDB-LV). The fits were perfect, and the numbers were reasonable. The data could not include alcohol or amine solvents because of reactions.

Polymers 372–389:

The data correlated for these 18 rubbers are from a RAPRA database.[19] The information used was satisfactory or unsatisfactory, all other information such as limited suitability was neglected. No precise weight gain or other information is available, just the general suitability or not. The values in parenthesis are (data fit/number of solvents).

ACM = Acrylate rubbers (0.981/55)
ECO = Epichlorohydrin rubbers (0.988/37)
CSM = Chlorosulphonated polyethylene rubber (0.906/53)
E = Ebonite (0.722/41)
EPM = Ethylene-propylene copolymer (0.987/47)
EPDM = Ethylene-propylene terpolymer (0.968/51)
FQ = Fluorosilicone rubber (0.844/40)
FKM = Hexafluoroprop.-Vinylidine fluoride copolymer (VITON) (0.769/50)
NR = Natural rubber (1.000/59)
NBR = Nitrile rubber (0.990/65)
FFKM = KALREZ® (Du Pont) too resistant to correlate
CR = Polychloroprene (0.877/54)
AU = Polyester polyurethane (0.959/63)
EU = Polyether polyurethane (0.959/63)
T = Polysulphide rubber (0.799/48)
Q = Silicone (0.748/53)
SBR = Styrene butadiene rubber (0.942/54)
TFP = Tetrafluoroethylene-propylene copolymer (0.744/26)

Polymers 390–412:

These correlations use data from the Rapra collection of data on chemical resistance for plastics.[20] Approach same as for RAPRA rubber data previously.

Polymers 413–450:

These data are from the collected report of the EC project on self-stratifying coatings reported in a full issue of *Progress in Organic Coatings*. The specific reference is Reference 21. The evaluations were made at different concentrations in many cases. Some alkyds were omitted here.

Polymers 451 and 452:

These data are for strong swelling of two different film samples of brominated butyl rubber.

Polymer 453:

The correlation is based on strong swelling of a film of polyisoprene.

REFERENCES (FOR TABLE A.2)

1. Saarnak, A., Hansen, C. M., and Wallström, E., Solubility Parameters — Characterization of Paints and Polymers, Report from Scandinavian Paint and Printing Ink Research Institute, January 1990, Hoersholm, Denmark.
2. Hansen, C. M., Solubility Parameters, in *Paint Testing Manual*, Manual 17, Koleske, J. V., Ed., American Society for Testing and Materials, Philadelphia, 1995, 383–404.
3. Hansen, C. M., Billing, C. B., and Bentz, A. P., Selection and Use of Molecular Parameters to Predict Permeation Through Fluoropolymer-Based Protective Clothing Materials, *The Performance of Protective Clothing, Fourth Volume, ASTM STP 1133,* McBriarty, J. P. and Henry, N. W., Eds., American Society for Testing and Materials, Philadelphia, 1992, 894–907.
4. Hansen, C. M., 25 Years with Solubility Parameters (in Danish: 25 År med Opløselighedsparametrene), *Dan. Kemi,* 73(8), 18–22, 1992.
5. Hansen, C.M. and Hansen, K.M., Solubility Parameter Prediction of the Barrier Properties of Chemical Protective Clothing, *Performance of Protective Clothing: Second Symposium, ASTM STP 989,* Mansdorf, S. Z., Sager, R., and Nielsen, A. P., Eds., American Society for Testing and Materials, Philadelphia, 1988, 197–208.
6. Wessling, R. A., The Solubility of Poly(vinylidene Chloride), *J. Appl. Polym. Sci.,* 14, 1531–1545, 1970.
7. Hansen, C. M., *Solubility Parameters for Polyphenylene Sulfide (PPS) and Polyether Sulphone (PES),* Centre for Polymer Composites (Denmark), Danish Technological Institute, Taastrup, 1991, 1–89. ISBN 87-7756-139-2
8. Wyzgoski, M. G., The Role of Solubility in Stress Cracking of Nylon 6,6, in *Macromolecular Solutions — Solvent Property Relationships in Polymers*, Seymour, R. B. and Stahl, G. A., Eds., Pergamon Press, New York, 1982, 41–60.
9. Anonymous, *Modern Plastics Encyclopedia,* McGraw-Hill, New York, 1984/1985, 482–555.
10. Fuchs, O., Tables of Solvents and Non-Solvents, *Polymer Handbook,* 3rd ed., Branderup, J. and Immergut, E. H., Eds., Wiley, New York, 1989, VII/379–VII/407.
11. Barton, A. F. M., *Handbook of Solubility Parameters and Other Cohesion Parameters,* CRC Press, Boca Raton, FL, 1983, 280–289.
12. Grulke, E. A., Table 3.4, Solubility Parameter Ranges of Commercial Polymers, *Polymer Handbook,* 3rd ed., Branderup, J. and Immergut, E. H., Eds., Wiley, New York, 1989, VII/544–VII/550.
13. Fuchs, O., Tables of Solvents and Non-Solvents, *Polymer Handbook,* 3rd ed., Branderup, J. and Immergut, E. H., Eds., Wiley, New York, 1989, VII/385.
14. Hansen, C. M. and Andersen, B. H., The Affinities of Organic Solvents in Biological Systems, *Am. Ind. Hyg. Assoc. J.,* 49(6), 301–308, 1988.
15. Robertson, A. A., Cellulose-Liquid Interactions, *Pulp and Mag. Can.*, 65(4), T-171–T-178, 1964.
16. Hansen, C. M., *Solvent Resistance of Polymer Composites — Glass Fibre Reinforced Polyether Sulfone (PES),* Centre for Polymer Composites (Denmark), Danish Technological Institute, Taastrup, 1994.
17. Hansen, C. M., *Conservation and Solubility Parameters*, Nordic Conservation Congress Preprints, Copenhagen, 1994, 1–13.
18. Pauly, S., Permeability and Diffusion Data, *Polymer Handbook,* 3rd ed., Branderup, J. and Immergut, E. H., Eds., Wiley, New York, 1989, VI/435–VI/449.
19. Anonymous, Chemical Resistance Data Sheets, Volume 2, Rubbers, New Edition — 1993, *Rapra Technology,* Shawbury, Shrewsbury, Shropshire, Great Britain, 1993.

20. Anonymous, Chemical Resistance Data Sheets, Volume 1, Plastics, New Edition — 1993, *Rapra Technology*, Shawbury, Shrewsbury, Shropshire, Great Britain, 1993.
21. Benjamin, S., Carr, C., and Walbridge, D. J., Self-Stratifying Coatings for Metallic Substrates, *Prog. Org. Coat.,* 28, 197–207, 1996.

Index

A

B

D

E

F

G

H

I

J

K

L

M

N

O

P

R

S

T

U

V

W

X

Z